Advances in Soil Science

GLOBAL CLIMATE CHANGE AND COLD REGIONS ECOSYSTEMS

Advances in Soil Science

Series Editor: B. A. Stewart

Other Titles on this Topic

Forthcoming Titles

Advances in Soil Science

GLOBAL CLIMATE CHANGE AND COLD REGIONS ECOSYSTEMS

Edited by

R. Lal
J.M. Kimble
B.A. Stewart

LEWIS PUBLISHERS

Boca Raton   London   New York   Washington, D.C.

Library of Congress Cataloging-in-Publication Data

Global climate change and cold regions ecosystems / edited by R. Lal, J.M. Kimble, B.A. Stewart.
    p. cm. -- (Advances in soil science)
    Based on papers from an international workshop held at Ohio State University, Mar. 31-Apr. 2, 1998.
    Includes bibliographical references.
    ISBN 1-56670-459-6 (alk. paper)
    1. Soil ecology--Cold regions--Congresses. 2. Climatic changes--Cold regions--Congresses. I. Lal, Rattan, 1945- II. Kimble, J. M. (John M.) III. Stewart, B. A. (Bobby Alton), 1932-
IV. Advances in soil science (Boca Raton, Fla.)
QH84.1.G55 2000
577.5'86276—dc21
                                          00-030439
                                             CIP

© 2000 by CRC Press LLC
Lewis Publishers is an imprint of CRC Press LLC

No claim to original U.S. Government works
International Standard Book Number 1-56670-459-6
Library of Congress Card Number 00-030439
Printed in the United States of America 1 2 3 4 5 6 7 8 9 0
Printed on acid-free paper

Preface

Cold ecosystems comprise arctic, subarctic, alpine, antarctic, boreal forests, and peatlands. Soils of these regions contain about 775 Pg (1 Pg = petagram = 10^{15} g) of organic carbon (SOC) and 275 Pg of inorganic carbon (SIC). The total C pool in soils of these regions represents 16.4% of the global soil C pool for the tundra region and 25.6% of the global soil C pool for the soils of the boreal forest ecoregion. Two soil orders, Gelisols and Histosols, combined account for about 75% of the total C pool in the Arctic regions. Soils of the tundra ecoregion are less developed than those of the boreal forest ecoregion, and have a very little horizon development. Cryoturbation, or soil movement due to frost action leading to patterned ground, is a principal process in soils of the tundra ecoregion. Usually, high SOC of arctic soils is partially due to cryoturbation which mixes organic matter into the subsoil. The boreal forest ecoregion comprises the great northern coniferous biome, bordered by the treeless lands of the Arctic tundra at about 68° N to about 58° N.

This ecoregion is circumpolar in extent, occupying a belt as wide as 1000 km in certain regions of North America. The alpine tundra comprises soils of the high altitude beyond the tree line. Important regions comprising alpine soils are the Himalayan-Tibetan ecoregion, Alps, Andes, and other high-altitude mountainous terrains. The Antarctic region, a very large land mass, represents a minor carbon pool but comprises an important component of the cold region soils. While these ecoregions have been a net sink of C in the past, they may become a major source due to anthropogenic activities in the region and elsewhere in other ecosystems. The database on total C pools in soils of these ecoregions is sketchy, and little is known about the C dynamics and its impact on the global C cycle.

In the event of global warming, these ecoregions are anticipated to undergo the most significant increases in mean annual temperature. This drastic increase in temperature could substantially increase the depth of the soil's active layer. Warmer soil temperature and more aeration porosity could have two significant but vastly different effects: (i) higher rates of SOC decomposition and CO_2 efflux to the atmosphere, and (ii) more photosynthesis and increase in SOC pool within the active soil layer. The increase in land area of more productive and warmer soils could lead to more C sequestration in the soil and the terrestrial and aquatic ecosystems.

The information on SOC pools for these ecosystems is limited and not available for all countries, especially for soils with permafrost. Potential environmental change is likely to influence this large C pool, and little is known about the net effects of two opposing scenarios on the global carbon cycle and agricultural productivity. It is with this background that an international workshop was held at the Ohio State University from March 31 to April 2, 1998. The specific objectives of the workshop included the following: (i) assess processes and the status of the total C pool in soils of these ecoregions, especially for Gelisols and Histosols, (ii) evaluate the impact of anthropogenic activities and potential global environmental changes on SOC and SIC pools and the global C cycle, (iii) collate available information on the C dynamics in soils of these ecoregions, (iv) identify knowledge gaps and suggest an action plan toward judicious management of these ecoregions, and (v) prioritize researchable issues.

This volume is based on papers presented at this workshop. A total of 17 chapters presented are organized into 4 thematic sections. Section I deals with soil C pools in different ecoregions and contains six chapters. These chapters deal with C pool in tundra and boreal soils of Russia, Canada, China, and Antarctic regions. Section 2 deals with the impact of natural and anthropogenic disturbances on soil C pool and other properties. A total of five chapters in the section deal with the impact of management and of political global warming on soil properties. Section III deals with the method of assessment of C and other properties of soils of the cold ecoregions. A total of five chapters presented in the section deal with the topic of spatial and temporal variations in soil properties, and the use of chronosequence and areal evaluation techniques to assess soil C pool. The last section contains a synthesis chapter and discusses the fate of C in soils of the cold ecoregions, and research and development priorities.

The organization of the symposium and publication of this volume was made possible by cooperation and funding of the U.S. Department of Agriculture, Natural Resources Conservation Service, and Ohio State University. The editors thank all authors for their outstanding efforts to document and present their information on the current understanding of soil processes and the carbon cycle in cold ecoregions in a timely fashion. Their efforts have contributed to enhancing the overall understanding of pedospheric processes in these important ecoregions with large reserves of soil C pool. These efforts have advanced the frontiers of soil science and improved the understanding of the pedosphere into the broader scientific arena of linking soils to the global carbon cycle.

Thanks are also due to the staff of Lewis Publishers and CRC Press LLC for their efforts in publishing this information in a timely fashion to make it available to the overall scientific community. Valuable contributions were made by numerous colleagues, graduate students, and staff of the Ohio State University. We especially thank Lynn Everett for her efforts in organizing the workshop and for handling the flow of chapters to and from the authors throughout the review process. Her tireless efforts, good humor, and good nature are greatly appreciated. We also offer special thanks to Brenda Swank for her help in preparing this material and for her assistance in all aspects of the symposium. The efforts of many others were also very important in getting this relevant and important scientific information out in a timely manner.

The Editorial Committee

About the Editors

Dr. R. Lal is a professor of soil science in the School of Natural Resources at Ohio State University. Prior to joining Ohio State in 1987, he served as a soil scientist for 18 years at the International Institute of Tropical Agriculture, Ibadan, Nigeria. Professor Lal is a fellow of the Soil Science Society of America, American Society of Agronomy, Third World Academy of Sciences, American Association for the Advancement of Science, Soil and Water Conservation Society, and the Indian Academy of Agricultural Sciences. He is a recipient of the International Soil Science Award, the Soil Science Applied Research Award of the Soil Science Society of America, the International Agronomy Award of the American Society of Agronomy, and the Hugh Hammond Bennett Award of the Soil and Water Conservation Society. He is past president of the World Association of the Soil and Water Conservation Society and the International Soil Tillage Research Organization. He is a member of the U.S. National Committee on Soil Science established by the National Academy of Sciences.

Dr. John Kimble is a Research Soil Scientist at the USDA Natural Resources Conservation Service National Soil Survey Laboratory in Lincoln, Nebraska. Dr. Kimble manages the Global Change Project of the Natural Resources Conservation Service, and has worked more than 15 years with the U.S. Agency for International Development on projects dealing with soils-related problems in more than 40 developing countries. He is a member of the American Society of Agronomy, the Soil Science Society of America, the International Soil Science Society, and the International Humic Substances Society.

Dr. B.A. Stewart is Distinguished Professor of Agriculture, and Director of the Dryland Agriculture Institute at West Texas A&M University. Prior to joining West Texas A&M University in 1993, he was director of the USDA Conservation and Production Research Laboratory, Bushland, Texas. Dr. Stewart is a past president of the Soil Science Society of America, and was a member of the 1990–93 Committee on Long Range Soil and Water Policy, National Research Council, National Academy of Sciences. He is a Fellow of the Soil Science Society of America, American Society of Agronomy, Soil and Water Conservation Society, a recipient of the U.S. Department of Agriculture Superior Service Award, and a recipient of the Hugh Hammond Bennett Award of the Soil and Water Conservation Society.

Contributors

M.J. Apps
Canadian Forest Service
Northern Forestry Centre
5320 122nd Street
Edmonton, Alberta T6H 3S5, Canada

R.M. Bajaracharya
The Ohio State University
School of Natural Resources
2021 Coffey Road
Columbus, OH 43210
Present Address:
P.O. Box 3411
Kathmandu, Nepal

J.S. Bhatti
Canadian Forest Service
Northern Forestry Centre
5320 122nd Street
Edmonton, Alberta T6H 3S5, Canada

P.W. Birkeland
University of Wisconsin at Madison
Soils Department
1525 Observatory Drive
Madison, WI 53706

W.L. Bland
University of Wisconsin at Madison
Soils Department
1525 Observatory Drive
Madison, WI 53706

J.G. Bockheim
Soils Department
University of Wisconsin at Madison
1525 Observatory Drive
Madison, WI 53706

G. Broll
Institute of Landscape Ecology
University of Muenster
Robert Koch Strasse 26
D-48149 Muenster, Germany

I.B. Campbell
New Zealand Antarctic Research Program
23 View Mount
Nelson 7001, New Zealand

A.E. Cherkinsky
Geochron Laboratories
Krueger Enterprises
711 Concord Avenue
Cambridge, MA 02138

O.A. Chichagova
Laboratory of Soil Geography and Evolution
Institute of Geography
Starnometny 29
Moscow 109017, Russia

G.G.C. Claridge
New Zealand Antarctic Research Program
23 View Mount
Nelson 7001, New Zealand

X.Y. Dai
University of Alaska at Fairbanks
Palmer Research Center
533 E. Fireweed Avenue
Palmer, AK 99645

L. Ericson
Swedish University of Agricultural Sciences
Northern District for Agricultural Experiments
Box 4097S-904 03
Umeå, Sweden

K.R. Everett (deceased)
Ohio State University
School of Natural Resources
2021 Coffey Road
Columbus, OH 43210

L.R. Everett
Ohio State University
Byrd Polar Research Center
108 Scott Hall
1090 Carmack Road
Columbus, OH 43210-1002

Z.T. Gong
Institute of Soil Sciences
Academia Sinica
Nanjing 210008, China

S.V. Goryachkin
Laboratory of Soil Geography and Evolution
Institute of Geography
Starnometny 29
Moscow 109017, Russia

D.H. Halliwell
Canadian Forest Service
Northern Forestry Centre
5320 122nd Street
Edmonton, Alberta T6H 3S5, Canada

K.M. Hinkel
University of Cincinnati
Department of Geography
Cincinnati, OH 45221-0131

H. Jiang
University of Alberta
Department of Renewable Resources
Edmonton, Alberta T6G 2H1, Canada

J.M. Kimble
National Soil Survey Center
USDA-NRCS
Federal Building, Room 152
100 Centennial Mall North
Lincoln, NE 68508-3866

R. Lal
Ohio State University
School of Natural Resources
2021 Coffey Road
Columbus, OH 43210

G.B. Luo
Institute of Soil Science Academia Sinica
P.O. Box 821
Nanjing, People's Republic of China

L. Mattsson
Swedish University of Agricultural Sciences
Box 7014
S-750 07, Uppsala, Sweden

G.J. Michaelson
University of Alaska at Fairbanks
Palmer Research Centre
533 E. Firewood Avenue
Palmer, AK 99645

L.L. Miller
University of Cincinnati
Department of Geography
Cincinnati, OH 45221-0131

F.E. Nelson
University of Delaware
Department of Geography
Newark, DE 19717-5267

T.E. Osterkamp
University of Alaska at Fairbanks
Geophysical Institute
Palmer, AK 99645

R.F. Paetzold
National Soil Survey Center
USDA-NRCS
Federal Building, Room 152
100 Centennial Mall North
Lincoln, NE 68508-3866

C.H. Peng
Canadian Forest Service
Northern Forestry Centre
5320 122nd Street
Edmontonk, Alberta T6H 3S5, Canada

C.L. Ping
University of Alaska at Fairbanks
Palmer Research Center
533 E. Firewood Avenue
Palmer, AK 99645

V.E. Romanovsky
University of Alaska at Fairbanks
Geophysical Institute
Palmer, AK 99645

D.S. Sheppard
New Zealand Antarctic Research Program
23 View Mount
Nelson 7001, New Zealand

V. Stolbovoi
International Institute for Applied Systems
  Analysis
A-2361 Laxenburg, Austria

C. Tarnocai
Eastern Cereal and Oilseed Research Centre
Agriculture Canada
K.W. Neatby Building
960 Carling Avenue
Ottawa, Ontario K1A 0C6, Canada

G.L. Zhang
Institute of Soil Science Academia Sinica
P.O. Box 821
Nanjing, People's Republic of China

Table of Contents

---

* Deceased

Section I

Soil C Pools of Different Ecoregions

1 Soil C Pool and Dynamics in Cold Ecoregions

R. Lal and J.M. Kimble

CONTENTS

INTRODUCTION

World soils contain about 1550 Pg of soil organic carbon (SOC) to a depth of 1 m (Eswaran et al., 1995). Four principal biomes which contain this global SOC pool are Tropical, Temperate, Boreal Forest, and Tundra, of which Tundra and Boreal Forest constitute the cold ecoregions. Both Tundra and Boreal Forest biomes are affected by permafrost to some extent and contain cryoturbated soils. Soils of the cold ecoregions, including Tundra and Taiga, may contain as much as 23 to 48% (350 to 750 Pg of C in the active soil layer and permafrost) of the global SOC pool (Tarnocai and Smith, 1992; Eswaran et al., 1995; Whalen et al., 1996). While some databases exist for SOC pools and dynamics of the Tropical (Lal et al., 1999) and Temperate (Lal et al., 1998) biomes, research information about the SOC pool and dynamics of the Boreal and Tundra biomes is scanty (Tarnocai et al., 1993; Kimble and Ahrens, 1994). Further, the SOC pool and dynamic of the biomes in the cold ecoregions are likely to be influenced more by the accelerated greenhouse effect than those of the Temperate and Tropical biomes (Oberbauer et al., 1991; Rivkin, 1997). If the projected global warmings were to occur, Tundra and Boreal Forest biomes may experience the greatest increase in mean annual temperature of up to 2 to 10°C (Maxwell, 1992; Oechel et al., 1993). In that scenario, the fate of the SOC pool of these biomes is not known. A high proportion of the SOC is stored in the upper permafrost layer which is likely to be subjected to thawing upon global warming. Potential thawing of permafrost could liberate a considerable quantity of greenhouse gases (CO_2 and CH_4), alter local hydrology, and affect NPP and landscapes in the circumpolar permafrost zone. Anthropogenic changes may also alter these ecoregions from a net carbon dioxide sink to a source.

Tundra is defined as the treeless land beyond the climatic limit of the Boreal Forest and woodlands. The Arctic ecoregions (e.g., Tundras, polar deserts, and the Antarctic) are characterized by permafrost, with only a shallow thaw layer of 30 to 75 cm in depth (Plate 1.1). In addition, there are also Alpine Tundras and Sub-Antarctic regions where the active thaw layer may occur to greater depths. The Boreal Forest ecoregion refers to the great northern coniferous biome, bordered by the treeless lands of the Arctic Tundra at about 68°N and the deciduous forest biome in the

PLATE 1.1 A permafrost soil of Alaska.

south. The Tundra-Boreal ecotone in the north is generally more sharply defined than the Boreal Forest boundary in the south. In Canada it occurs near the 75 to 80 KJ cm^{-2} isoline. The Boreal forest is circumpolar in extent, occupying a belt as wide as 1000 km in certain regions of North America (Larsen, 1980).

The objective of this chapter is to describe general characteristics of the two biomes of the cold ecoregions, briefly outline predominant soil types and their SOC pool, and discuss the potential impact of anthropogenic and natural perturbations on soil C pool and fluxes.

CLIMATE AND VEGETATION OF THE TUNDRA

The Tundra biome constitutes three geographic/ecologic regions: Alpine, Antarctic, and Arctic. The Alpine Tundra comprises soils of high altitude beyond the tree line. Important regions comprising Alpine soils are the Himalayan-Tibetan ecoregion, Alps, Andeas, and other high-altitude mountainous terrain. The Antarctic region constitutes a very large land mass, and over 97% is ice covered at this time. Nevertheless, it represents an important component of the cold region soils.

Principal climates of the Tundra include the Arctic, Sub-Arctic, Sub-Antarctic, and Mountain and maritime Tundras (Figure 1.1). These climates are characterized by the warmest month average temperature of <10°C (Köppen, 1918). The winter temperatures may range from –20 to –30°C (Table 1.1a), with days with snow cover ranging from 0 to 300, and mean annual precipitation of 100 to 1500 mm/yr (Barry et al., 1981). The Arctic receives much less solar radiation than lower latitudes and experiences higher annual variation (Table 1.1b). The annual total solar radiation incident at the ground surface in North America decreases from about 4200 MJ m^{-2} at 50° N to 3100 MJ m^{-2} at 75° N (Barry et al., 1981), with corresponding values for summer ranging from 15 to 25 MJ m^{-2} day^{-1} (Table 1.1b). The radiation during winter months may range from 0 to 7 MJ m^{-2} day^{-1} (Table 1.1b). The Arctic is characterized by continuous darkness during several of the winter months and continuous daylight during summer. These extremes are altered by several

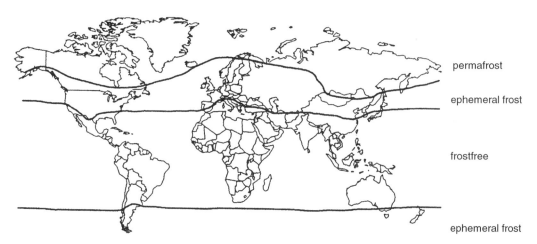

permafrost

ephemeral frost

frostfree

ephemeral frost

FIGURE 1.1 A map of the cold ecoregions.

site-specific factors. The albedo can be as high as 0.8 during October to May with a low of 0.2 during summer. Soil temperature is another important feature of Tundra climate and it governs the depth of the active layer above the permafrost. Soil temperature regime is a complex function of air temperature, SOC content, moisture content, or drainage and aspect.

The growing season of plants is climatically defined by the total number of degree-days above 5°C. The nature of principal vegetation can be defined by the annual growing degree-day totals. The boundaries between the low and mid-Arctic and between mid- and high-Arctic coincide roughly with 6 to 8°C and 5 to 6°C mean July isotherms, respectively (Maxwell, 1992). Principal vegetation of most Tundra are sedges, forbes, and grasses, and drastically stunted pine, birch, alders, and other species (Tieszen, 1978; Webber et al., 1980; Bliss, 1981;Bliss and Matveyeva, 1992; Reynolds and Tenhunen, 1996) (Plate 1.2). Bliss and Matveyeva (1992) estimated C reserves in the major Arctic vegetation types. In the low Arctic region, the C reserves are estimated at about 24 Pg, comprising 0.48 Pg in tall shrub, 2.32 Pg in low shrub, 3.41 Pg in tussock sedge-dwarf shrub, 17.13 Pg in mire, and 0.65 Pg in semi-desert. In the high-Arctic region, the C reserves are estimated at 2.33 Pg, comprising 1.36 Pg in mire, 1 Pg in semi-desert, and 0.02 Pg in polar desert. Thus the total C reserves in major Arctic vegetation types is about 26.33 Pg (Bliss and Matveyeva, 1992). Swanson (1994) described landscape (including soils and vegetation) characteristics of two sites in Upper Kobuk Valley, Alaska. The predominant vegetation of the flood plains comprised balsam poplar, white and black spruce, and alder shrubs. In contrast, soils of the older moraines include black spruce with moss and lichens. Vegetation of the Tundra and Boreal biomes is described in Table 1.2. The net primary production is low due to limitations of the climate (Wielgolaski et al., 1981). The topography ranges from flat, rolling, coastal plains and hills near sea level (Plate 1.3) to rugged mountain ridges rising up to 3500 m above sea level (Plate 1.4).

SOILS OF THE TUNDRA

Soils of the Tundra (continuous zone of permafrost) are predominately Gelisols (soil having permafrost and/or cryoturbation within the top meter of the soil profile). These soils generally have an active layer of 25 to 75 cm and the depth of the active layer may depend on the relief, the normalized difference vegetation index, and the interaction between them (McMichael et al., 1997). The active layer contains over 70% of the living biomass of the Tundra biome (Gersper et al., 1980). It is in this layer that roots grow, nutrients are recycled, invertebrates and other fauna burrow and mix the soil, and soil organic carbon changes. Many soils have histic epipedons (organic surface layers). These soils are less developed, have little horizon development, and are derived from a

TABLE 1.1a
Monthly Temperature and Precipitation of the Tundra Biome
Near Bettles, Alaska

Month	Mean temperature (°C)	Mean precipitation (mm)
January	−25	1.8
February	−23	1.6
March	−16	1.7
April	−6	1.6
May	6	1.6
June	14	3.7
July	15	4.9
August	12	5.8
September	5	4.4
October	−8	3.1
November	−19	2.3
December	−23	2.3
Annual	−6	34.9

TABLE 1.1b
Wind Speed, Solar Radiation and Day Length of Barrow, Alaska

Month	Wind speed (m s^{-1})	Solar radiation (MJ m^{-2} day $^{-1}$)	Day length (hr)
January	5.0	0	0.7
February	4.9	1.6	6.8
March	5.0	7.4	11.7
April	5.2	15.5	16.7
May	5.2	21.9	23.1
June	5.1	23.0	24.0
July	5.2	18.5	24.0
August	5.5	10.8	19.0
September	5.9	5.0	13.4
October	6.0	1.7	8.6
November	5.6	0.2	2.4
December	5.0	0	0

Adapted from Swanson, D.K., Proc. Meet. Classification, Correlation, Manage. Permafrost-Affected Soils, USDA-SCS, NSSC, Lincoln, NE, 1994.

wide range of bedrock types overlain by sediments. Principal types of sediments include marine, lacustrine, glacial, fluvial, and aeolian (Plate 1.5). Some soils are developed on loess, rubbly colluvial, and reworked (colluvial/alluvial) materials (Everett et al., 1981). The intensity of soil-forming processes (e.g., podzolization, calcification, alkalization, and SOC deposition) changes along the bioclimatic gradient (Tedrow, 1968). The genesis of permafrost-affected soils is mostly affected by freeze-thaw processes and faunal activity. Soil moisture regime (drainage) is not a major factor since most soils have high moisture content. This is a result of water perched above the permafrost in the active layer, even in areas receiving a very low input of precipitation and a very low rate of evapotranspiration. In most soils, morphological separations are neither parallel to the surface nor are they continuous horizontally. In addition to the parent material, soil temperature is

PLATE 1.2a Flat and rolling landscape near sea level; a typical meandering river in the coastal plain on the North Slope of Alaska.

PLATE 1.2b Flat and rolling landscape near sea level; Wright Valley in Antarctica area is completely underlain by permafrost.

TABLE 1.2
Vegetation and Evapotranspiration of Tundra-Boreal Biomes

Ecoregion	Evapotranspiration (mm)	Dominant vegetation
Tundra	30.5 to 31.7	Tundra
Boreal-Tundra ecotone	35.5 to 36.8	Tundra and lichen woodland
Open Boreal woodland	41.8 to 43.1	Lichen woodland
Boreal forest	47.0 to 48.2	Spruce forest, spruce-fir association
Boreal-mixed forest ecotone	50.6	Closed forest with white and red pine, yellow birch, etc.

Source: Adapted from Larsen, J.A., *The Boreal Ecosystem*, Academic Press, San Diego, 1980.

the predominant climatic factor that influences pedogenic processes affecting the rate and depth of weathering. Cold soil temperatures are the primary factor driving the cryogenic process. Mean monthly soil temperature during winter may range from -12 to $-20°C$ at 50 to 150 cm of depth (Tarnocai, 1994). The thinner snow cover on Tundra soil causes winter soil temperatures to be much colder than in forest soils. The number of freeze-thaw cycles in these soils may range from 20 to 40 per year.

The prevailing cold soil temperature regimes influence soil characteristics in two ways: the depth of active soil layer above the permafrost, and cryoturbation.

1. **Permafrost:** Permafrost is defined as a thermal condition in which soils and rocks remain at or below $0°C$ for at least two consecutive years. Consequently, water occurs as ice throughout the year in at least a portion of the soil within the control section. There are about 1.8 billion ha of soil characterized with permafrost, covering as much as 13% of earth's land area without including Antarctica (Plate 1.6), which is mostly concentrated in Alaska, Canada, and Russia (Table 1.3). Permafrost soils comprise 65% of the land area of Russia, 40% of the land area of Canada, and 85% of the world's high-altitude lands.

2. **Cryoturbation:** It refers to soil movement due to frost action, leading to formation of patterned ground or polygons (Plate 1.7a). The freeze-thaw process in soil results from the movement of two freezing fronts, one moving downward from the soil surface and the other moving upward from the permafrost table (Tarnocai, 1994). The upward-moving front, called freeze-back, occurs during the spring and fall. When these two freezing fronts move towards each other, unfrozen soil materials clamped between them generate cryostatic pressure. The latter is not uniform because freezing and moisture contents are highly variable. The cryoturbation is caused by squeezing of the unfrozen soil material, especially that with high moisture content, from areas of high pressure to those of low pressures. Because of the high moisture content, the unfrozen material moves readily. The unusually high SOC content of Arctic soils is partially due to cryoturbation that mixes organic matter into the subsoil. It is also due to simple thawing, that affects some of the buildings with improperly constructed foundations (Plate 1.7b).

These soils behave like entirely organic soils for at least two reasons. One, the mineral horizon underneath is part of the permafrost and is below the active thaw layer. Two, the organic horizon above the mineral soil is so thick that plant roots are no longer able to reach it (Everett, 1983). Another feature of soils with permafrost is the formation of Pingos (Plate 1.8).

PLATE 1.3a Rugged mountains and steep slopes of Tundra: Brooks Range in Alaska.

PLATE 1.3b Rugged mountains and steep slopes of Tundra: permafrost-affected soils at high elevations in the Brooks Range.

PLATE 1.4a General vegetation of Arctic Tundra near Prudhoe Bay, Alaska.

PLATE 1.4b General vegetation of black spruce in northern Alaska.

PLATE 1.4c General vegetation of cotton grass on the North Slope of Alaska.

PLATE 1.4d General vegetation at the interface between Tundra and black and white spruce forests on the Seward Peninsula of Alaska.

PLATE 1.4e General vegetation of spruce in Alaska at the interface of the tree line and Tundra.

PLATE 1.4f General vegetation of mixed forests in Alaska in an area of discontinuous permafrost.

PLATE 1.5a Some soils of Tundra: stratified loess deposit with permafrost at a depth of about 1 m.

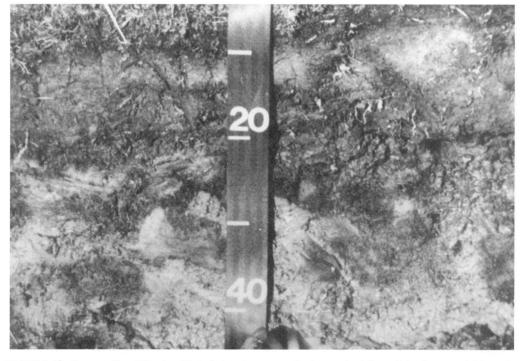

PLATE 1.5b Some soils of Tundra: Turbel showing cryoturbation and mixing of the Oa material into the C-horizon.

PLATE 1.5c Some soils of Tundra: Turbel on the North Slope of Alaska, Histic epipedon with cryoturbation and O material concentrated along the top of the permafrost at 50 to 60 cm.

PLATE 1.6a Permafrost-affected soils: Turbel with permafrost at 40 to 60 cm.

PLATE 1.6b Permafrost-affected soils: Turbel on the North Slope of Alaska with permafrost at a depth of 50 to 60 cm.

TABLE 1.3
Land Area Under Permafrost-Affected Soils

Region	Area (10^6 km²)	Relative area
Earth	18	13% of Earth's land surface
Russia	11	65% of Russia's land area
Canada	3.7	40% of Canada's land area
Quinghai-Tibet	—	85% of world's high-altitudinal permafrost

Source: Adapted from Bockheim and Tarnocai, 1998; Huijun et al., 1997.

PLATE 1.7a Effects of cryoturbation on patterned ground in far-eastern Russia near the Upper Kolyma River.

PLATE 1.7b Effects of cryoturbation on raised polygons with large frost cracks on the coastal plains of the North Slope of Alaska.

PLATE 1.7c Effects of cryoturbation on frost boils on the North Slope of Alaska.

PLATE 1.7d Cryoturbation effects of frost action on a building.

PLATE 1.8 Pingo formation in permafrost soils: the Pingo near Dead Horse, Alaska.

SOC CONTENT OF SOILS OF THE TUNDRA

Most soils of the Tundra biome have high SOC content. The decomposition rate is extremely low due to continuously low temperatures. Because the rate of addition of biomass greatly exceeds that of the decomposition, it leads to buildup of SOC content in these environments. More than 90% of the total C in the terrestrial Tundra ecosystem is below the ground surface, and one-third is in the upper 10 cm of soil where biological activity is concentrated (Gersper et al., 1980). The large amount of SOC determines the specific soil structure and fluxes of water, heat, and oxygen. The C and N, with a mean C:N ratio of 20:1, make up from 10 to 40% of total soil weight (Gersper et al., 1980). The SOC in the surface horizon is mostly fibric, and the degree of decomposition increases with depth. This material includes biomass whose generic characteristics can be recognized and which is interlaced by an abundance of living roots and rhizome. The top 10-cm layer may contain 600 to 700 g m^{-2} of below-ground plant parts. Soil bulk density of the layer may range from 0.05 to 1.5 Mg m^{-3} depending on the SOC content. The bulk density reflects mineral content, water-holding capacity, and depth of active layer. Oechel and Billings (1992) estimated the C reserves in biomass and dead organic matter at 61.1 Pg in Tundra and 283.4 Pg in the Boreal Forest biome.

Estimating the SOC pool of Tundra soils is difficult. No detailed determinations have been made of the SOC content of the highly cryoturbated soils, nor has the carbon content of the active layer been separated from that in the permafrost layer (Kimble et al., 1993). The SOC pool of several Tundra soils of Alaska ranges from 3 to 100 kg m^{-3}. The SOC pool is usually less (< 10 kg m^{-3}) in comparatively warm and nutrient-rich pergelic soils. The fraction of total SOC pool in the permafrost layer may range from 0 to 84%, depending on the location and site characteristics (Kimble et al., 1993). The data in Table 1.4 from a pedon in Mackenze Valley, Canada has an SOC pool of 71 kg m^{-3} of which 39% is located in the permafrost layer. The data in Table 1.5 of a pedon in the Yukon Peninsula show the SOC pool of 73 kg m^{-3}, of which 54% is located in the permafrost layer. The data of a deep profile from Siberia (Table 1.6) show that a

TABLE 1.4
The SOC Pool of a Pedon (Y-39) in Mackenze Valley, Canada
(Latitude 65°10' N, Longitude 127°27' W

Horizon	SOC (%)	Bulk density (Mg m⁻³)	SOC pool[a] (kg m⁻³)
Of	51.8	0.40	29.82
Bmy	3.5	1.22	3.87
Bmyzl	2.6	1.22	5.33
Ofz	51.8	0.40	1.06
Bmyz2	2.6	1.22	1.36
Ahzy	11.3	1.00	12.75
Cz	1.2	1.89	7.69
Wz	0	0.90	0
Total			70.88

[a] Adjusted for 1 m³ of soil; 61% SOC in the active layer and 39% in the permafrost layer.

Source: Adapted from Kimble, J.M., Tarnocai, C., Ping, C.L., Ahrens, R., Smith, C.A.S., Moore, J., and Lynn, W., *Proc. Joint Russ.-U.S. Semin. Cryopedol. Global Change*, Russian Academic of Sciences, Puschino, 1993.

TABLE 1.5
SOC Pool of Pedon Y66 (Latitude 68°55' N, Longitude 137°50' W)
on the Coast of Yukon, Canada

Horizon	SOC[a] (%)	Bulk density (Mg m⁻³)	SOC pool[a] (kg m⁻³)
L-H	46.9	0.74	11.08
Om	51.8	0.98	14.94
Bmgy1	1.7	1.90	3.11
Bmgy2	3.8	1.36	4.64
Bcgy1	3.8	1.22	2.69
Bcgy2	3.8	0.10	1.10
Bcgyz1	1.5	1.89	4.84
Bcgyz2	4.3	1.89	4.67
Cz1	3.2	1.89	13.89
Cz2	3.0	1.89	11.67
Total	—	—	72.63

[a] SOC pool is 46% in the active layer, and 54% in the permafrost layer.

Source: Adapted from Kimble, J.M., Tarnocai, C., Ping, C.L., Ahrens, R., Smith, C.A.S., Moore, J., and Lynn, W., *Proc. Russ-U.S. Joint Semin. Cryopedol. Global Change*, Russian Academy of Sciences, Pushchino, 1993.

permafrost layer can have a high SOC content (and low C/N ratio) to a depth of more than 10 m. The permafrost layers are indeed a major pool of soil C.

The data in Tables 1.7a to d show the SOC content of some soils of Tundra biome. The SOC content is as much as 50 to 80% for some soils of Tareyo, Russia; 20 to 25% for soils of Barrow, Alaska; and 10 to 20% for those of Prudhoe Bay, Alaska (Table 1.7a). Not all soils, however, have a very high SOC content. It is only 2 to 9% for a soil of Queen Elizabeth Island in Canada, and less than 1% for a soil in Ingelfield Land, Greenland (Table 1.7a). The data in Table 1.7b show the

TABLE 1.6
SOC Content of Some Horizons of a Permafrost Site
of the Alas Depression in Kolyma-Intigirka-Lowland,
Northeast Siberia

Depth (m)	SOC content (%)	C/N
Active Layer		
0.0 to 0.05	5.66	15
0.10 to 0.15	6.87	13
0.25 to 0.30	4.80	12
0.40 to 0.45	4.27	14
Permafrost Layer		
0.60 to 0.70	12.22	13
1.20 to 1.30	13.18	16
2.90 to 3.0	2.87	10
4.2 to 4.3	7.73	14
5.0 to 5.1	1.48	7
6.2 to 6.3	2.55	17
7.2 to 7.3	1.38	9
8.2 to 8.3	1.38	13
9.7 to 9.8	1.44	6
10.2 to 10.3	1.08	12

Source: Adapted from Pfeiffer, E.M. and Janssen, H., Proc. Meet. Classification, Correlation, Manage. Permafrost-Affected Soils, Alaska, U.S. and
Northwest Territories, Canada, USDA-SCS, NSSC, Lincoln, NE, 1994.

SOC content of three Sub-Antarctic soils that range from 38 to 42% for O_i and O_e horizons. It is only in the AC horizon of a Typic Cryaquept that the SOC content is low.

The data in Table 1.7c show SOC contents of some bog soils in Scandinavia and England. While the SOC content of the mineral horizons (A2 and B21) is low, between 0.4 and 0.8%, that of organic horizons is between 42 to 53%. In fact, the organic horizon can be several meters thick (Plate 1.9). Organic soils of cold regions constitute a major part of the global C pool.

Similar to the Arctic Tundra, soils of the Alpine Tundra also have high SOC content in the O horizon. The data in Table 1.7d show that the SOC content of O horizon of some soils of Mt. Patscherkofel, Austria is about 45%, while that of E and Bh horizons is 3 to 10%. The SOC content of the A horizon of Alpine Tundra soils of Niwot Ridge, Colorado is 7 to 11% for A1 horizons and 7% for B2 horizon (Table 1.7d).

Precise estimates of the global SOC pool of the soils of Tundra biome (comprising Arctic, Antarctic, and Alpine) are not known. Soils of Tundra biome probably constitute 14% (Oberbauer et al., 1991) to 16% (Eswaran et al., 1995) of the global soil C pool. The SOC pool to a 1-m depth comprises 393 Pg SOC and 17 Pg SIC pool (Table 1.8). Some argue that the SOC pool of Tundra soils has been grossly underestimated (Ping, 1997). Soils of Tundra biome have been a net sink of C in the past. An increase in mean annual temperature due to potential global warming may, however, render them a large source of C due to an increase in decomposition rate.

SOC CONTENT OF SOILS IN THE BOREAL FOREST ZONE

The predominant vegetation of the Boreal biome is coniferous forest (Plate 1.10). Soils of the Boreal zone vary with large areas of Gelisol in the zone of discontinuous permafrost mixed with

TABLE 1.7a
SOC Content of Some Selected Horizons of Soils of Subpolar Tundra

Soil	Horizon	SOC (%)
Queen Elizabeth Island, Canada		
Gleysolic static Cryosol	A	9
Pergelic Cryaquept	B	2
Upland Tundra	C	2
Tareva, Russia		
Gleysolic static Cryosol	Oa1 (At)	78
Histic Pergelic Cryaquept	Oi1 (At)	48
Bog Tundra, peaty raw humus gley	Bgh (Gh)	1.5
Ingelfield Land, Greenland		
Regosolic static Cryosol	A (cl)	0.9
Pergelic Cryoorthent	B (cl)	0.3
Arctic desert	C	0.1
Barrow, Alaska		
Gleysolic, turbic, Cryosol	Oa	24.4
Upland Tundra, polygon rim	Oe1	22.3
Arctic Tundra, weakly gleyed humus	Oe2	24.6
Prudhoe Bay, Alaska		
Brunisolic turbic Cryosol	A	9.6
Pergelic Cryaquoll	O2	19.9
Tundra, peaty gley	IIC	2.1

Source: Adapted from Everett, K.R., Vassiljevskaya, V.D., Brown, J., and Walker, B.D., *Tundra Ecosystems: A Comparative Analysis*, Bliss, L.C., Heal, O.W., and Moore, J.J., Eds., Cambridge University Press, Cambridge, U.K., 1981.

TABLE 1.7b
SOC Content of Some Horizons of Sub-Antarctic Soils

Soil	Horizon	SOC (%)
Livingston, Island		
Fibric organo Cryosol	Oi1	42
Lithic Cryofibrist	Oi2	38
Bog soil	Oi3	39
Acid peat soil	Oi4	39
MacQuarie Island		
Typic Cryaquept	AC	1
Typic Cryosaprist	Oe	39
South Georgia		
Typic Cryumbrept	Oe	8
Typic Cryaquept	Oi/C	38
Typic Cryofibrist	Oe	38

Source: Adapted from Everett, K.R., Vassiljevskaya, V.D., Brown, J., and Walker, B.D., *Tundra Ecosystems: A Comparative Analysis*, Bliss, L.C., Heal, O.W., and Moore, J.J., Eds., Cambridge University Press, Cambridge, U.K., 1981.

TABLE 1.7c
SOC Content of Some Horizons of Bog Soils

Soil	Horizon	SOC (%)
Kevo, Finland		
Humo-ferric podzol	A_1	44
Entic Cryorthod	A_2	0.4
Podzol	B_{21}	0.5
Stordalen, Sweden		
Cryic fibrisol	Oi1(T)	45
Bog soil	Oi2(T)	45
High moor (peat)	Oi3(T)	44
Hardangervidda, Norway		
Typic Cryohemist	Oe1(T)	44
Bog soil	Oe2(T)	42
Low moor	Oe3(T)	53
Moorehouse, England		
Lithic brunisol	(A1)[A]	9.8
Lithic haploboroll	(A12)[A]	0.8

Source: Adapted from Everett, K.R., Vassiljevskaya, V.D., Brown, J., and Walker, B.D., *Tundra Ecosystems: A Comparative Analysis*, Bliss, L.C., Heal, O.W., and Moore, J.J., Eds., Cambridge University Press, Cambridge, U.K., 1981.

TABLE 1.7d
SOC Content of Some Horizons of Alpine Tundra Soils

Soil	Horizon	SOC (%)
Mt. Patscherkofel, Austria		
Humic ferric podzol	O1(L)	45
Humic Cryothent	Oe(O_m)	44
Terro-humic podzol	E(Ae)	3
	Bh	10
Niwot Ridge, Colorado		
Brunisolic static Cryosol	A1	11
Tundra illuvial-humus	B2	7
Meadow Tundra	A1	7

Source: Adapted from Everett, K.R., Vassiljevskaya, V.D., Brown, J., and Walker, B.D., *Tundra Ecosystems: A Comparative Analysis*, Bliss, L.C., Heal, O.W., and Moore, J.J., Eds., Cambridge University Press, Cambridge, U.K., 1981.

Histosols, Spodosols, and Inceptisols. In the more southern zones of the Boreal biome, the soils are predominately Histosols, Spodosols, and Inceptisols. Soils of the northern Boreal biome have permafrost thaw during the summer to a depth of at least 50 cm. In the parts of the biome with no permafrost, all of the past season's frost is gone in the summer. During summer, soils of the Boreal Forest tend to be slightly cooler at the 50-cm depth than soils of the Tundra because of the thick organic layer covering the forested soil. In winter, however, soil temperature at depths of 50 to 150 cm may range from –1 to –2°C (Tarnocai, 1994).

PLATE 1.9 Typic Hemistel found on the North Slope of Alaska near Dead Horse.

TABLE 1.8
Carbon Stocks in Tundra and Boreal Soils

Ecoregion	SOC	SIC	Total	% of global SOC stock
		Pg		
Tundra	393	17	410	16.4
Boreal	382	258	640	25.6
World total	1550	947	2497	100.0

Source: Adapted from Eswaran, H.E. et al., *Soils and Global Change*, Lal, R. et al., Eds., CRC Press LLC, Boca Raton, FL.

The SOC content of soils of the Boreal Forest biomes is also high, but lower than that of the Tundra soils. The SOC density may range from 3 to 100 kg m^{-2} with a mean of about 20 kg m^{-2} (Kimble et al., 1993). The total C pool of soils of this biome exceeds that of Tundra and represents 25.6% of the global C pool (Eswaran et al., 1995). The SOC pool to a 1-m depth comprises 382 Pg of SOC and 258 Pg of SIC (Table 1.8). Analysis conducted in forest frozen soils in the Barkal region of Russia showed that SOC content is related to high clay content, high CEC, and high concentration of exchangeable cations (Kelnov, 1997). The accumulation of SOC is attributed to the higher rate of addition of biomass and lower rate of its decomposition. Similar to the pool of the Tundra biome, soils of this biome have also been a net sink in the past. Potential global warming may, however, render them a large source of C due to an increase in the decomposition rate. Road construction and similar activities can lead to thermokrast formation that often leads to increased erosion and oxidation of the exposed soil material. Physics and chemistry of seasonally frozen soils are well known (Gundelwein and Pfeiffer, 1997; Iskander et al., 1997; Sharratt et al., 1997; Polubesova and Shirshova, 1997).

PLATE 1.10 Gold mining in far eastern Russia by pumping water to thaw the permafrost.

C FLUX DUE TO NATURAL AND ANTHROPOGENIC PERTURBATIONS

Soils of Tundra and cold ecoregions are currently a net sink for atmospheric CO_2. The rate of C accumulation generally exceeds that of the decomposition. Vourlitis and Oechel (1997) reported that coastal Tundra landscapes were net sources of CO_2 at the rate of 0.02 to 0.05 mol m^{-2} day^{-1} during the early season snowmelt period. During the midseason, from mid July to mid August, these landscapes were net sink at 0.09 mol m^{-2} day^{-1}. Over both seasons, Tundra was currently net sink at the rate of 1.7 mol m^{-2}. At present Tundra soils also show a high potential for CH_4 oxidation that might result in a negative feedback to atmospheric CH_4 increases (Whalen et al., 1996). Anthropogenic changes, however, may change this to a net source (Oechel et al., 1993). Important anthropogenic perturbations that may enhance C flux from soil to the atmosphere include soil disturbance due to exploration and mining for minerals and oils (Plate 1.11), and the attendant land use change. Crude oil and diesel spills can also create drastic ecological disturbance leading to C efflux from soil. An increase in acid precipitation may also lead to an increase in the gaseous flux. Warmer soils and an unchanged water table may cause an increase in CH_4 emissions.

The Boreal Forest is undergoing a rapid change in land use due to deforestation and biomass burning with attendant disturbance to the soil. The rate of deforestation is especially high in Russian Siberia. In contrast to anthropogenic perturbations, there are also natural or wild fires in the biome which cause drastic ecological perturbations. Experiments conducted by O'Neill et al. (1997) have shown that loss of the vegetative cover by wild fires leads to: (i) increased soil temperature, (ii) thawing of the permafrost layer, (iii) change (usually a decrease) of soil moisture, (iv) an increase in microbial activity, and (v) high efflux of CO_2 and CH_4 from soil to the atmosphere. This chain of events leads to depletion of the SOC pool making the affected soils of the Boreal biome a net source of CO_2. Vedrova (1997) studied the impact of soil disturbance on the air temperature and mineralization rate of SOC in Siberia. He observed that when undisturbed, soils under the 80-year-old pine plantation were a net sink of carbon. However, disturbance increased the atmospheric

PLATE 1.11a Oil field in the North Slope of Alaska: general view of the oil field.

PLATE 1.11b Oil field in the North Slope of Alaska: wellhead in the foreground, and other wells.

temperature by 1.2 to 3.4°C in the summer. The alteration in temperature increased the mineral-ization rate of the humus layer in 0 to 10-cm depth. The net result was the efflux of C at the rate of 20 Mg ha^{-1} yr^{-1}. Verdova observed that the impact of a change in air temperature on the mineralization rate under an 80-year-old plantation in middle Siberia was more on soils with high rather than low SOC content. The destruction of litter and grass cover caused the loss of 18% of humus supply in 11 years. Similar to the Boreal Forest, disturbance of Alpine grasslands can also lead to emission of radiatively active gases from soils to the atmosphere. The potential response of Arctic carbon budgets to climate change will depend strongly on changes in water and nutrient availability and on time scales. All other factors remaining the same, the combination of rising temperatures and rising CO_2 would increase photosynthesis, growth, and net primary productivity (NPP), thereby increasing the C sequestration in Tundra ecosystem (Oechel and Billings, 1992). Increase in soil mineralization rate would also enhance the NPP. In contrast, warmer soils could deepen the active layer, lower the watertable and lead to thermokrast erosion and eventual loss of permafrost over much of the Arctic and Boreal Forest (Oechel and Billings, 1992). This trend would accelerate decomposition rates and increase C loss to the atmosphere. If the temperature rises and the precipitation remains the same (which is unlikely), vegetation zones may move northward over a period of several hundred years. The extent of change may depend on soil types, landscape, and precipitation, which introduce considerable uncertainties. Because of numerous uncertainties, it is difficult to say at present whether increases in NPP will offset high decompo-sition rates under a possible climate change. These uncertainties can be minimized through long-term monitoring of pools and dynamics of C reserves in vegetation and soil, and changes in soil properties and hydrology.

CONCLUSIONS

Four principal biomes comprising global pools include Tundra, Boreal, Temperate, and Tropical. Soils of the Tundra biome, treeless lands beyond the climatic limit of Boreal Forest or woodlands, contain an estimated 393 Pg of SOC (25.4% of the world total) and 17 Pg of SIC (1.8% of the world total). The total C pool in soils of the Tundra biome is estimated to be 410 Pg or 16.4% of the world soil C pool. The Boreal Forest ecoregion comprises the northern coniferous biome bordered by the treeless lands of the Arctic Tundra. It is circumpolar in extent occupying a belt as wide as 1000 km in certain regions of North America. The Forest-Tundra ecotone is more sharply defined than the southern limit of the Boreal Forest. Soils of the Boreal biome contain 382 Pg of SOC (24.7% of the world total) and 258 Pg of SIC (27.2% of the world total). The total C pool in soils of the Boreal biome is estimated at 640 Pg or 25.6% of the world soil C pool. Together, soils of Tundra and Boreal biomes comprise 772 Pg of SOC (50.1% of the world total), 275 Pg of SIC (29.0% of the world total), and 1050 Pg of total C (42.0% of the world total). Numerous anthro-pogenic activities (e.g., agriculture, biomass burning, mining, urbanization) and the potential green-house effect may have drastic effects on the C pool in soils of these biomes. Histosols and Cryosols, two principal soils of these biomes, have been major C sinks in the historic past. However, anthropogenic activities and the potential greenhouse effect are disturbing the C cycle, and may render these soils a major source of atmospheric CO_2, CH_4, and other radiatively active gases. Research information is needed to obtain reliable estimates of SOC and SIC pools of these biomes, and on the impact of anthropogenic activities and the potential greenhouse effect on soil C dynamics.

REFERENCES

Barry, R.G., Courtin, G.M., and Labine, C., Tundra climates, in *Tundra Ecosystems: A Comparative Analysis*, Bliss, L.C., Heal, O.W., and More, J.J., Eds., Cambridge University Press, Cambridge, U.K., 1981, pp. 81-114.

Bliss, L.C., The evolution of tundra ecosystem, in *Tundra Ecosystems: A Comparative Analysis,* Bliss, L.C., Heal, O.W., and More, J.J., Eds., Cambridge University Press, Cambridge, U.K., 1981, pp. 1-46.

Bliss, L.C. and Matveyeva, N.V., Circumpolar Arctic vegetation, in *Arctic Ecosystems in a Changing Climate: An Ecophysiological Perspective,* Chapin, F.S., III, Jefferies, R.L., Reynolds, J.F., Shaver, G.R., and Svoboda, J., Eds., Academic Press, San Diego, 1992.

Bockheim, J.G. and Tarnocai, C., Recognition of cryoturbation for classifying permafrost affected soils, *Geoderma,* 81, 281, 1998.

Brown, J., Everett, K.R., Webber, P.J., Maclean, S.F., Jr., and Murray, D.F., The coast tundra at Barrow, in *An Arctic Ecosystem: The Coast Tundra at Barrow Alaska,* Brown, J., Miller, P.C., Tieszen, L.L., and Bunnell, F.L., Eds., Dowden, Hutchinson and Ross, Stroudsburg, PA, 1980.

Bunnell, F.L., MacLean, S.F., Jr., and Brown, J., Structure and functions of tundra ecosystems. Papers presented at the IBP Tundra-Biome Vth Int. Meet. Biological Productivity of Tundra, Bisko, Sweden, April 1974. Ecological Bull. 20, Swedish Natural Science Research Council, Stockholm.

Eswaran, H., Vanden Berg, E., Reich, P., and Kimble, J., Global soil carbon resources, in *Soils and Global Change,* Lal, R., Kimble, J., Levine, E., and Stewart, B.A., Eds., CRC Press, Boca Raton, FL, 1995, pp. 27-43.

Everett, K.R., Histosols, in *Pedogenesis and Soil Taxonomy. II. Soil Orders,* Developments in Soil Science, Wilding, L.P., Smeck, N.E., and Hall, G.F., Eds., Elsevier, Amsterdam, 1983.

Everett, K.R., Vassiljevskaya, V.D., Brown, J., and Walker, B.D., Tundra and analogous soils, in *Tundra Ecosystems: A Comparative Analysis,* Bliss, L.C., Heal, O.W., and Moore, J.J., Eds., Cambridge University Press, Cambridge, U.K., 1981, pp. 139-186.

Gersper, P.L., Alexander, V., Barkely, S.A., Barsdate, R.J., and Flint, P.S., The soils and their nutrients, in *An Arctic Ecosystem: The Coast Tundra at Barrow Alaska,* Brown, J., Miller, P.C., Tieszen, L.L., and Bunnell, F.L., Eds., Dowden, Hutchinson and Ross, Stroudsburg, PA, 1980.

Gundelwein, A. and Pfeiffer, E.-M., *Int. Symp. Phys. Chem. Ecol. Seasonally Frozen Soils,* Iskander, I.K., Wright, E.A., Radke, J.K., Sharatt, B.S., Groenevelt, P.H., and Hinzman, L.D., Eds., Fairbanks, AK, June 10-12, 1997.

Huijun, J., Guodong, C., and Qing, L., Early winter CO_2 and CH_4 emissions from Alpine grassland soils at Qingshuihe, Quinghai-Tibet Plateau, in *Int. Symp. Phys. Chem. Ecol. Seasonally Frozen Soils,* Iskander, I.K., Wright, E.A., Radke, J.K., Sharatt, B.S., Groenevelt, P.H., and Hinzman, L.D., Eds., Fairbanks, AK, June 10-12, 1997.

Iskander, I.K., Wright, E.A., Radke, J.K., Sharatt, B.S., Groenevelt, P.H., and Hinzman, L.D., Eds., *Int. Symp. Phys. Chem. Ecol. Seasonally Frozen Soils,* Fairbanks, AK, June 10-12, 1997.

Kelnov, B.M., Humus formation in frozen forest soils in the Barkal region, in *Int. Symp. Phys. Chem. Ecol. Seasonally Frozen Soils,* Iskander, I.K., Wright, E.A., Radke, J.K., Sharatt, B.S., Groenevelt, P.H., and Hinzman, L.D., Eds., Fairbanks, AK, June 10-12, 1997.

Kimble, J.M., Tarnocai, C., Ping, C.L., Ahrens, R., Smith, C.A.S., Moore, J., and Lynn, W., Determination of the amount of carbon in highly cryoturbated soils, in *Proc. Joint Russ.-Am. Semin. Cryopedol. and Global Change,* Russian Acadamy of Sciences, Pushchino, Russia, 1993.

Kimble, J.M. and Ahrens, R.J., Eds., Proc. Meet. Classification, Correlation, Manage. of Permafrost-Affected Soils, Alaska U.S. and Yukon and Northwest Territories, Canada, USDA-SCS, NSSC, Lincoln NE, 1994.

Köppen, W., Klassifikation der Klimate nach Temperatur, Niederschlag und Jahreslauf, *Petermanns Geograph. Mitt.,* 64, 193, 1918.

Lal, R., Kimble, J.M., Follett, R.F., and Stewart, B.A., Eds., *Soil Processes and the Global C Cycle,* CRC Press, Boca Raton, FL, 1998.

Lal, R., Kimble, J.M., Eswaran, H., and Stewart, B.A., Eds., *Global Climate Change and Tropical Ecosystems,* CRC Press, Boca Raton, FL, 1999.

Larsen, J.A., *The Boreal Ecosystem,* Academic Press, San Diego, 1980.

Maxwell, B., Arctic climate: potential for change under global warming, in *Arctic Ecosystems in a Changing Climate: An Ecophysiological Perspective,* Chapin, F.S. III, Jefferies, R.L., Reynolds, J.F., Shaver, G.R., and Svoboda, J., Eds., Academic Press, San Diego, 1992.

McMichael, C.E., Hope, A.S., Stow, D.A., and Fleming, J.B., The relation between active layer depth and a spectral vegetation index in Arctic Tundra landscape of the North Slope of Alaska, *Int. J. Remote Sensing,* 18, 2371, 1997.

Oberbauer, S.F., Tenhuen, J.D., and Reynolds, J.F., Environmental effects on CO_2 efflux from water track and tussock Tundra in Arctic Alaska, U.S., *Arctic Alpine Res.,* 23,162, 1991.

Oechel, W.C. and Billings, W.D., Effects of global change on the carbon balance of Arctic plants and ecosystems, in *Arctic Ecosystems in a Changing Climate: An Ecophysiological Perspective,* Chapin, F.S., III, Jefferies, R.L., Reynolds, J.F., Shaver, G.R., and Svoboda, J., Eds., Academic Press, San Diego, 1992, p. 139-168.

Oechel, W.C., Hastings, S.J., Vourlitis, G., Jenkins, M., Riechers, G., and Grulke, N., Recent change of Arctic tundra ecosystems from a net carbon dioxide sink to a source, *Nature,* 361, 520,1993.

O'Neill, K.P., Kasischke, E.S., Richter, D.D., and Krasovic, V., Effects of fire on temperature, moisture and CO_2 emissions from soils near Tok, AK: an initial assessment, in *Int. Symp. Phys. Chem. Ecol. Seasonally Frozen Soils,* Iskander, I.K., Wright, E.A., Radke, J.K., Sharatt, B.S., Groenevelt, P.H., and Hinzman, L.D., Eds., Fairbanks, AK, June 10-12, 1997.

Pfeiffer, E.M. and Janssen, H., Characterization of organic carbon using the $^{\delta 13}C$ value of a permafrost site in the Kolyme-Indigirka-Lowland, Northeast Siberia, in Proc. Meet. Classification, Correlation, Manage of Permafrost-Affected Soils, Alaska U.S. and Yukon and Northwest Territories, Canada, USDA-SCS, NSSC, Lincoln, NE, 1994.

Ping, C.-L., Characteristics of permafrost soils along a latitudinal transact in Arctic Alaska, *AgroBorealis,* 29, 35,1997.

Polubesova, T. and Shirshova, L., Exchangeable cations and composition of organic matter in soils as affected by acidification and freezing, in *Int. Symp. Phys. Chem. Ecol. Seasonally Frozen Soils,* Iskander, I.K., Wright, E.A., Radke, J.K., Sharatt, B.S., Groenevelt, P.H., and Hinzman, L.D., Eds., Fairbanks, AK, June 10-12, 1997.

Reynolds, J.F. and Tenhunen, J.D., Eds., *Landscape Function and Disturbance in Arctic Tundra,* Springer-Verlag, New York, 1996.

Rivkin, F.M., Additional methane emissions from thawing Cryosols, Yamal Peninsula, Russia, in *Int. Symp. Phys., Chem., Ecol. Seasonally Frozen Soils,* Iskander, I.K., Wright, E.A., Radke, J.K., Sharrat, B.S., Groenevelt, P.H., and Hinzman, L.D., Eds., Fairbanks, AK, June 10-12, 1997.

Sharratt, B.S., Radke, J.K., Hinzman, L.D., Iskander, I.K., and Groenevelt, P.H., Physics, chemistry and ecology of frozen soils in managed ecosystems: an introduction, *Int. Symp. Phys., Chem. Ecol. Seasonally Frozen Soils,* Iskander, I.K., Wright, E.A., Radke, J.K., Sharrat, B.S., Groenevelt, P.H., and Hinzman, L.D., Eds., Fairbanks, AK, June 10-12, 1997.

Swanson, D.K., Soils and landscape of the upper Kobuk Valley, Alaska, in Proc. Meet. Classification Correlation Manage. of Permafrost-Affected Soils, Alaska U.S. and Yukon and Northwest Territories, Canada, Kimble, J.M. and Ahrens, R.J., Eds., USDA-SCS, NSSC, Lincoln, NE, 1994, pp. 134-142.

Tarnocai, C., Smith, C.A.S., and Fox, C.A., International tour of permafrost-affected soils: the Yukon and Northwest Territories of Canada, Center for Land and Biological Research, Agriculture Canada, 1993.

Tarnocai, C.S., Genesis of permafrost-affected soils, in Proc. Meet. Classification, Correlation, Manage. of Permafrost-Affected Soils, Alaska U.S. and Yukon and Northwest Territories, Canada, Kimble, J.M. and Ahrens, R.J., Eds., USDA-SCS, NSSC, Lincoln, NE, 1994, pp. 143-154.

Tarnocai, C.S. and Smith, C.A.S., The formation and properties of soils in the permafrost regions of Canada, First Int. Conf. Cryopedal. Workshop Proc., Puschino, Russia, 1992, pp. 21-42.

Tedrow, J.C.F., Pedogenic gradients of the polar regions, *Arctic,* 11, 166, 1968.

Tieszen, L.L., Ed., *Vegetation and Production Ecology of an Alaskan Arctic Tundra,* Springer-Verlag, New York, 1978.

Vedrova, E.F., Humus formation in frozen forest soils, in *Tundra Ecosystems: A Comparative Analysis,* Bliss, L.C., Heal, O.W., and Moore, J.J., Eds., Cambridge University Press, Cambridge, U.K., 1997, pp. 403-407.

Vourlitis, G.L. and Oechel, W.C., Landscape-scale CO_2, H_2O vapour energy flux of moist-wet coastal tundra ecosystems over two growing seasons, *J. Ecol.,* 87, 575, 1997.

Webber, P.J., Miller, P.C., Chapin, F.S., III, and McCown, B.H., The vegetation: pattern and succession, in *An Arctic Ecosystem: The Coast Tundra at Barrow Alaska,* Brown, J., Miller, P.C., Tieszen, L.L., and Bunnell, F.L., Eds., Dowden, Hutchinson & Ross, Stroudsburg, PA, 1980, pp. 186-217.

Whalen, S.C., Reeburg, W.S., and Reimers, C.E., Control of Tundra methane emission by microbial oxidation, in *Landscape Function and Disturbance in Arctic Tundra,* Reynolds, J.F. and Tenhunen, J.D., Eds., Springer-Verlag, Berlin, 1996, pp. 257-275.

Wielgolaski, F.E., Bliss, L.C., Svoboda, J., and Doyle, G., Primary production of tundra, in *Tundra Ecosystems: A Comparative Analysis,* Bliss, L.C., Heal, O.W., and Moore, J.J., Eds., Cambridge University Press, Cambridge, U.K., 1981, pp. 187-225.

2 Bioavailability of Organic Matter in Tundra Soils

X.Y. Dai, C.L. Ping, and G.J. Michaelson

CONTENTS

INTRODUCTION

Northern ecosystems may play a key role in future climate change because of their potential feedback to the global climate (Chapin and Matthews, 1993). Tundra ecosystems alone contain approximately $191*10^{15}$ g of soil C which represent 12% of the global soil C pool (Billings, 1987). Recent measurement of CO_2 flux in Arctic Alaska indicates that tussock and wet sedge tundra are now sources of atmospheric CO_2. If the current efflux of CO_2 from the Arctic tundra continues, these ecosystems could represent a strong positive feedback to global atmospheric CO_2 concentration and concomitant climate change (Oechel et al., 1993).

The quantity and quality of soil organic matter could play an important role in the terrestrial C cycle which would have a direct influence on global climate change. Soil organic matter (SOM) includes a multitude of different organic components ranging from plant residues and microbial biomass to highly stable humus (Cheng and Molina, 1995). Since not all SOM participates equally in the transformation processes, it is necessary to identify the specific soil organic carbon fractions (bioreactive SOC) which are actively involved in the carbon cycling processes when studying the influence of climate change on the dynamics of SOM transformations in soils.

Soil organic matter is a complex series of related but different molecules with varying physical and chemical structures and different functional groups. SOC can be associated with itself, clays, and microorganisms, and it is within and between aggregates (Paul et al., 1995). Therefore, there is not a readily available method for identifying and characterizing the bioreactive organic carbon fractions in soils. Of the many that have been tried, including chemical fractionation, physical fractionation, and modeling approaches, radioisotope fractionation approaches are most commonly used to identify and characterize SOM pools (Cheng and Molina, 1995).

The classical chemical fractionation scheme uses NaOH as an extractant, followed by acidification of the extract to separate fulvic acid, humic acid, and humins (Stevenson, 1965). Such schemes are not designed to separate or differentiate the bioreactive SOC from other SOC (Cheng and Molina, 1995). Physical fractionation reduces the chemical alteration of organic materials, but SOM may be redistributed among fractions during the disruption of the soil (Elliott and Cambardella, 1991). A combined chemical and resin scheme using dilute base to extract soil organic matter (Malcolm et al., 1995), followed by a tandem XAD-8 and XAD-4 resin technique to isolate the dissolved organic matter (Malcolm and MacCarthy, 1992), was conducted. The scheme was further modified in which soil organic matter was isolated into six fractions: humic acid (HA), fulvic acid (FA), low molecular weight acid (LMA), hydrophobic neutral (HON), hydrophilic neutral (HIN), and low molecular weight neutral (LMN). The objectives of this study were to test the role of the active fractions in CO_2 production in the arctic ecosystems and to study the bioavailability of SOM fractions in Arctic tundra soils.

MATERIALS AND METHODS

MATERIALS

Soil samples were taken from four soils including three Gelisols from Alaska (Figure 2.1) and one Mollisol from Nebraska. The description of sampling sites and soil classification are presented in Table 2.1 and selected properties of these soil samples are presented in Table 2.2. All soil samples were placed in a cooler in the field and then frozen before shipping. The soil samples were thawed and visible roots were removed. Then they were homogenized by hand before extraction and fractionation. Soil subsamples were homogenized and air-dried before total C and N analysis.

METHODOLOGY

Extraction with 0.1 *M* NaOH and Fractionation by Tandem XAD Resins

All soil samples were extracted with 0.1 *M* NaOH under N_2 at a solution-to-soil ratio of about 15:1. The extractants were separated by centrifugation and pressure-filtered under N_2 through a 0.45-µm polysulfanone membrane filter. The extraction was repeated four times (Ping et al., 1995). Each filtrate was acidified with 6 *M* HCl to pH 1 and left overnight in an ice bath to allow precipitation of humic acids. The humic acids were then separated by centrifugation and filtration. The supernatant solutions were passed through 10-l columns of XAD8/XAD4 resins, following the technique of Thurman and Malcolm (1981). The nonextractable fraction which was not dissolved in NaOH was rinsed with DI water to remove salts and kept frozen.

Incubation of Whole Soil and Nonextractable Fractions (NEF)

A 5 g (dry weight) fresh sample of soil or NEF was placed in a 250 ml jar, the NEF samples were inoculated with a 2 ml suspension of their original fresh soils (Catroux and Schnitzer, 1987), and the pH was adjusted to 7 with phosphate buffer. A slurry was made by addition of 25 ml DI water and samples were incubated in the dark on orbital shakers rotating at 60 rpm at 4 and 25°C for 4 months. CO_2 was determined by Perkin-Elmer 8500 gas chromatography once a week in the first month and twice a month in the last three months.

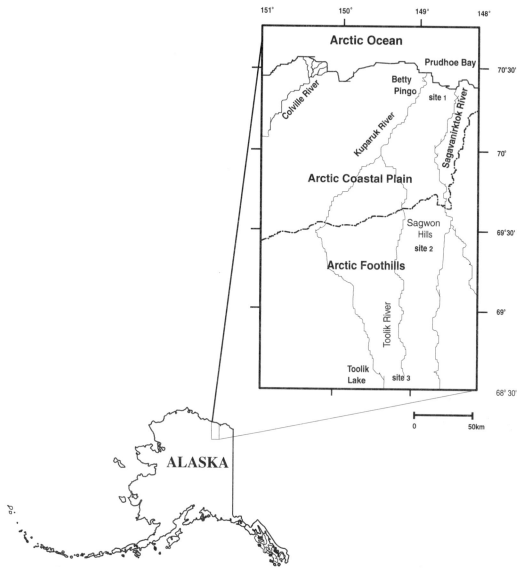

FIGURE 2.1 Sites in Alaska where soil samples were obtained.

Incubation of Extractable Fractions (EF)

A 55 ml solution of each extractable fraction isolated from the Prudhoe Bay Oa1 horizon, containing 30 mg C, was placed in a wide-mouth 250 ml septum-incubation jar, inoculated with a 2 ml suspension of the original fresh soil (same as above), and the pH was adjusted to 7 with phosphate buffer. The jars were fitted with sponge stoppers and placed on an incubator shaker rotating at 60 rpm in the dark at 25°C for 12 months. CO_2 was measured by a Perkin-Elmer 8500 gas chromatography once a week in the first month, twice a month in the next five months, and once a month in the last six months.

TABLE 2.1

Location, Classification, and Site Characteristic of Soils Used in the Study

Site ID	Area	Land cover class[a] Soil Classification[b]	Landform microrelief	Active layer (cm)
Site 1	Prudhoe Bay, AK	Moist nonacidic tundra	Coastal Plain,	35
		Euic Sapric Glacistels	Flat-polygon center	
Site 2	Sagwon Hills, AK	Moist acidic tundra	Hills	40
		Fine-loamy, mixed Ruptic-Histic Aquiturbel	Tussock tundra	
Site 3	Toolik Lake, AK	Moist acidic tundra	Hills	35
		Loamy, mixed Ruptic-Histic Aquiturbel	Tussocks, mid-slope	
Site 4	Nebraska	Great Plains	Grassland	—
		Fine silty, mesic Calcic Hapludoll	Plain	

[a] Adapted from Auerbach and Walker, 1995; [b] Adapted from Soil Survey Staff, 1994.

TABLE 2.2

Some Properties of Whole Soils and Nonextractable Fractions

Site ID	Horizon	Depth (cm)	pH 1:1	Ext-C %[b]	TOC %	TN %	C/N	HA	FA	HON	LMA	LMN	HIN
										% of total extractable C			
Site 1	Oa1	10–22	5.7	30	14.2	0.9	15.4	45.1	18.9	2.2	11.1	4.4	18.3
Prudhoe Bay	Oa1-non		5.9[a]	24	9.42	0.6	16.7	44.2	17.4	2.0	12.4	4.5	19.5
	Oaf	22–50	6.6	47	19.9	1.3	15.8	41.1	9.2	12.1	9.9	3.1	24.6
	Oaf-non		6.9[a]	31	15.6	0.97	16.2	35.5	5.0	21.2	12.6	3.4	22.3
Site 2	Oe	8–16	5.2	32	36.0	1.4	25.0	37.2	15.1	8.6	8.8	7.0	23.3
Sagwon Hills	Oe-non		5.3[a]	24	30.3	0.95	31.9	32.9	6.1	10.8	9.0	14.5	26.7
	O/A	37–50	5.8		11.9	0.75	15.8						
	O/A-non		5.9[a]		7.4	0.46	16.3						
Site 3	Cf	30–100	5.4		6.95	0.39	18.1						
Toolik Lake	Cf-non		5.4[a]		4.4	0.25	17.7						
Site 4	Ap3	10–26	6.1		1.4	0.15	9.6						
Nebraska	Ap3-non		6.0[a]		1.1	0.12	9.2						

TOC — total organic, C; TN — total nitrogen; Oa1 — non residue of 0.1 *M* NaOH; HA — humic acid; FA — fulvic acid; HON — hydrophobic neutral; LMA — low molecular weight acid; LMN — low molecular weight neutral; HIN — hydrophilic neutral.

[a] Phosphate adjusted pH.
[b] 0.1 M NaOH extractable SOC of the total SOC.

Carbon, Nitrogen, and Hydrogen Analysis

Total carbon and nitrogen were determined on a LECO 1000 CHN analyzer. Inorganic carbon was determined from the carbon dioxide evolution after treating the samples with 1 N HCl (CO_2 was determined on a Perkin-Elmer 8500 gas chromatograph).

RESULTS AND DISCUSSION

CO_2 EVOLUTION FROM WHOLE SOIL AND THE NONEXTRACTABLE FRACTION

Nonextractable Fraction and the Whole Soil

Nonacidic tundra and acidic tundra samples had higher CO_2 evolution in the nonextractable fractions than in whole soils at both 4 and 25°C. The Mollisol Ap3 showed a different trend; the whole soil

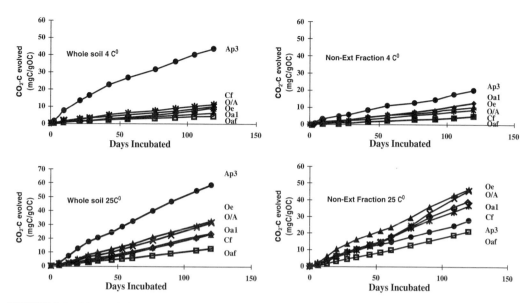

FIGURE 2.2 CO_2 evolution of whole soils and nonextractable fractions incubated at 4 and 25°C (Non-Ext Fraction: residue not disolved in 0.1 M NaOH).

generated more CO_2 than the nonextractable fraction (Figure 2.2). In the tundra soils, the nonextractable fractions contain about 50 to 80% hemicellulose and cellulose (Ping et al., 1998), which are readily decomposed by a large number of microorganisms and are used as both a source of energy and carbon for the build-up of the microbial cell substance (Waksman and Starkey, 1951). The extractable organic carbon of the tundra soils is 24 to 47% of the total soil organic carbon (Table 2.2). After extraction with alkali solution, hemicellulose and cellulose remain and dominate in the nonextractable fraction. In the Mollisol, extractable carbon was 24% of the total SOC. It is likely that the NEF of Mollisol contains little hemicellulose and cellulose compared with that of tundra soils but more humin, which is very resistant to degradation by microbes.

Whole Soils

In the four-month incubations, the Mollisol Ap3 had 43.4 ± 0.9 and 58.8 ± 8.3 mg C/g OC evolved as CO_2 at 4°C and 25°C, respectively, which was the highest CO_2 evolution at both temperatures. Nonacidic tundra Oaf had the lowest CO_2 evolution of 4.3 ± 0.4 and 12.8 ± 0.9 mg C/g OC at 4°C and 25°C, respectively (Table 2.3, Figure 2.2). The low CO_2 evolution of Oaf might be due to the higher degree of humification and recalcitrant nature of the soil organic matter. The radiocarbon dates from similar samples (Ping et al., 1997) showed that Oa1 was 700 ± 175 years B.P. and "modern" for both the whole soil and NEF, Oaf was 4580 ± 295 and 5075 ± 585 years B.P. for the whole soil and NEF, respectively. The organic matter from Oaf was much older than that of Oa1, hence, age may be an important difference influencing the activity of the SOM from the two horizons.

Nonextractable Fractions

At 4°C incubation, the nonextractable fraction of Mollisol Ap3 had the highest CO_2 evolution of 15.4 ± 1.3 mg C/g OC and nonacidic tundra Oaf had the lowest CO_2 evolution of 3.6 ± 0.6 mg C/g OC, which showed the same trends as the whole soils. Whereas, at 25°C, acidic tundra Oe and O/A horizons had the highest CO_2 evolution, which were 31.5 ± 1.6 and 31.8 ± 6.6 mg C/g OC, the Mollisol Ap3 was the second lowest; it evolved 23 ± 4.4 mg C/g OC which may be due to the

TABLE 2.3
Total CO$_2$ Evolution of Whole Soil, Nonextractable Fractions, and
Extractable Fractions, and Incubated at 4 and 25°C for 120 Days

Site	Sample ID	CO$_2$-C mg C/g OC 4°C	CO$_2$-C mg C/g OC[b] 25°C	CO$_2$-C/OC (%)[c] 4°C	CO$_2$-C/OC (%) 25°C
Prudhoe Bay	Whole soil Oa1	6.3 ± 0.6	23.3 ± 2.3	0.6	2.3
	Non-Ext[a] Oa1	8.1 ± 0.5	30.5 ± 1.1	0.8	3.1
	Whole soil Oaf	4.3 ± 0.4	12.8 ± 0.4	0.4	1.3
	Non-Ext Oaf	3.6 ± 0.6	16.1 ± 10.6	0.4	1.6
Sawgon Hills	Whole soil Oe	10.1 ± 1.9	32.8 ± 6.9	1.0	3.3
	Non-Ext Oe	11.8 ± 1.5	31.0 ± 5.9	1.2	3.1
	Whole soil O/A	10.3 ± 2.8	31.5 ± 1.6	1.0	3.2
	Non-Ext Cf	6.4 ± 1.6	31.8 ± 6.6	0.6	3.2
Toolik Lake	Whole soil Cf	11.0 ± 3.8	23.3 ± 1.3	1.1	2.3
	Non-Ext Cf	5.4 ± 2.4	27.8 ± 3.1	0.5	2.8
Nebraska	Whole soil Ap3	43.4 ± 0.9	58.8 ± 8.3	4.3	5.9
	Non-Ext Ap3	15.4 ± 1.3	23.0 ± 4.4	1.5	2.3

[a] Residue of 0.1 *M* NaOH.
[b] mg C of the total g of organic C.
[c] CO$_2$-C percentage of the total organic C.

explanation of the cellulose and hemicellulose contents discussed above. The Oaf was the lowest at 16.1 ± 10.6 mg C/g OC (Table 2.3, Figure 2.2), which is consistent with the explanations of the radiocarbon age mentioned above.

Temperature Effects on CO$_2$ Evolution

Increasing the temperature had a positive effect on the CO$_2$ evolution of both the whole soil and NEF of all the samples. Among tundra soils, the acidic tundra was influenced by temperature more than the nonacidic tundra, and all tundra soils were affected more than the Mollisol Ap3. The Mollisol Ap3 nonextractable fraction was not influenced by temperature (Figure 2.3), indicating its recalcitrant nature.

CO$_2$ Evolution from Extractable Organic Fractions

At 25°C, HIN had the highest CO$_2$ evolution at 289 ± 18 mg C/g EFC. LMN and HON had the second highest at 211 ± 18 and 185 ± 13 mg C/g EFC, respectively. LMA was the third highest at about 84 mg C/g OC CO$_2$ evolution, and HA and FA were the lowest at 23 ± 4 and 34 ± 3 mg C/g OC, respectively (Figure 2.4, Table 2.4). The distribution of the HA, FA, HON, LMA, LMN, and HIN in the extractable C is 45, 19, 2, 11, 4, and 18%, respectively. The contribution of CO$_2$ from HA and FA accounted for 7% of the total CO$_2$-C evolved from the extractable OC, LMA accounted for 10%, and neutrals (LMN, HON, and HIN) accounted for 77%. Hydrophilic neutrals accounted for 18% of the total extractable C, and contributed 35% of the CO$_2$-C evolved from the extractable C (Table 2.4). Therefore, the neutrals, especially HIN, are the major bioreactive C pool in Tundra soils. The results are consistent with those of Michaelson et al. (1998) and Qualls and Haines (1992). The hydrophilic neutral fraction contains free carbohydrates which are not bound to humic substances (Leenheer, 1981), and has been generally regarded as one of the most biodegradable components of SOM. Qualls and Haines (1992) found that there was a positive correlation between the percentage of DOC lost during the incubation period and the initial content of hydrophilic

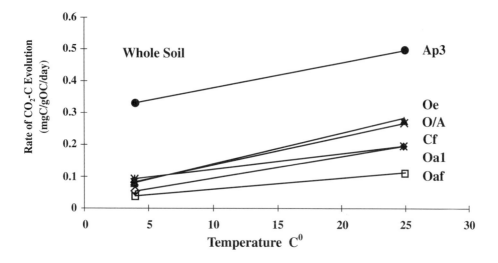

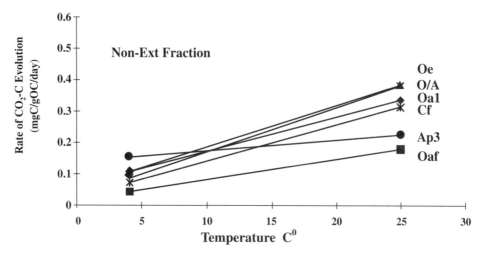

FIGURE 2.3 The effect of temperature on the CO_2 evolution rate (mg CO_2-C/g OC/day).

neutral substances (r = 0.83). In this study, the positive correlation between the CO_2 evolution and the initial content of HIN of soils is also found at both temperatures (r_{4C} = .81, r_{25C} = .91).

CONCLUSIONS

In tundra soils, the nonextractable fraction had higher CO_2 evolution than the whole soils, whereas the Mollisol was the opposite. The nonextractable fraction of tundra soils comprises 60 to 80% of the total organic C. It responded to temperature more than the Mollisol. This might be due to the fact that the tundra soils are less humified than the Mollisol; they contain more hemicellulose and cellulose in the nonextractable fraction than the Mollisol. Hemicellulose and cellulose are relatively easily decomposed by microorganisms. They are likely the bioreactive fractions in the Tundra soils in addition to the readily decomposed simple acids and sugars in the hydrophilic neutrals (Ping et al., 1998). With climate warming, the tundra soils may have a greater potential to contribute to greenhouse gas emissions and could become the net source of greenhouse gases. In the extractable fractions of Prudhoe Bay Oa1 horizon, the neutrals (LMN, HIN, and HON) are the most bioreactive fractions and represent 25% of the total extractable C. However, they account for 77% of the CO_2

TABLE 2.4
Distribution and Total CO$_2$ Evolution of Extractable Organic
Fractions Incubated at 25°C for 111 days (Prudhoe Bay)

Fraction ID	CO$_2$-C mg/g EFC[a]	CO$_2$-C mg/g OC[b]	C/Ec[c] %	CO$_2$-C/EC[d] %
Oa1-HA	23 ± 4	4.5	45.1	3
Oa1-FA	34 ± 3	0.9	18.9	4
Oa1-LMA	84 ± 7	1.7	11.1	10
Oa1-LMN	211 ± 18	4.3	4.4	26
Oa1-HIN	289 ± 18	11.8	18.3	35
Oa1-HON	185 ± 13	7.5	2.2	22
Oaf-HA	25 ± 2	2.9	44.2	3
Oaf-FA	34 ± 1	1.5	17.4	4
Oaf-LMA	100 ± 9	0.9	12.4	11

[a] 0.1 *M* NaOH extractable organic carbon in each fraction.
[b] Total soil organic carbon.
[c] Extractable organic C in each fraction of the total extractable C.
[d] CO$_2$-C evolved from each fraction of the total extractable C.

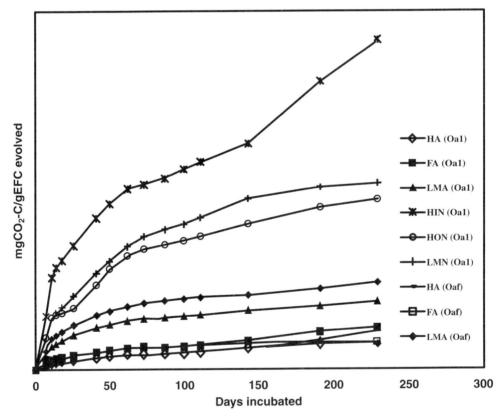

FIGURE 2.4 CO$_2$ evolution (mg CO$_2$-C/g EFC) from different extractable fractions of Prudhoe Bay Oa1 and Oaf horizons (HA: humic acid, FA: fulvic acid, LMA: low molecular weight acid, HIN: hydrophilic neutral, HON: hydrophobic neutral, LMN: low molecular weight neutral).

evolution of the total extractable fraction. Humic and fulvic acids are the most stable fractions. They account for 64% of the total extractable C, but contributed 7% of the CO_2 evolution of the total extractable fraction. LMA is in between, with 11% of the total extractable C and producing 10% of the CO_2. Hence, in the extractable fraction, neutrals are the most important contributors to the labile pool. This agrees with the previous postulation based on chemical composition.

REFERENCES

Catroux, G. and Schnitzer, M., Chemical, spectroscopic, and biological characteristics of the organic matter in particle size fractions separated from an aquoll, *Soil Sci. Soc. Am. J.,* 51, 1200, 1987.

Billings, W.D., Carbon balance of Alaska tundra and taiga ecosystems: past, present, and future, *Q. Sci. Rev.,* 6, 165, 1987.

Chapin, F.S., III and Matthews, E., Boreal carbon pools: approaches and constraints in global extrapolation, *Carbon Cycling in Boreal Forest and Sub-Arctic Ecosystems: Biospheric Responses and Feedbacks to Global Climate Change,* Vinson, T.S. and Kolchugina, T., Eds., Oregon State University, Corvallis, 1993, pp. 9-20.

Cheng, H.H. and Molina, J.A.E., In search of the bioreactive soil organic carbon: The fractionation approaches, in *Advances in Soil Science — Soil and Global Changes,* Lal, R., Kimble, J., Levine, E., and Stewart, B.A., Eds., CRC Press, Boca Raton, FL, 1995, p. 343.

Elliott, E.T. and Cambardella, C.A., Physical separation of soil organic matter, *Agric. Ecosyst. Environ.,* 34, 407, 1991.

Leenheer, J.A., Comprehensive approach to prepare isolation and fractionation of dissolved organic carbon from natural waters and waste waters, *Environ. Sci. Technol.,* 15, 578, 1981.

Malcolm, R.L., Kennedy, K., Ping, C.L., and Michaelson, G.J., Fractionation, characterization, and comparison of bulk soil organic substances and water-soluble soil interstitial organic constituents in selected cryosols of Alaska, in *Soil Processes and the Carbon Cycle,* Lal, R., Kimble, J., Levine, E., and Stewart, B.A., Eds., CRC Press, Boca Raton, FL, 1995, p. 343.

Malcolm, R.L. and MacCarthy, P., Quantitative evaluation of XAD-8 and XAD-4 resins used in tandem for removing organic solutes from water, *Environ. Int.,* 18, 597, 1992.

Michaelson, G.J., Ping, C.L., Kling, G.W., and Hobbie, J.E., The character and bioactivity of dissolved organic matter during thaw and spring runoff in waters of the arctic tundra north slope, Alaska, *J. Geophys. Res.,* 103(D-22), 28939, 1998.

Oechel, W.C., Hastings, S.J., Bourlitis, G., Jenkins, M., Riechers, G., and Grulke, N., Recent change of Arctic tundra ecosystems from a net carbon dioxide sink to a source, *Nature,* 361, 520, 1993.

Paul, E.A., Horwath, W.R., Harris, D., Follett, R., Leavitt, S.W., Kimball, B.A., and Pregitzer, K., Establishing the pool sizes and fluxes in CO_2 emissions from soil organic matter turover, in *Advances in Soil Science — Soil and Global Change,* Lal, R., Kimble, J., Levine, E., and Stewart, B.A., Eds., CRC Press, Boca Raton, FL, 1995, p. 297.

Ping, C.L., Michaelson, G.J., Loya, W.M., Chandler, R.J., and Malcolm, R.L., Characteristics of soil organic matter in arctic ecosystems of Alaska, in *Soil Processes and the Carbon Cycle,* Lal, R., Kimble, J., Follett, R. F., and Stewart, B.A., Eds., CRC Press, Boca Raton, FL, 1998, p. 157.

Ping, C.L., Michaelson, G.J., and Malcolm, R.L., Fractionation and carbon balance of soil organic matter in selected Cryic soils in Alaska, in *Advances in Soil Science — Soil and Global Change,* Lal, R., Kimble, J., Levine, E., and Stewart, B.A., Eds., Lewis Publishers, Boca Raton, FL, 1995, p. 307.

Ping, C.L., Michaelson, G.J., Cherkinsky, A., and Malcolm, R.L., Characterization of soil organic matter by stable isotopes and radiocarbon ages of selected soils in Arctic Alaska, in *The Role of Humic Substances in the Ecosystems and in Environmental Protection,* Drozd, J., Gonet, S.S., Senesi, N., and Weber, J., Eds., PTSH, Wroclaw, Poland, 1997, p. 475.

Qualls, R.G. and Haines, B.L., Biodegradability of dissolved organic matter in forest throughfall, soil solution and stream water, *Soil Sci. Soc. Am. J.,* 56, 578, 1992.

Stevenson, F.J., Gross chemical fractionation of organic matter, *Methods of Soil Analysis — Part 2,* Black, C.A., Evans, D.D., White, J.L., Ensminger, L.E., and Clark, F.E., Eds., American Society of Agronomy, Madison, WI, 1965, p. 1409.

Thurman, E.M. and Malcolm, R.L., Preparative isolation of aquatic humic substances, *Environ. Sci. Technol.,* 15, 463, 1981.
Waksman, S.A. and Starkey, R.L., Role of microbes in the decomposition of organic substances in the soil, in *The Soil and the Microbe,* John Wiley & Sons, New York, 1951, p. 75.

3   Carbon Pools in Tundra Soils of Russia: Improving Data Reliability

V. Stolbovoi

CONTENTS

INTRODUCTION

Soil organic carbon (SOC) has been the subject of examination for more than 200 years (Orlov et al., 1996). During this period, research has been carried out to observe the organic carbon content in various soils, to analyze its distribution within soil profiles, and to determine its chemical composition, mechanisms, and biochemical processes driving humus formation, etc. The results have shown that SOC greatly affects physical, chemical, and biological soil parameters, and controls soil biological activity, fertility, and productivity. The significance of SOC for many practical applications has stimulated intensive investigations, which has resulted in much scientific literature. In turn, the achievements gained in this branch of knowledge led to the attitude that this type of research would no longer be necessary. This opinion resulted in the loss of scientific interest in a field that was, until only a few decades ago, very strong (Orlov et al., 1996).

Policymakers recognized the problem, which resulted in intensified SOC investigations at the end of the 1980s. This initiative was associated with the growing social awareness of global climate change, and the recognition that soils are of particular importance in the atmospheric CO_2 budget. Containing a large reserve of carbon, which exceeds by several times the atmospheric CO_2–C pool, soils can significantly affect atmospheric CO_2 concentrations. Approximately 10% of the atmosphere's CO_2 passes through terrestrial soils each year (Raich and Potter, 1995). In addition, the expected climate warming might strongly influence soil carbon pools (SCP) and consequently increase the rate of soil CO_2 efflux, as this depends significantly on temperature. A scenario was pictured where climate warming would unbalance the processes of plant photosynthesis, respiration, and soil organic accumulation and mineralization, which in turn would increase the greenhouse gas

(CO_2, CH_4, N_2O) fluxes to the atmosphere. The necessity for assessing the magnitude of these effects, as well as understanding the global carbon balance, formulated an urgent need for new estimates of SOC reserves.

Numerous attempts have been made to calculate SCP at the global and regional scales. This research has led to a considerable improvement in knowledge about soil organic matter; however, it has also had negative implications. Many drastically different figures were produced in the SCP estimates at various scales. This resulted in much confusion, particularly because no single figure value for calculation of the global carbon balance could be found. Greenland (1995) concluded that all estimates must be accepted with considerable reservation. Although this observation is correct, it does not make the situation easier. Indeed, the disparity of the carbon pool estimates for North America and Russia varies by about 200% (Apps et al., 1993). The latest global estimates of SOC range from 1220 (Sombroek et al., 1993) to 1555 Pg C (Eswaran et al., 1995), or have a variation of about 22%. The estimates of the SCP for Russia vary from 296 (Orlov et al., 1996) to 342 Pg C (Rozhkov et al., 1996), or a difference of about 14%. This uncertainty does not come anywhere near the commitments proposed in the Kyoto negotiations (Berg, 1998) where the developed countries will have a 5.2% reduction of carbon emissions (compared to the 1990 level) and a carbon market will be introduced. Clearly, there is a need to develop methods for making carbon sink and flux calculations more reliable.

One of the biggest gaps in the current SCP estimates is that they are based on the SOC of virgin or natural soils. The fact has been overlooked that nearly all the agriculturally suitable soils in the world have been cultivated and the actual organic content in these soils is very different from that of natural soils. Numerous publications show that due to cultivation and improper technology, the humus content in Chernozems of Russia has been reduced by nearly twice (Orlov et al., 1996). This leads to the recognition that the carbon balance of terrestrial ecosystems is markedly dependent upon the direct impact of human activities. Therefore, in order to make the SCP assessments more reliable, land-use effects also have to be incorporated into estimates.

The overall goal of our study is to illustrate major data discrepancies in estimations for the tundra soils in Russia and indicate ways for improvement of the data reliability. More specifically, our study is devoted to the following issues:

- To define soil extent.
- To delineate and characterize tundra soils.
- To view soil degradation as a correction factor of the SCP estimate for the Russian tundra.

MATERIALS AND METHODS

Most of this study presents results from the analysis of the database developed for the territory of Russia, which includes a set of digital maps handled by the geographical information system (GIS) Arc/Info. This database has been widely implemented in various analyses within the Sustainable Boreal Forest Resources project carried out by the International Institute for Applied Systems Analysis (IIASA), Austria. GIS tools have been taken to create all original figures presented in the research. Russian soil terminology was taken from the legend of the soil map of Russia at the scale of 1:2,500,000 (Fridland, 1988). FAO soil names were used in accordance with the revised legend (FAO, 1988).

SOIL EXTENT DEFINITION

SOIL CONTINUITY AND ITS EFFECT ON ESTIMATE DISCREPANCIES

The model used for the SCP estimate can be restricted to a numeric integration by a rectangular method based on a few variables: soil unit area, humus content for the referenced soil profile for

TABLE 3.1
Tundra Zone Area of Russia According to Different Authors

Author	Source map (scale)	Area (million ha)
Karelin et al. (1994)	Landscape map of the U.S.S.R., 1988 (1:4,000,000)	203.7
Kolchugina et al. (1995)	Geographical belts and zonal types of landscape of the world, 1988 (1:15,000,000)	234.1
Orlov et al. (1996)	Natural-agricultural regionalization of the U.S.S.R. 1975 (1:8,000,000)	346.9
Stolbovoi (current study)	Vegetation map of the U.S.S.R., 1990 (1:4,000,000)	252.8

the unit, soil bulk density, and coarse fragments content. In general, the approach is so simple that sometimes it seems strange that the results of the calculations are so drastically different. There are a number of publications (Bouwman, 1990; Batjes, 1995, 1996; Greenland, 1995) that consider various sources of errors of the SCP estimates, i.e., due to determination of parameters, differences in analytical methods, etc. It is believed that all these factors affect the increase of uncertainties. However, the discrepancies of the results are much larger than should be expected. The laboratory errors are generally systematic and thus assumed to be attributed to all calculations. This should lead to the leveling of the total number of mistakes independently, shifting it toward overestimation or underestimation. Thus, other explanations for the discrepancies have to be determined. The major source for the disparity of the estimates has to be attributed to the soil, which can be investigated in terms of its nature and general methodology. This has been applied for basic entry definitions, especially for SCP calculations.

Basic soil science assumes that soil belongs to a continuum body or pedosphere. Applying various concepts, pedologists artificially divide the pedosphere into soil horizons, profiles, taxonomic classes, corresponding spatial units, and patterns. For example, at the highest level of the soil classification hierarchy, and according to different theories, 28 major soil groups have been established in the FAO revised legend (FAO, 1988), 19 soil divisions (Shishov, 1989), and 11 soil orders (USDA, 1994). A similar approach has been applied to the laboratory methods. For example, the amount of defined matter (exchange cations, humus content, amount of soluble salts, etc.) is not an absolute value but a relative one and varies depending on the extraction, its concentration, temperature, etc. This pedological postulate leads to the general statement that the results of SCP estimates based on different concepts will always be drastically incompatible if they are not based on consensus on a common methodology.

In reality, the situation is much more diverse and confusing. Obviously, researchers are not very critical in the selection of source maps, which have been used for providing soil areas essential for SCP calculations. For example, and as shown in Table 3.1, they apply to soil area definition, e.g., the map of soil-vegetation types of ecoregions (Kolchugina et al., 1995), the map of natural-agricultural zones (Orlov et al., 1996), a landscape map (Karelin et al., 1994), and a vegetation map (Stolbovoi, current study). This simple listing identifies a great diversity in mapping materials. All these source maps differ thematically, conceptually, by scale, age, etc. It is important to note that these maps are intended to delineate the area of other objects, such as landscape, agroregions, or vegetation communities. These map sources do not indicate the geographical soil distribution, which means that soil spatial reality has been replaced in these SCP estimates by other objects. Indeed, the reliability of these calculations is very limited and the resulting figures vary considerably depending on the objects being investigated. However, the most alarming message that these studies pose is that each of them claims to be equally efficient for SCP calculations. Undoubtedly, a specialist could easily assess the value of the research results with a specific interpretation and limits for its extrapolation. However, it should be considered that these results have been addressed to a wide audience, including decision-policymakers. A lay person is not obliged or experienced

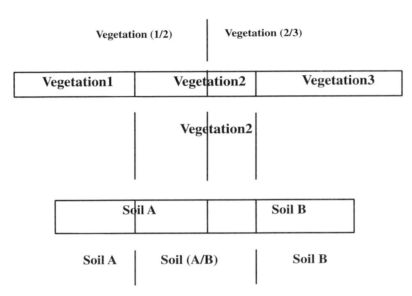

FIGURE 3.1 Spatial delineation of soil patterns related to three vegetation units (—) and vegetation patterns related to two soil units (— — —).

enough to filter these scientific findings and recommendations; therefore, we adopt this function. This generates the necessity to review the application of non-soil maps for SCP estimates.

APPLICABILITY OF NON-SOIL MAPS FOR SOIL CARBON POOL ESTIMATES

SCP estimates are firmly linked with soil classes and soil extent; therefore it is logical to apply soil maps for the calculations. However, soil maps are not always available for the estimates. As a result, researchers are often forced to replace soil maps with other cartographic material. This approach is based on the concept of close soil-landscape interrelations and most likely derives from the Russian pedological school, which is founded on Dokuchaev's factor-genetic idea of nature units.

More often, the investigation aims to have SCP related to other objects, such as administrative units or geographical elements, i.e., major soil regions of the world (Eswaran et al., 1995), the map of Holdridge life-zones (Batjes, 1995). This approach is popular in the global change community. The idea is mainly to link SCP with other natural features which, for instance, are believed to be more sensitive to environmental change. Figure 3.1 abstractly illustrates the logistics of this operation.

Two patterns are shown in Figure 3.1: (i) vegetation with spatial units 1, 2, and 3; and (ii) soil represented by the polygons A and B. The intention is to delineate soils according to vegetation patterns or, alternatively, vegetation, which is related to soil spatial patterns. Applying vegetation spatial units (vegetation units 1, 2, and 3) for the delineation of soils, the soil coverage (soils A, A/B, and B) is formed. Using soil spatial units (soils A and B) for the delineation of vegetation patterns, the resulting vegetation polygons would be 1/2 and 2/3. In both cases, there is no difficulty in estimating the soil or vegetation areas within the new polygons, where the internal spatial structure of coverages is known. In this case, soil extents within vegetation polygons or vegetation within soil patterns could easily be incorporated. Properly selected and statistically approved samples/profiles could characterize each of them.

However, let us assume that the internal structure of objects occurring in a new complex polygon is unknown. In this situation, the researcher should establish some value for soil or vegetation (in our case), which should be nominated as a representative for the polygon. This procedure has been applied for soil area definition by non-soil maps. In this approach, the area of the polygon replaces soil extents within the vegetation polygon. The soil value (sample/profile/carbon content, etc.) is established on the basis of subjective reasoning, i.e., expert opinion or published data.

Clearly, the result of SCP estimates will depend heavily upon the soil values which have been introduced. It is likely that there are very few possibilities to improve the calculation based on such an approach. Nevertheless, it seems that some room for improvement in the estimate may be found if SOC variability within a complex mapping unit can be defined. Thus, the problem of the implementation of soil variability arises when establishing soil carbon values for non-soil mapping units. SOC variability can be reasonably predicted if additional data on maximum and minimum carbon contents and their approximate extent can be applied. For this purpose some basic soil geography concepts have to be implemented. It could be suggested that SOC variability depends on the degree of qualitative differentiation of soil combinations or their contrast. According to Fridland (1972), the contrast reflects the closeness of genetic interrelations between components of the soil combination, which suggest that it is higher where it is more intense. Fridland (1972) notes that intensive surface redistribution of the precipitation due to runoff leads to the formation of the very contrasting dry soils in the interfluve and wet soils in depressions. Interpreting this soil combination from the perspective of SOC, one can assume that soil in the interfluve will have less organic content compared to that in the depression. Thus, relief heterogeneity could be used to predict soil contrasts. A basic topographic map can be applied for assessment of the areas of relief forms. Soil profiles have to be separated by relief forms, subsequently.

The contrast will also be high where the occurrence of soil combinations has been driven by a variety of different soil parent materials (texture, chemical characteristics, mineralogical composition, etc.) i.e., sandy and clay soils with a drastically different absorption capacity, calcareous and noncalcareous soils with different organic coagulation potentials, variation of ground water levels, and their qualities. These examples clearly show that application of the non-soil maps to SCP estimates requires the involvement of various additional sources of information, which are able to indicate regional soil heterogeneity. However, this task is very delicate and can be performed only by an experienced soil geographer. It is important to note that a methodological consensus for this type of analysis does not exist at present. Nevertheless, the SCP estimates based on this approach would be more realistic.

APPLICABILITY OF SOIL MAPS: CONCEPTUAL AND AGE CONSISTENCY

Very few maps have been compiled and designed to show the spatial distribution of the SOC. For instance, as illustrated in Dokuchaev's classical map of "isohumus bands" (Dokuchaev, 1948), the procedure for the soil organic mapping (including sampling size, sampling unit, sampling dispersion, contour delineation, standards on the map accuracy, scale specification, etc.) has not been fully developed. However, the very intention to compile such special soil organic maps indicates that this idea is not new. This leads to two questions: Is there anything rational in this intention? and, Why are general soil maps not sufficiently reliable for the delineation of spatial soil carbon patterns?

Unfortunately these very important aspects have not been seriously discussed in the scientific literature. This may be explained by the strong belief that soil maps suit the task. With respect to this opinion, it is interesting to consider the following example. Soil aggregations in Russia within a soil polygon at the small, middle, and observational scales are based on the principles of the genetic unity or the uniformity of the soil cover patterns. These generalizations do not take into account organic profile peculiarities. In both types of aggregation, the Gley-podzolic soil is combined with Gley-podzolic peat and muck soils that are genetically close and form single soil toposequences (soil catena). Eventually, these soils have considerably different soil organic profiles and thus the aggregated soil map loses its important detail for SCP estimates. This example illustrates the basic cartographic rule that any soil map generalization should always be thematically orientated. It can therefore be concluded that soil maps intended to be applied for SCP estimates have to be specially prepared.

Agrochemical surveys have developed their own methods of soil sampling for mapping of the nutrient elements. These dramatically different methods are motivated by the high prices of fertil-

TABLE 3.2
Soil Maps for Soil Carbon Pool Estimates for Russia

Geographic scope and year of compilation	Scale	Format	Availability and accessibility
Soil and terrain database of Russia (FAO revised legend) 1995	1:5,000,000	Digitized	Dokuchaev Soil Institute, Moscow, Russia. International Institute for Applied Systems Analysis, Laxenburg, Austria. FAO, Rome, Italy. Public domain.
Soil map of Russia Federative Social Republics, 1988	1:2,500,000	Paper, digitized in 1997	Dokuchaev Soil Institute, Moscow, Russia. International Institute for Applied Systems Analysis, Laxenburg, Austria Public domain.
State soil map, completed in 1995	1:1,000,000	Paper	Dokuchaev Soil Institute, Moscow, Russia. Public domain.
Soil maps of administrative regions, mostly updated at the end of the 1980s	1:300,000	Paper	Goskomzem, Moscow, Russia. Public domain.
Soil map of administrative districts	1:100,000 1:50,000	Paper	Regional Roskomzem departments located at the administrative cities. Internal use.
Soil maps of agricultural enterprises, updated every 5 years	1:10,000	Paper	Regional agricultural administration, and agricultural enterprises. Internal use.
Soil maps of the forest enterprises, updated every 15 years	Vary from 1:100,000 to 1:25,000	Paper	Regional forest administration, and forest enterprises. Internal use.

izers and their careful application due to the negative effects on the environment, which is also costly. Thus, the economy and market are believed to be driving forces in agrochemistry. A similar development for the soil carbon account can be easily predicted if the Kyoto agreement is implemented. In the near future special cartographic methods for SCP estimates have to be developed; however, at present, the only way to improve data reliability is to reach an agreement on the source soil maps which are used for such calculations.

Table 3.2 lists the soil cartographic data available in Russia that could be recommended for the implementation. This table shows that Russia has a high extent of mapped soil data allowing estimations of both inherent and global SCP. It is also important to note that soil maps at a local scale represent a valuable source for such calculations, because they combine contemporary and historical knowledge on the diversity of SOC from the whole country. Undoubtedly, most countries have problems integrating local data into a regional and global context. The development of a common methodology would solve this, otherwise the final results might be inconsistent and it would take much more effort to harmonize the estimates.

Selection of the source soil maps should take into account their conceptual and age consistency. Most of the soil maps at an observational scale are based on theoretical assumptions on the soil distribution. Before publication of the soil map at the scale of 1:2,500,000 in 1988, the basic source for Russia was the soil map published in 1964 (Rozov, 1964). At that time, the soil–geographic concept was based on knowledge obtained mainly from the European part of Russia, which has soil distributions strictly following the bioclimatic zonality. Two-thirds of the country area, including the territory of the European North, the West, and East Siberia and the Far East, were practically unknown from a pedological point of view. Thus, these territories were shown on the soil map in accordance with the author's hypotheses. That is, soil cartographers extrapolated knowledge on soil geographical distributions which were obtained for one region of Russia and applied it to other regions by creating abstract soil–geographical space. Recent soil cartographic knowledge indicates that such extrapolation was incorrect. Currently, soil geographers in Russia are recognizing the necessity of revising a general soil geographical concept of the country in order to maintain consistency with the actual geographic distribution of soils (Stolbovoi and Sheremet, 1997). This revision is not a criticism of the basic soil–geographic concepts produced by generations of soil scientists and geographers, which, however, should not be dogmatized. The revision intends to demonstrate that conceptual soil maps are matters of permanent revisions. They are strongly dependent upon the author's preferences and temporal (age) consensus. This means that application of soil maps for SCP estimates should be very delicate, and a concept analysis should be requested which is always subjective and informal. From this point of view, it seems that conceptual soil maps will be impossible to introduce when moving to the carbon market.

Another important feature of soil maps is that the soil condition is fixed at the time of the map compilation. Thus, they do not consider the soil organic temporal dynamics. In particular, the results of calculations may change considerably when soils have been converted from one land-use practice to another. There is evidence that organic matter will rapidly decline when previously virgin soil has been cropped, etc. Based on this assumption, the age consistency of the SOC data might be critical for many sources of information.

DELINEATION OF TUNDRA SOILS

As mentioned above, most of the SCP estimates have usually been related to other geographical elements (climate zones, biomes, living zones, ecosystems, etc.). Table 3.1 shows that the extent of tundra and polar zones in Russia will be different depending on the concepts of their delineation, the scale of the source map, its age, etc. In turn, the figures of the SCP calculations will be drastically different based on the fact that the area estimates of the tundra and polar deserts range from 200 to 350 million ha (Table 3.1). This disparity indicates that the procedure of delineating the basic geographical elements, which have been selected to create a spatial base for SCP estimates, has to be precisely specified.

However, this problem is not as simple as it seems. In the case of the tundra zone,* the question is how should it be delineated? Should one use relief, vegetation, climate, soils, or combinations of them? Similar questions could be asked for the delineation of all spatial patterns based on complex geographical concepts.

In order to delineate the tundra zone related to SCP estimates, the first issue to be addressed is soil classification. Could any soil class/classes be applied to identify this zone? It should be noted that there have been few attempts to distinguish the specific class of tundra soils in Russia (i.e., Ignatenko, 1979; Vasilevskay et al., 1986), assuming uniform soil-forming factors (cold climate, dwarf vegetation, etc.) and soil morphological features (permafrost, cryoturbations, redoximorphism, etc.). None of the external (soil-forming factors) or internal (soil characteristics) criteria fulfilled the task completely. Thus, currently there is no soil that is purely tundra specific that could be used as an indicator for the tundra zone. This therefore leads to the necessity of identifing the tundra zone through some other characteristic features.

Permafrost is one of the key criteria that could be used for distinguishing the tundra zone. It refers to the complex phenomena of rocks having negative (less than 0°C) temperature values (or soils if these temperature conditions are found within the upper 2-m layer), their formation, behavior in time and space, and lithogenetic effects, such as rock diagenesis. The permafrost zone covers more than 60% (more than 1,000 million ha) of Russian territory (Figure 3.2). The southern boundary of the permafrost zone goes through the northeastern corner of the European part of the country, crosses north of the Ural mountains, continues over the West Siberian plain (at about the

---

* The term tundra originated in Finland and refers to unforested, open highland.

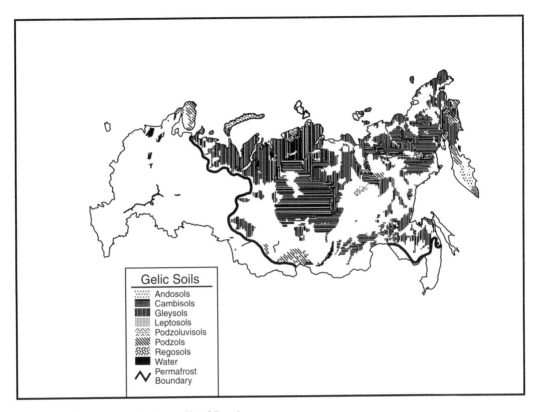

FIGURE 3.2 Permafrost-affected soils of Russia.

62nd parallel), turns south at the 90th meridian, and crosses the country's border at the 96th meridian. Most of East Siberia and the Far East have permafrost. In some parts of Kamchatka, Sakhalin, and southern Siberia the permafrost is distributed in the form of islands.

Table 3.3 represents a correlation of the Gelic soil units (permafrost-affected soils) of the FAO Soil Map of the World (1988), Soil Taxonomy (ST) subgroups, and the soil classes from the soil map of Russia (Fridland, 1988). It is important to note that 20 soil classes of the permafrost-affected soils from the soil map of Russia have been aggregated into 7 FAO Gelic soil units and 10 ST subgroups For instance, the FAO soil unit of Gleysols Gelic combines 3 ST subgroups (Pergelic Cryaquepts, Histic Pergelic Cryaquepts, and Humic Pergelic Cryaquepts) and 8 classes of Gleyzems and tundra meadow soils of the Russian system. This suggests that the SCP estimates based on the Russian classification will be more sophisticated.

As can be seen from Table 3.3, half of the area (about 154 million ha) of the permafrost-affected soils is represented by excessively moist Gelic Gleysols, Cambisol (Sod-brownzems gleyic and gley), and Podzoluvisols (Podzolic over-permafrost-gleyic soils). All these soils and Taiga peaty-muck high-humic non-gleyic soils (total area 210 million ha) have a predominantly fine texture and ice permafrost. Coarse-textured Gelic soils (Regosols, Leptosols, Andosols, and Podzols) very often have dry permafrost, which is difficult to use as a field-measured diagnostic characteristic.

The soil correlation from Table 3.3 illustrates the well-known fact that permafrost is not ultimately attributable to tundra soils and can be found in a wide range of bioclimatic conditions, varying from arctic desert to taiga. The tundra and arctic soils comprise about 206 million ha. This is considerably less than the area of the permafrost-affected soils (307 million ha) and far less than the total area of the permafrost zone (more than 1,000 million ha). The latter figure includes mountains and swamps (i.e., territories where permafrost has not been measured or indicated by

TABLE 3.3

Correlation and Area of Gelic Soil Units (FAO, 1988) and Soil Taxonomy Subgroups (1994) with the Legend of the Soil Map of Russia (Fridland, 1988)

Gelic soil units (index)	Soil Taxonomy subgroups	Legend of the soil map of Russia	Extent		
			Million ha	% of soil unit	% of total Gelic soils
Gleysols Gelic (GL i)	Pergelic Cryaquepts	Gleyzems arctic	1.47	<1	
	Histic Pergelic Cryaquepts	Gleyzems arctotundra muck-gley	25.40	17	
	Histic Pergelic Cryaquepts	Gleyzems and weak-gley humic tundra	16.22	11	
	Histic Pergelic Cryaquepts	Gleyzems tundra shallow and deep peat	58.44	38	
	Histic Pergelic Cryaquepts	Gleyzems tundra differentiated peaty-muck and peat	9.22	6	
	Histic Pergelic Cryaquepts	Tundra meadows	<0.01		
	Histic Pergelic Cryaquepts	Gleyzems weak-gley peaty-humic taiga	10.49	7	
	Histic Pergelic Cryaquepts	Gleyzems peaty-muck taiga	32.00	21	
Total			153.24		50
Regosols Gelic (Rgi)	Pergelic Cryorthents	Arctic desert	<0.01		
	Pergelic Cryorthents	Arctic cryozems	4.35	100	
	Oxyaguic Cryorthents	Arctic hydromorphic nongleyic	<0.01		
	Total		4.35		<1
Leptosols Gelic (Lpi)	Pergelic Cryorthents	Soils of spots (saline, arctic, and tundra)	6.41	49	
	Histic Pergelic Cryaquepts	Muck-calcareous tundra	6.62	51	
Total			13.03		4
Andosols Gelic (ANi)	Typic Gelicryands	Volcanics illuvial-humic tundra	1.73	100	
Total			1.73		
Cambisols Gelic (Cmi)	Oxyaguic Cryumbrepts	Taiga peaty-muck high-humic nongleyic	56.26	99	
	Humic Pergelic Cryaquepts	Sod-brownzems gleyic and gley	0.39		
Total			56.65		18
Podzoluvisols Gelic (Pdi)	Aguic Cryoboralfs	Podzolics over-permafrost-gleyic	0.13	<1	
Total			0.13		<1
Podzols Gelic (Pzi)	Pergelic Humicryods	Podburs dark tundra	16.42	21	
	Pergelic Haplocriods	Podburs light tundra	2.99	4	
	Pergelic Haplocriods	Podburs tundra (without subdivision)	58.24	75	
Total			77.65		25
Total Gelic			306.78		

TABLE 3.4
Vegetation Communities of the Polar Deserts and the Tundra Zone of Russia and Their Extent

Bioclimatic zones and vegetation communities	Million ha	% of subzones within zone and communities within subzone	% of plain and alpine from total
Polar deserts	2.26		100
Open (unclosed) aggregations of lichen (*Pertusaria, Ochrolechia*), moss (*Ditrichum flexicaule, Bryum, Pohlia*) and arctic species of flowering plants	2.26	100	
Plain tundra	124.95		49
Arctic tundra	20.52	10	
Grass-moss and low bush-grass-moss	20.52	100	
Northern tundra	50.36	47	
Grass-moss and low bush-moss with *Carex ensifolia* ssp. *arctisibiria*; *Betula* spp., *Salix glauca, S. lanata*	21.40	42	
Low bush-moss (*Dryas punctata, Cassiope tetragone, Aulacomnium* spp., *Tomenthypnum nitens, Hylocomium splendens* var. *alaskanum* with *Betula exilis, Salix pulchra, S. lanata*)	6.77	13	
Small willow stand (*Salix glauca, S. Reptans*)	15.55	31	
Cotton grass and moss (*Aulacomnium* spp., *Hylocomium splendens* var. *alaskanum, Eriophorum vaginatum*) hummocky	6.65	14	
Southern tundra	54.08	43	
Shrubbery grass-low bus-moss	38.09	88	
Low bush-cotton grass-moss (*Ledum decumbens, Eriophorum vaginatum, Sphagnum* spp. *Aulacomnium* spp.) together with *Betula exilis, Salix pulchra*, in some places *Duschekia fruticosa*	15.99	12	
Alpine tundra	127.80		51
Open (unclosed) aggregations of crustaceous and foliose lichen (species such as *Rhizocarpon, Lecanora, Lecidea, Umbilicaria, Gyrophora*), moss (species of *Rhacomitrium*), arctic-alpine flowering plants	61.92	48	
Low bush-moss, grass-low bush-moss and lichen (*Novosieversia glacialis, Dryas* Spp.)	13.21	10	
Low bush-lichen and low bush-moss in combination with shrubs and sparse vegetation among rock streams	52.66	42	
Total plain and alpine tundra	252.75		100

Source: Modified from Isachenko et al., 1990.

pedologists). In view of this information, permafrost cannot be designated as the criterion for the delineation of the tundra zone. Therefore, the tundra zone must be delineated by other external geographical elements. Vegetation could be an ideal criterion, especially as it has classes directly indicating tundra territories (Table 3.4). Table 3.4 shows 11 tundra vegetation communities (including polar desert) with a total area of about 253 million ha. Tundra is represented by two nearly equal parts (Figure 3a,b) that have different reliefs: plain (about 125 million ha) and mountain or alpine (about 128 million ha).

The tundra plain zone occupies the extreme north of the Kola Peninsula, Kanin Peninsula, the north of the European plain, Yamal, Tazovskey, Giydan, and Taimyr Peninsulas, the north of the East Siberia and Yano-Indigirskaja plain, and plains on the North-East and the northern Kamchatka Peninsula (Figure 3.3a). Three vegetational subzones have been distinguished in the plain tundra

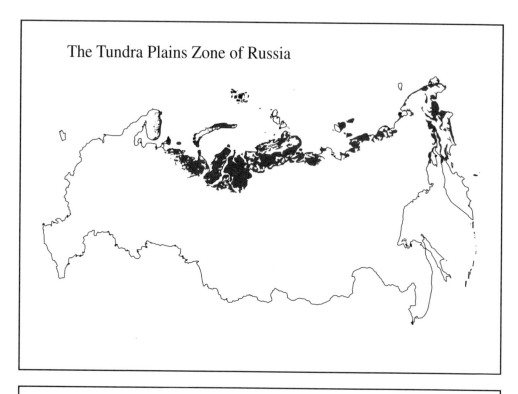

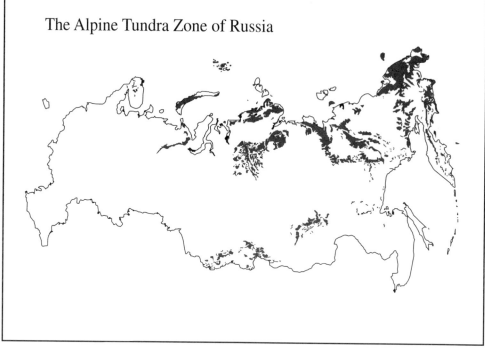

FIGURE 3.3 The tundra zone of Russia: plain and Alpine.

with respect to bioclimatic conditions and enrichment of vegetation communities: arctic, northern, and southern tundra. The arctic tundra has the most severe climatic conditions and is to be found only on islands. It covers a total area of more than 20 million ha. The vegetation is represented by grass-moss and low bush-grass-moss communities with a total phytomass stock of about 5 t ha^{-1} (in this and further examples the phytomass stock is given after Bazilevich, 1993). The northern tundra has a much milder climatic condition due to its continental location. However, the arctic influence on these plains is extremely strong. The total extent of the northern tundra is about 50 million ha. It is covered by four types of vegetational communities: (i) grass-moss and low bush-moss with *Carex ensifolia* ssp. *arctisibirica*, *Betula* spp., *Salix glauca*, and *S. lanata* (about 21 million ha, phytomass stock about 13 t ha^{-1}); (ii) low bush-moss with *Dryas punctata*, *Cassiope tetragone*, *Aulacomnium* spp., *Tomenthypnum nitens*, *Hylocomium splendens* var. *alaskanum*, with *Betula exilis*, *Salix pulchra*, and *S. lanata* (about 7 million ha, phytomass stock about 22 t ha^{-1}); (iii) small willow stand with *Salix glauca* and *S. reptans* (about 16 million ha, phytomass stock about 22 t ha^{-1}); and (iv) cotton grass and moss with *Aulacomnium* spp., *Hylocomium splendens* var. *alaskanum*, and *Eriophorum vaginatum* (about 7 million ha, phytomass stock about 11 t ha^{-1}). The southern tundra occupies locations that are protected from the arctic by mountains. Therefore, it has a relatively mild climate. It covers about 54 million ha in total and is represented by two types of vegetational communities: (i) shrubbery grass-low bush-moss (about 38 million ha, phytomass stock about 14 t ha^{-1}); and (ii) low bush-cotton grass-moss with *Ledum decumbens*, *Eriophorum vaginatum*, *Sphagnum* spp., and *Aulacomnium* spp., together with *Betula exilis*, *Salix pulchra*, and in some places *Duschekia fruticosa* (about 16 million ha, phytomass stock 23 t ha^{-1}).

Alpine tundra (Figure 3.3b) occupies the mountain territories of Spitsbergen Island, the northern part of the Ural mountains, the northwestern part of the Middle-Siberian plateau, Verhojanskey mountain range, the mountains of the Northwest, and the Kamchatka Peninsula. It is also found in the West and East Sayans and the southern Siberian mountains. The alpine tundra has not been subdivided into vegetational subzones. It has a total extent of about 128 million ha. The alpine tundra comprises three types of vegetational communities that are differentiated by the amount of species, their density, and phytomass stock: (i) open (unclosed) aggregations of crustaceous and foliose lichen (species such as *Rhizocarpon*, *Lecanora*, *Lecidea*, *Umbilicaria*, *Gyrophora*), moss (species of *Rhacomitrium*), arctic–alpine species of flowering plants (about 62 million ha, phytomass stock about 1.5 to 2 t ha^{-1}); (ii) low bush-moss, grass-low bush-moss and lichen with *Novosieversia glacialis*, *Dryas* spp. (about 13 million ha, phytomass stock about 15–16 t ha^{-1}); and (iii) low bush-lichen and low bush-moss in combination with shrubs and sparse vegetation among rock streams (about 53 million ha, phytomass stock about 15 to 16 t ha^{-1}).

The vegetation map and table (Figure 3.3a,b; Table 3.4) show reasonable tundra diversity reflecting the variety of phytomass growing conditions, which could meet the requirements for SCP estimates. A more detailed analysis is needed to identify soils related to the tundra vegetation. Such analysis should answer two questions: What spectrum of soils is related to the tundra vegetation zone; and, Is there any geographical differentiation of soils within the tundra vegetational zone, mainly regarding reliefs of the territory?

The spectrum of the FAO major soil groupings within the tundra zone of Russia is shown in Figure 3.4. As this figure shows, the variety of soils within the tundra zone is relatively high. The total of 19 soils that have been found on the tundra territory of Russia (Stolbovoi and Sheremet, 1995) can be classified according to 9 FAO major soil groupings. From the soil genetic point of view, it should be noted that the cold and humid climatic conditions influence the main soil features within the tundra vegetation zone. The cold climate slows down the intensity of soil-forming processes and brings about the development of shallow soils (Regosols, Leptosols, and Cambisols). It also restricts biomass growth, which in turn reduces the organic accumulation, limiting peat formation and the development of Histosols. The climate humidity, together with poor surface drainage, generate an expansion of soil redoximorphism, which leads to the formation of huge areas of Gleysols.

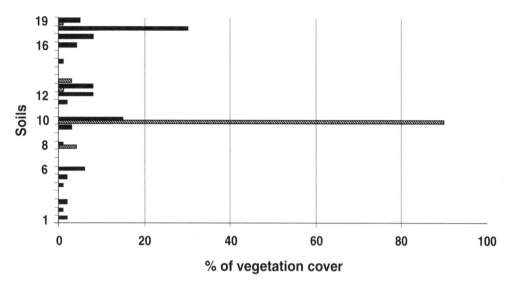

FIGURE 3.4 Soil spectrum within a low bush-moss vegetation community: Andosols: 1. Haplic; 2. Vitric; 3. Gelic, Cambisols; 4. Eutric; 5. Dystric; 6. Gelic, Fluvisols; 7. Dystric; 8. Umbric, Gleysols; 9. Dystric; 10. Gelic, Leptosols; 11. Dystric; 12. Rendzic; 13. Gleyic; 14. Gelic, Podzoluvisols; 15. Gelic, Podzols; 16. Haplic; 17. Cambic; 18. Ferric; 19. Gelic.

Great dissimilarities can be found between soil spectrums for the plain (northern tundra) and mountain (alpine) regions (Figure 3.4). The plain territories are characterized by the development of poorly drained Gleysols and Histosols, which do not occur in mountain regions. Andosols and Cambisols are not found on the plains and are widespread in mountains. Mountain regions have many soils formed from coarse-textured and internally well-drained deposits, mainly eluvium (Podzols, Leptosols, Cambisols, and Andosols).

Thus, the general finding is that the tundra vegetation zone incorporates many different soils that have substantially different soil characteristics. Table 3.5 shows the internal soil composition within the major tundra vegetational communities.

Three items of interest are shown in Table 3.5. First, it might be found that the soil extent does not correspond to the area of vegetation patterns. For example, soils occupy only 27% of the polar desert area. The rest is covered by non-soil formations (glaciers, rock outcrops, and sands). The average plain tundra has about 90% of the surface covered by soils. In mountain regions the share of soil formations is 80 to 85% on average. These figures demonstrate that the area of tundra vegetation cannot replace the soil extent itself in the SCP estimates.

Second, each tundra vegetation pattern has its own specific soil spectrum. In general, it can be noted that the soil variety in the tundra vegetation zone depends greatly upon the diversity of soil-forming factors. This regularity is very common and has also been found in other soil-natural zones. In particular, the soil diversity is driven by the amount of parent materials where the drainage condition will not deplete their specific features. This can be illustrated by the soil collection which is formed under a low bush-lichen and low bush-moss vegetation pattern in the mountain relief (Table 3.5). The soils are formed from different parent materials varying from volcanic deposit (Andosols), to coarse- and fine-textured alumosilicates (Podzols and Podsoluvisols) and carbonate (Leptosols rendzic) eluviums.

Finally, Table 3.5 shows that differences in soil spectrums within various tundra vegetation patterns are so great that none of the soils can be regarded as typical. Soil spectrums vary by a number of soils, by dominant soils, and by soil percentage. For instance, the share of Gleysols

TABLE 3.5
Distribution (Million ha/% of Vegetation Pattern) of Soil Units Within Tundra Vegetation in Russia

Soil units	Polar deserts[a]	Arctic tundra[b]	Northern tundra A[c]	B[d]	C[e]	D[f]	Southern tundra A[g]	B[h]	Alpine tundra A[i]	B[j]	C[k]
Andosols											
Halpic											0.8/2
Vitric											0.3/1
Gelic											0.7/2
Cambisols											
Eutric										0.9/2	0.5/1
Dystric											1.0/2
Calcaric											
Gleyic											
Gelic								0.1/1		2.2/5	2.5/6
Fluvisols											
Eutric		1.1/6				0.2/3			0.1/1		
Dystric							0.3/1	0.1/1		0.1/0	0.1/0
Umbric			1.2/7	0.3/4	0.3/2		1.1/3	0.4/3		0.3/1	0.6/1
Tionic	<0.1/0	<0.1/0			0.3/2	0.1/2	0.1/0				
Gleysols											
Dystric		0.7/4	1.5/8		0.9/6	0.5/9	1.6/4	2.9/19	0.6/7	1.0/2	1.4/3
Umbric											
Gelic		9.0/51	11.0/62	6.2/9.0	12.0/84	4.3/74	23.4/64	7.8/52	3.3/40	4.6/10	6.7/15
Histisols											
Terric			0.4/2		<0.1/0	0.3/5	2.8/8	1.0/7			0.1/0
Fibric							0.5/1	1.0/7			0.1/0
Without Subdivision							0.1/0				
Leptosols											
Dystric										1.5/3	1.1/2
Rendzic		0.1/1	0.1/1	<0.1/0						1.3/3	3.6/8
Mollic											0.2/0
Umbric											0.2/0
Gleyic	0.1/17	0.7/4	0.7/4	0.1/1	<0.1/0	0.3/5	0.2/1	0.1/1	1.9/23	9.9/22	3.6/8
Gelic		3.0/17	0.9/5	0.2/3						1.6/4	
Podzoluvisols											
Eutric											<0.1/0
Gleyic							0.4/1			0.3/1	0.3/1
Podzols											
Haplic							1.3/4	0.5/3		2.6/6	1.8/4
Cambic			0.1/1		<0.1/0		1.6/4	0.5/3	0.4/5	7.9/17	3.5/8
Ferric		0.7/4	0.3/2		0.8/6	0.1/2	1.9/5	0.2/1	1.6/20	6.6/15	13.2/30
Gleyic							0.3/1				
Gelic		0.5/3	1.2/7	0.1/1			1.1/3	0.5/3	0.1/1	3.8/8	2.0/5
Regosols											
Gelic	0.5/83	1.9/11	0.3/2						0.2/2	0.6/1	
Total Soils	0.6/27	17.7/86	17.7/83	6.9/100	14.3/92	5.8/87	36.7/94	15.1/94	45.2/73	8.2/62	44.3/84

[a] Open (unclosed) aggregations of lichen (*Pertusaria Ochrolechia*), moss (*Ditrichum flexicaule, Bryum, Pohlia*), and arctic species of flowering plants.

[b] Grass-moss and low bush-grass-moss.

[c] Grass-moss and low bush-moss with *Carex ensifolia* ssp., *arctisibiria Betula* spp., *Salix glauca, S. lanata*.

TABLE 3.5 (continued)

[d] Low bush-moss (*Dryas punctata, Cassiope tetragone, Aulacomnium* spp., *Tomenthypnum nitens, Hylocomium splendens* var. *alaskanum* with *Betula exilis, Salix pulchra, S. lanata*).

[e] Small willow stand (*Salix glauca, S. Reptans*).

[f] Cotton grass and moss (*Aulacomnium* spp., *Hylocomium splendens* var. *alaskanum, Eriophorum vaginatum*).

[g] Shrubbery grass-low bus-moss.

[h] Low bush-cotton grass-moss (*Ledum decumbens, Eriophorum vaginatum*, species of *Sphagnum Aulacomnium*) together with *Betula exilis, Salix pulchra*, in some places *Duschekia fruticosa*.

[i] Open (unclosed) aggregations of crustaceous and foliose lichen (species such as *Rhizocarpon, Lecanora, Lecidea, Umbilicaria, Gyrophora*), moss *Rhacomitrium* spp., arctic-alpine species of flowering plants.

[j] Low bush-moss, grass-low bush-moss and lichen (*Novosieversia glacialis, Dryas* spp.).

[k] Low bush-lichen and low bush-moss in combination with shrubs and sparse vegetation among rock streams.

Gelic ranges from 0% in polar deserts to 51% in the grass-moss and low bush-grass-moss pattern, and 90% in the low bush-moss pattern (northern tundra). Histosols occupy from 9 to 14% of the southern tundra and do not exist in some other vegetation patterns. These facts clearly demonstrate that the concept of the zonal soil cannot be applied for SCP estimates.

Thus, the following conclusions should be made regarding delineation of the tundra zone as far as SCP estimates are concerned:

- The tundra zone cannot be distinguished based on direct soil criteria. Additional cartographic data must be used for this task.
- A vegetation map can provide essential data for delineation of the tundra zone into smaller tundra vegetation patterns.
- When the tundra zone and its most detailed patterns have been spatially detected, the relevant internal soil cover structures must be considered.
- Soil diversity within tundra vegetation patterns is very high. Therefore, detailed analyses should be carried out to indicate the spectrum of soils that can be applied for establishing the spectrum of reference soil profiles for SCP calculations.

CHARACTERIZATION OF ORGANIC CONTENT OF TUNDRA SOILS

The selection of representative (reference) soil profiles is a critical and delicate task affecting the SCP estimates. Common problems arise when SCP estimates are made based on the literature. Some of these problems are illustrated in Table 3.6 by three randomly selected referenced soil profiles. The profiles represent tundra soils and they have been classified as a single soil unit under the FAO revised legend (Gleysols Gelic) and the soil classification of Russia (Gleyzems muck or peaty). A simple comparison of the soil profiles shows that all are very different. This can be divided intro three areas.

First, the use of nonstandard symbols to describe the soil profiles does not allow for any substantial analysis. For example, the horizon symbols are not practical for the diagnosis of the soils. On the one hand, this partly reflects the fact that the soil diagnostic system is not well developed in Russian soil surveys and authors follow their preferences rather than any standards. On the other hand, the diagnostic system is well established for agricultural and forests soils. The diagnosis and classification of tundra soils needs to be considerably improved in order to meet requirements for accurate SCP estimates.

Second, soil-sampling depths vary drastically. Some important soil horizons for SCP estimate, depth 0 to 3 cm (Tonkonogov, 1979), are missing. Some soil profiles (Vasilevskay et al., 1986) have been sampled continuously, e.g., 0 to 14 cm, 14 to 19 cm, 19 to 33 cm, while others (Ignatenko,

TABLE 3.6
Reference Soil Profiles for Gleysols Gelic

Name, sample location	Horizon	Depth (cm)	Humus (after Turin), in %	Source
1. Gleezems tundra muck, Yamal Peninsula	O1	0–3	Not defined	Tonkonogov (1979)
	A1Bh	3–5	5.3	
	Bg (ochre)	8–10	1.2	
	G1	15–20	1	
	G2	20–25	0.8	
	Cg	40–50	0.8	
	[a]	70–90	0.7	
		150–170	Not defined	
2. Gleyzems tundra peaty, Yamal Peninsula	O2A0	0–14	28.7[b]	Vasilevskay et al. (1986)
	B1g	14–19	3.1	
	B2g	19–33	2.5	
	G2	33–56	2.1	
	Bclg	56–71	2.1	
	BC2g	71–95	2.5	
	[a]Cg	95–100	0.1	
3. Gleyzems tundra peaty, Kolva Basin	Aov	0–8	96.9[b]	Ignatenko (1979)
	At	10–20	88.6[b]	
	Bgh	22–29	1.9	
	Ghx1	29–37	2.7	
	Ghx2	37–43	6.4	
	[a]Ghx3	45–55	3.6	
	[a]Cg	60–70	1.2	
	[a]Cg	90–100	Not defined	
	[a]Cg	100–120	Not defined	

[a] Permafrost.
[b] Ignition losses (%).

1979; Tonkonogov, 1979) only discretely, e.g., 0 to 8 cm, 10 to 20 cm, 22 to 29 cm. It is difficult to unify such soil profile samplings for SCP estimates.

Finally, soil organic content in the soil profiles is also very different, despite the fact that, as mentioned previously, all soils have been classified identically. For example, at the 15 to 20 cm profile depth of Tonkonogov (1979), the humus content is 1%, while at practically the same depth (14 to 19 cm) of the Vasilevskay et al. (1986) data, it is about 3%. Another example is the ignition losses from the topsoil horizons of the Vasilevskay et al. (1986) reference soil profile (about 29%) and those of Ignatenko (1979), which are about 97%. These examples illustrate that soil classes, which have been established for general soil classification, are rather broad as far as SCP is concerned. In order to convert these classes into carbon-related groups, special methods have to be employed. This finding is similar to the issue of the applicability of the soil maps to the SCP estimates.

In summary, three factors must be considered to deal with the problems of the soil reference profiles:

- First, efforts should be made to improve knowledge of tundra soils, their diagnosis, and classification, making them relevant for SCP estimates.
- Second, special attention should be paid to the standardization of soil sampling.
- Third, serious investigations should be carried out to establish methods to harmonize existing data for their incorporation into SCP estimates.

TABLE 3.7
Soil Degradation in the Tundra Vegetational Zone in Russia

Degradation types	Plain Million ha	Plain % of plain soils	Alpine Million ha	Alpine % of alpine soils	Total Million ha	Total % of total degraded soil
Thermocarst	11.4	48	1.4	18	12.8	40
Surface corrosion	10.8	45	1.2	15	12	38
Compaction	1.7	7	4.9	61	6.6	21
Disturbance due to fire			0.4	5	0.4	1
Water erosion			0.1	1	0.1	
Total plain and alpine	23.9		8.0		31.9	
Percent of degraded soil		19		6		13

DEGRADATION OF TUNDRA SOILS IN RUSSIA

Most SCP estimates are based on the assumption that soils are undisturbed or natural. However, results from the GLASOD study (Oldeman et al., 1990) on the status of human-induced soil degradation show that a vast area comprising 1,965 million ha of the world's soils — or 15% of the total global land area — is degraded. The development of these negative processes varies from continent to continent, reaching 23% in Europe and 5% in North America. These alarming facts have been supported by further detailed investigations (e.g., Government [National] Report, 1993; Land of Russia, 1995; Lynden, 1995). Soil degradation has been suggested as one of the indicators for sustainable land use (Pieri et al., 1995). Stolbovoi and Fischer (1998) have shown that soil degradation is widespread in Russia, affecting 14.5% of the country's soil cover. Furthermore, it has also been recognized that soil degradation is not just limited to agricultural land (cropland and pastures) on which the GLASOD study focused. Soil degradation has also strongly affected forest land, where about 9% has been degraded (Stolbovoi, 1997).

In a broad sense soil degradation refers to "a process that describes human-induced phenomena which lower the current and/or future capacity of the soil to support human life" (Lynden, 1995). About 10 to 15 years ago it became a subject of intensive investigation due to the expanding population, economic development, and the growing demand for various land-based products. It was recognized that an increasing pressure on soils, water resources, and plants in many developing and developed countries exceeded critical thresholds, with people facing problems of deteriorating land resources, declining productivity, and consequently, reduced income. However, soil degradation affects not only the economy, it also has very strong environmental consequences due to the impact on the functioning of the ecosystem.

Table 3.7 illustrates soil degradation within the tundra vegetation zone of Russia. The total area of degraded soils is about 32 million ha, or 13% of the total tundra area. Most prevailing types of soil degradation are thermocarst (e.g., land subsidence, cryogenic lakes, caverns), comprising 12.8 million ha (40% of degraded soils), and surface corrosion (e.g., frost mounds, frost mudboils, earth hummocks solifluction, and landslides) comprising 12 million ha (38% of degraded soils). Thus, in total, about 80% of soil degradation in the tundra zone deals with permafrost. These soil degradation types have mostly developed in plain tundra because of the occurrence of deep, loose, and fine-textured deposits with an ice content of 30 to 70% (by volume). Mountain regions, with shallow coarse-textured soils, do not have this type of soil degradation. Unfortunately, there are no data allowing for the separation of human-induced and natural thermocarst and surface corrosion. Nevertheless, it is well known that permafrost soils are very sensitive to any disturbances of the top organic layer, which controls the temperature balance and protects permafrost from melting during the summer. The huge extent of the permafrost-related types of soil degradation definitely indicates that any intensification of human activities in the tundra zone should be accompanied by sensitively designed technologies in order to keep the topsoil layer undisturbed.

Compaction is the next most widespread type of soil degradation in the tundra. Soil compaction refers to soil bulk density 1.2 times higher than in natural soils. About 6.6 million ha is affected, alpine tundra in particular. Overgrazing by reindeer on tundra pastures causes soil compaction. However, the mechanism of this process is not clear, although it is closely linked with the soil organic balance on tundra pastures. An official report (Government [National] Report, 1993) shows that during the last three decades the moss phytomass stock in tundra pastures has declined two to three times. It is likely that this can explain an increase in soil bulk density (compaction). Reindeer use mountainous uplands in summer when winds protect them from mosquitoes. Therefore, mosses are affected from both overgrazing and intensive trampling when unfrozen. In addition, the decrease of biomass residuals leads to a soil organic misbalance, which results in a decline of the soil organic content. The latter has also been further driven by changes of soil hydrothermal regimes due to removal of vegetation and exposure of a bare soil surface. Additional soil warming initiates biochemical processes of organic decomposition and mineralization. The decline of soil organics, in turn, leads to the deterioration of humus adhesion capacity and affects the soil structure water resistance. It causes detachment of soil macroaggregates, decreases soil macroporosity, and increases soil compaction.

About 0.4 million ha of tundra soils in the alpine zone (accounted area for 10 years) are influenced by fires. For the most part, fires affect vegetation and the soil surface organic horizon, and are predominantly (95%) caused by human carelessness.

Overgrazing and fires contribute to the development of water erosion (0.1 million ha). This soil degradation type is found in alpine tundra.

There are no data presently available assessing the impact of soil degradation on soil organic content in the tundra soils. Nevertheless, all the above-mentioned types of soil degradation (i.e., thermocarst, surface corrosion, compaction, disturbances of the organic topsoil layer due to fires, erosion, etc.) lead to changes in productivity and reduce the extent of the organic carbon accumulation and circulation in the ecosystem. The latter development increases CO_2 fluxes to the atmosphere and provides a positive feedback to climate warming. There is no doubt that soil degradation leads to great misbalances within the terrain-atmospheric carbon cycling. Considering the huge extent of soil degradation in the tundra zone of Russia (13% of the tundra zone), this phenomenon might be instrumental in changing the figures on SCP estimates based on traditional natural soils. Thus, in order to improve the reliability of the SCP calculations soil degradation has to be taken into account.

CONCLUSIONS

- Large uncertainties regarding SCP assessments in the tundra zone of Russia originate from the lack of a convention on methods for the estimates. The existing assessments are based on thematically different maps, distinct concepts, and inconsistent expert views.
- A consensus has to be reached in order to harmonize the procedure of SCP estimates, thus making data compatible and more reliable. Current flexibility in the identification of the parameters for the calculations has to be replaced by standard measurements, which should be introduced at the national and international level. An essential part of the standards has to be devoted to the uniform procedure of incorporation of regional knowledge in the global context.
- SCP estimates on the tundra soils can be assessed on the basis of two sources: (i) a vegetation map, which provides delineation of the tundra zone and its most detailed vegetation patterns; and (ii) a soil map showing the spectrum of soils, their extent, and corresponding soil reference profiles. Estimates based on the selected (zonal) soil profiles should be regarded as unsatisfactory.

- The extent of soil degradation in the tundra of Russia comprises 13% of the tundra zone. This degradation has strong environmental consequences, affecting ecosystem functioning and imbalancing the land-atmospheric carbon cycling process. The SCP estimates will be more reliable based on the actual soil status of tundra soils.

REFERENCES

Apps, M.J., Kurtz, W.A., Luxmoore, R.J., Nilsson, L.O., Sedjo, R.A., Schmidt, R., Simpson, L.G., and Vinson, T.S., Boreal forests and tundra, *Water, Air, Soil Pollut.,* 70, 39, 1993.

Batjes, N.H., World soil carbon stocks and global change. Preprint 95/11, International Soil Reference and Information Center, Wageningen, The Netherlands, 1995.

Batjes, N.H., Total carbon and nitrogen in the soils of the world, *Eur. J. Soil Sci.,* 47, 151, 1996.

Bazilevich, N.I., *Biological Productivity of the Ecosystems of Northern Eurasia,* Nauka, Moscow, 1993 (in Russian).

Berg, M., The Kyoto Protocol — next step on a long, bumpy road, *Change,* 40, 4, 1998.

Bouwman, A.F., Ed., *Soils and the Greenhouse Effect,* John Wiley & Sons, Chichester, U.K., 1990.

Dokuchaev, V.V., *Chernozems of Russia,* Vol. I, Selected papers, Selhozgiz, Moscow, 1948 (in Russian).

Eswaran, H., Van den Berg, E., Reich, P., and Kimble, J., Global soil carbon reserves, in *Soils and Global Change,* Lal, R., Kimble, J., Levine, E., and Stewart, B.A., Eds., CRC Press, Boca Raton, FL, 1995.

FAO-Unesco, Soil Map of the World — Revised Legend, World Resources Report No. 60, Food and Agricultural Organization of the United Nations, Rome, 1988.

Fridland, V.M., *Structure of Soil Cover,* Mysl, Moscow, 1972 (in Russian).

Fridland, V.M., Ed., Soil map of the Russian Soviet Federative Socialist Republic, scale 1:2,500,000, Government Administration on Geodesy and Cartography, Moscow, 1988 (in Russian).

Government [National] report on the Status and Use of Land in the Russian Federation, Publications Committee of the Russian Federation on Land Resources and Land-Use Planning, Moscow, 1993 (in Russian).

Greenland, D.J., Land use and soil carbon in different agroecological zones, *Soil Management and the Greenhouse Effect,* Lal, R., Kimble, J., Levine, E., and Stewart, B.A., Eds., CRC Press, Boca Raton, FL, 1995, p. 9.

Ignatenko, I.V., *Soils of the East-European Tundra and Forest Tundra,* Nauka, Moscow, 1979 (in Russian).

Isachenko, T.I. et al., Vegetation of the U.S.S.R. Map at the scale of 1:4,000,000, Government Administration on Geodesy and Cartography, Moscow, 1990.

Karelin, D.V., Gilmanov, T.G., and Zamolodchikov, D.G., An estimate of the carbon stock in the terrain ecosystems of the tundra and forest zones of the Russian North: Phytomass and primary production. Reports of the Academy of Science, *Gen. Biol.,* 335, 530, 1994 (in Russian).

Kolchugina, T.P., Vinson, T.S., Gaston, G.G., Rozhkov, V.A., and Shvidenko, A.Z., Carbon pools, fluxes, and sequestration potential in soils of the former Soviet Union, in *Soil Management and the Greenhouse Effect,* Lal, R., Kimble, J., Levine, E., and Stewart, B.A., Eds., Lewis Publishers, Boca Raton, FL, 1995, p. 25.

Land of Russia — 1995, Problems, Figures, Commentaries, RUSSLIT, Moscow, 1996 (in Russian).

Lynden, G.W.J. van, Ed., Guidelines for the Assessment of the Status of Human-Induced Soil Degradation in South and Southeast Asia (ASSOD), International Soil Reference Information Center (ISRIC), Wageningen, The Netherlands, 1995.

Oldeman, L.R., Hakkeling, R.T.A., and Sombroek, W.G., World map of the Status of Human-Induced Soil Degradation, Global Assessment of Soil Degradation, GLASOD, October, 1990, Food and Agricultural Organization of the United Nations, Rome, 1990.

Orlov, D.S., Biryukova, O.N., and Sakhanova, N.I., *Soil Organic Matter of Russia,* Nauka, Moscow, 1996 (in Russian).

Pieri, C., Dumanski, J., Hamblin, A., and Young, A., Land Quality Indicators, Discussion Paper #315, World Bank, Washington, D.C., 1995.

Raich, J.W., and Potter, C.S., Global patterns of carbon dioxide emissions from soils, *Global Biogeochem. Cycles,* 9, 23, 1995.

Rozov, N.N., Ed., Soil Map of the U.S.S.R., Government Administration on Geodesy and Cartography, Moscow, 1964.

Rozhkov, V.A., Wagner, V.B., Kogut, B.M., Konyushkov, D.E., Nilsson, S., Sheremet, V.B., and Shvidenko, A.Z., Soil carbon estimates for Russia, WP-96-60, IIASA, Laxenburg, Austria, 1996.

Shishov, L.L. and Sokolov, I.A., A new version of soil classification in the Soviet Union, *Pochvovedenie,* 4, 112, 1989 (in Russian).

Sombroek, M.G., Nachtergaele, F.O., and Hebel, A., Amounts, dynamics, and sequestration of carbon in tropical and subtropical soils, *Ambio,* 22, 417,1993.

Stolbovoi, V. and Sheremet, B., A new soil map of Russia compiled in the FAO system, *Pochvovedenie,* 2, 149, 1995 (in Russian).

Stolbovoi, V. and Sheremet, B., On the soil fund of Russia, *Eurasian Soil Sci.,* 30, 1278, 1997.

Stolbovoi, V.S., Degradation of forestland in land-use/cover patterns of Russia, IR-97-070, IIASA, Laxenburg, Austria, 1997.

Stolbovoi, V.S. and Fischer, G., A new digital georeferenced database of soil degradation in Russia, in *Towards Sustainable Land Use: Furthering Cooperation Between People and Institutions, Vol. 1, Advances in GeoEcology,* 31, Blume, H.P., Ed., Catena-Verlag, Reiskirchen, Germany, 1998.

Tonkonogov, V.D., Gleyzems specific to the Yamal-Gydan tundra, in *Peculiarities of Soil Forming in the Siberia,* Acadamy of Sciences, Novosibirsk, 1979 (in Russian).

Vasilevskay, V.D., Ivanova, V.V., and Bogatirev, L.G., Soils of the northern part of West Siberia, Moscow State University, Moscow, 1986 (in Russian).

4 Carbon Pools in Antarctica and Their Significance for Global Climate Change

G.G.C. Claridge, I.B. Campbell, and D.S. Sheppard

CONTENTS

INTRODUCTION

It is now generally accepted that the consequence of an ever-increasing global population and its associated human activities is a parallel increase in impacts on the earth's environment and eco-systems (Houghton et al., 1996). The most widely reported concerns arise from increased global warming and the subsequent melting of polar ice, depletion of upper atmosphere ozone through the release of harmful substances into the air, acid rain, diminishing water quality, and increasing atmospheric CO_2 content resulting very largely from the burning of fossil fuels (Bouma et al., 1996).

The increase in the greenhouse effect, with a postulated marked increase in global temperatures due to increased CO_2 content, is now considered by many as being responsible for the climatic extremes presently being experienced in various parts of the globe. While much attention has focused on the magnitude of atmospheric CO_2 changes and associated effects, there is now a greater interest and need to consider the implications of these changes.

The importance of the Arctic region in relation to global climate change and associated processes has been previously documented (Harrison, 1991; Koster et al., 1994) and climatic records from a number of stations in the Arctic suggest a steady warming trend over the last few decades. The Arctic and Sub-Arctic regions are also of special interest in terms of the global carbon budget. For example, Post et al. (1982) have estimated the total global organic carbon mass at 1550 Gt while Tarnocai and Lacelle (1996) have estimated the total pool of organic carbon in the upper 30 cm of Canadian soils to be 72.8 Gt, and that in the upper meter as 262.3 Gt, with the highest values in the Sub-Arctic forest tundra zones. With a significant proportion of global organic carbon occurring within the Arctic region, any long-term shift in Arctic climate zones is therefore likely to have an impact on associated vegetation and soil patterns and the associated carbon exchange processes.

The Arctic region owes its particular vegetation and soil characteristics largely to the relative distribution of land and ocean. The Arctic region consists of several land masses which surround an ocean, the Arctic Sea. The Antarctic, on the other hand, has a very different configuration and essentially comprises a landmass which is surrounded by several oceans (Figure 4.1). Because of the existence of this polar landmass, the Antarctic is much colder and therefore much drier than the Arctic, and this profoundly influences terrestrial biological productivity and the associated carbon exchange processes.

In this chapter we consider the carbon pools in Antarctica and their significance in respect to global climate change and polar ecosystems.

THE ANTARCTIC ENVIRONMENT

PHYSICAL CONDITIONS

Antarctica is the world's largest continent, with an area of approximately 14 million square kilometers. It is completely surrounded by the Southern Ocean, which extends from about the 40th parallel to the Antarctic Circle at 60° S. The continent is shaped like an upturned saucer, rising steeply from the coast to a vast interior plateau, and divides easily into two topographic units. East Antarctica, the largest, lies between 30° W and 150° E longitude, and rises to an altitude of >4000 m. The smaller part, West Antarctica, rises to about 1500 m while the Antarctic Peninsula projects toward South America.

Since the ice sheet reaches the sea almost everywhere around the perimeter, floating ice shelves are found in many places, especially where they are buttressed by rock outcrops on either side or protected by shallow water. More particularly, large ice shelves occupy the embayments between East and West Antarctica: the Ross Ice Shelf, south of New Zealand and the Ronne and Filchner Ice Shelves south of the Weddell sea. Smaller ice shelves are found in other embayments around the coast and along the Antarctic Peninsula.

Various estimates of the area of the continent, of the floating ice shelves, and of the ice-free ground have been made, with increasing accuracy as more accurate survey data become available. The most recent compilation, based on accurate mapping of many areas, coupled with satellite imaging and radio-echo sounding information on ice thickness is that of Fox and Cooper (1994), replacing earlier estimates by Bardin and Suvetova (1967) and Shumskiy (1967). According to the latest estimates, the total area of Antarctica, including ice shelves, is 13.946×10^6 km², of which 12.348×10^6 km² is grounded ice-sheet, 1.555×10^6 km² is floating ice shelves, and only 0.033×10^6 km² is ice-free ground. The greatest change in these figures from those derived by

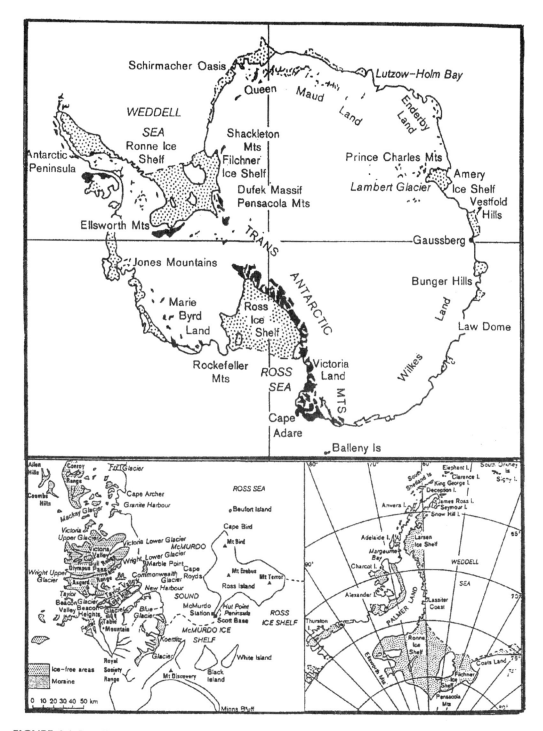

FIGURE 4.1 Locality map, showing places mentioned in the text and the sites of some of the national bases around Antarctica.

Bardin and Suvetova (1967) is the increase in the area of the ice shelves and decrease in the area of ice-free ground.

Estimates of the total volume of ice in Antarctica depend largely on the estimate on the ice profile and thickness. Bull (1971) quotes volumes of between 24×10^6 km^3 and 30×10^6 km^3. Since the estimate of the total area of Antarctica has not changed much with more accurate data, we will assume that the volume of permanent ice in Antarctica, including the ice shelves is 24×10^6 km^3. If all the ice, estimated to be about 24.5×10^6 km^3, were to melt, sea level would rise about 50 to 60 m.

The area of frozen sea varies in size from about 2.65×10^6 km^2 in summer to 19×10^6 km^2 in winter. Only a very small area of the continent is free of ice with exposed areas of bare ground.

CLIMATE

Antarctica is now cold because of the small amounts of solar radiation that it receives — about 16% of that reaching equatorial regions, and because of the high average elevation. Since the outgoing terrestrial radiation exceeds the absorbed solar radiation during the course of a year, the Antarctic continent is one of the Earth's major heat sinks.

Temperatures of –89°C have been recorded in inland regions with a mean of –49°C at the South Pole. Temperatures increase northwards however, with means of –20°C at Vanda Station in the McMurdo Oasis and –18°C at McMurdo Station. Because of the cold, precipitation is almost entirely as snow, but much of the precipitation which falls is unavailable for soil or biological processes due to its removal by ablation and wind.

The mean annual precipitation over Antarctica is about 50 mm y^{-1}, with least falling in inland regions and the most in coastal areas. In the Dry Valley region of McMurdo Sound, precipitation over 20 years averaged 13 mm y^{-1} at a valley floor site and 100 mm y^{-1} in nearby upland mountains (Chinn, personal communication). Precipitation is almost always snow.

Further north in coastal areas of East Antarctica, warmer climates are experienced (Burton and Campbell, 1980). At Davis Station in the Vestfold Hills, for example, mean annual temperature is –10.2°C but may rise to +13°C in January. At Molodezhnaya, the temperature range is similar (–11°C to +9°C) and precipitation of around 650 mm, mostly as snow, has been recorded.

The Antarctic Peninsula protrudes from the Antarctic Continent into the cyclonic belt and as a consequence the climate is quite different (Burdecki, 1967; Holdgate, 1970) and is not representative of the rest of the continent. Mean temperatures range from –11°C at Marguerite Bay to –4°C at Signy Island with maxima of +9°C and +12°C, respectively. Mean annual precipitation of >100 mm has been recorded at some stations in this region, which may fall as rain or sleet at times.

The thermal regimes and the water content of soils in the McMurdo Dry Valley region have been investigated by Balks et al. (1995), Campbell et al. (1997a, 1997b, 1997c), and MacCulloch (1996). These studies have shown that in this region, water contents of the active zone, often less than 10 cm deep, may range from 5% to less than 1% during the brief 6-week summer. The cryosols formed in this region of Antarctica have developed under the most extreme conditions for soil formation, with cold temperatures and severe aridity modulating the main weathering processes of oxidation and salinization. As a consequence, over most of the scattered and exposed bare ground areas, such as the Dry Valley region which is the most extensive area of bare ground in Antarctica, vegetation is absent or is restricted to isolated occurrences of moss or lichen assemblages. The organic regime of these soils, to say the least, is minimal.

Around the coast of East Antarctica, where the climate is less extreme, plant cover, mainly mosses and lichens, is more extensive, and may cover up to 10% of the surface. Under these vegetation patches, Histosols associated with Leptosols, and Podsols can be found (Blume and Bolter, 1994; Smith, 1990).

In the maritime Antarctic zone of the Antarctic Peninsula and associated islands, organic soils formed mainly by debris of mosses and algae are widespread. Under some bryophyte communities there is considerable peat formation (Holdgate et al., 1967). Because of the much warmer prevailing temperatures the soils are much more moist. Moisture contents of up to 15% were measured in the active layer of soils on King George Island, in the South Shetland Islands (Zhu et al., 1991.)

CARBON POOLS IN THE ANTARCTIC

OCEANS AND ATMOSPHERES

The most abundant carbon compound in the atmosphere is carbon dioxide, with a current concentration close to 360 ppm, and currently increasing due to the release of carbon from the burning of fossil fuel and will probably exceed 400 ppm early in the next century (Raynaud et al., 1993; Holland, 1978). The atmosphere also contains about 1.7 ppm of methane and 0.02 to 0.05 ppm of carbon monoxide. Other gases are present in very much lower concentrations. Measurements of CO_2 and CH_4 in air extracted from polar ice cores have shown that the concentration of these gases in the atmosphere was considerably lower in preindustrial times: 270 ppm CO_2 and 0.7 ppm CH_4 at 3700 years B.P. (Nakazawa et al., 1993).

These gases are in equilibrium with the oceans and other parts of the terrestrial biosphere. Carbon dioxide dissolves in sea water, where it may be taken up as insoluble carbonates of calcium and magnesium. The oceans cover 70% of the earth's surface and act as vast reservoirs of CO_2. The mass of CO_2 in the atmosphere is estimated to be 0.68×10^6 Mt, while the ocean basins contain more than 50 times this amount and more than 20 times that in the terrestrial biosphere (Lucas and Lindesay, 1991). Only a small amount of this carbon is involved in rapid interchange with the atmosphere, as there is only 3.8×10^3 Mt above the thermocline in the world's oceans. The remainder is contained in the deep water basins and undergoes very slow interchange with the ocean surface.

The organic carbon content in the oceans is estimated to be about 1.5 mg l^{-1}. However in closed seas and in highly productive coastal waters the content of organic carbon can be four to five times higher. Even in Antarctica, higher values are observed in coastal waters. For example Pecherzewski (1980) found values of up to 3 mg l^{-1} in Admiralty Bay, King George Island and up to 4 mg l^{-1} in Bransfield Strait. Probably similar values can be found in other highly productive waters around the continent. The Southern Ocean, south of the Antarctic Convergence, makes up about 12% of the global ocean surface, with an area of approximately 38×10^6 km^2 and a volume of the order of 150×10^6 km^3 (Carmack, 1990). With an organic carbon content of 1.5 mg l^{-1}, the Southern Ocean contains about 225×10^3 Mt. However, because the Southern Ocean is in continual exchange with more northerly waters, we will not consider the carbon content of the Southern Ocean as part of the carbon pool of Antarctica.

THE ICE SHEETS

Organic Carbon Content of the Ice Sheet

Estimates of the carbon content of the Antarctic ice sheet are rare. Because of the prevailing low temperatures organic life is absent or almost completely nonexistent in Antarctic snows of inland regions. Organic carbon does not appear to have been measured in ice core samples where the main interest has been in the isotopic or gas content of the ice.

In coastal regions, such as on the Antarctic Peninsula and near Davis Station, algal blooms giving rise to red and green snows are sometimes observed (Longton, 1985; Fogg, 1967; Heatwode et al., 1989), where the rapid appearance during summer is associated with accumulation of algal cells during snow melt and ablation, rather than by high productivity. Heatwode et al. (1989) reported organic carbon concentrations of 400 g m^{-3} in red snow near Casey Station, but these were very localized and would probably not contribute very much to the total carbon pool.

Some organic constituents may be trapped in falling snow and held within the ice mass, however. Downes et al. (1986) report carbon values of between 1 and 3 g m^{-3} in some samples of snow and glacial ice from the McMurdo Sound region. The organic matter was considered to be windblown material from lacustrine and marine sediments or possibly aerosol materials from the nutrient-rich Ross Sea.

One possible source of organic carbon for the Antarctic ice sheet may be protein-rich material blown inland from adjoining seas, in a similar manner to that described by Wilson (1959) for the origin of similar material in snows from New Zealand, and which was considered to be the source of nitrate salts in Antarctic soils by Claridge and Campbell (1968). However they considered that organic carbon compounds would have been fully oxidized during atmospheric transport, and therefore snow falling on the icecap would not contain appreciable amounts of carbon derived from this source. If the organic carbon found by Downes et al.(1986) is aerosolic in origin then it must have come from close to the continent and would not have blown far inland, so that only snows close to the coast would contain much organic carbon.

If the whole of the continental ice cap contains 1 g m^{-3} of carbon, then the total organic carbon contained in the ice cap would be 24×10^3 Mt. However, we consider that the organic matter brought in by winds off the sea will not penetrate far inland, and that carbon contents of the order of 1 g m^{-3} will not be found more than 100 km from the coast. We therefore consider that the total contribution of carbon to the pool from this source could not be more than 1% of the calculated figure, of the order of 250 Mt.

Organic Carbon Content in Floating Ice

According to Fox and Cooper (1994) the area of floating ice shelves is 1.555×10^6 km^2. In winter, the seas surrounding Antarctica freeze over and the area of floating ice surrounding the continent is greater than the continent itself. The sea ice provides an extensive but transient habitat for a wide variety of organisms, which live in the sheltered situations that it provides, especially on the underside. Microbial production associated with sea ice is a very significant contributor to the biological carbon pump. The thin layer at the underside of floating ice contributes about a quarter of the net primary production of the ice-covered ocean (Knox, 1990). Much of the fixed carbon is however released again to the oceans as the ice melts again and does not form part of the fixed carbon pool of the Antarctic.

However the floating ice shelves and associated fast ice around the coastal margins of the continent do not melt and, although there is little information on the organic carbon content at the base of the thick floating ice shelves, we may assume as a first approximation that the situation there is somewhat similar to that under the seasonal sea ice.

According to Archer et al. (1996), the bottom 5 cm of ice contains more than 20 g m^{-3} organic carbon at the time of maximum growth, while in the ice above values are very much lower, up to 0.1 g m^{-3}: very much less than the organic carbon figures found by Downes et al. (1986) for glacial ice. If this situation prevails under permanent floating ice then the contribution to the pool of carbon from this source would be 1.555×10^3 Mt.

Dissolved Carbon Dioxide in the Ice Sheet

As stated above, the atmosphere contains about 350 ppm of carbon dioxide, 1.7 ppm of methane, and 0.02 to 0.05 ppm of carbon monoxide. When snow falls it is relatively porous and contains trapped air, which has free interchange with the atmosphere, until the snow becomes sufficiently compact for gases to be trapped in discrete bubbles (Figure 4.2). At this stage further exchange with the atmosphere is not possible and the proportion of CO_2 and other gases in the trapped air remains unchanged during further compression.

Much work has been carried out on the composition of gases in ice cores from the Antarctic, particularly the cores from Vostok (Figure 4.1), where the oldest ice sampled was found to be up to 200,000 years old. Many measurements of the proportion of CO_2 in the gases extracted from

FIGURE 4.2 Determining the annual snow accumulation from a snow pit. Air within the snow retains CO_2 and other gases and is trapped within sealed pores. Increased precipitation from climate change may increase the CO_2 trapped in ice sheets, but this may be offset by accelerated ice sheet decay.

the ice cores have been made with a view to obtaining an indication of past atmospheric composition (Nakazawa et al., 1993; Barnola et al., 1991, 1995; Lorius et al., 1985, etc). These studies have shown that the CO_2 content of the atmosphere has increased markedly since preindustrial times as part of a general warming trend since the last glacial maximum, but was again higher during the last interglacial. These applications have considerable implications for climatic studies.

Most of the studies, however, concentrate on the proportion of CO_2 and other gases in the trapped air, but do not determine the CO_2 content of the ice itself. The only figure available is given by Van der Wal et al. (1994) who give a value of 16 µg kg^{-1} C as CO_2 in ice. If we assume that this value is representative of the whole of the ice contained in the Antarctic ice sheets, neglecting that which is in unsealed pores and available for exchange with the atmosphere, then the amount of carbon trapped as CO_2 is approximately 400 Mt. The other carbon-containing constituents of the atmosphere are present at much lower concentrations. The methane content is about 200 times less than that of carbon dioxide and the contribution of trapped methane to the carbon pool would be about 2 Mt.

Although there is no other measurement of the CO_2 content of Antarctic ice available, it is possible to calculate a maximum value. According to Barnola et al. (1991), air contained within firn is effectively sealed off from the atmosphere at a density of 0.8 g cm^{-3}. If all this air remains within the ice as it is compressed to its maximum density, then at a CO_2 content of 300 ppm, the maximum carbon content due to trapped CO_2 would be 40 µg kg^{-1}, giving a maximum possible value for the carbon content of the Antarctic ice sheet of 1000 Mt. We consider that some air would be lost during compression and transition from firn to glacial ice, and therefore that the value given by Van der Wal (1994) is reasonable.

At temperatures below –78.5°C carbon dioxide should precipitate in the solid form from the atmosphere in the central part of the ice cap during the winter when temperatures are very low. However, since the mean temperature of the ice sheet is above the freezing point of CO_2, it should not accumulate in this form within the ice. We know of no observations of the precipitation of solid CO_2 on the ice cap, however.

TERRESTRIAL CARBON POOLS

Extent and Distribution of Ice-Free Areas

As described above, recent estimates of the proportion of ice-free ground in Antarctica have been drastically reduced downwards in recent years from the 2% estimated by Shumskiy (1967) and accepted uncritically by many authors since then, to the very much lower figure of 0.33% or 46,000 km², derived by Fox and Cooper (1994). This figure is not a true measure of the extent of ice-free ground, but a measure of the extent shown on the best available maps. If the whole continent were to be accurately mapped at a larger scale, the calculated figure for ice-free ground would probably be even lower. Much of the ice-free ground mapped in Antarctica is bare rock, with little or no soil development, and it is difficult to quantify this amount. The exposures of ice-free ground are scattered throughout the Transantarctic Mountains, around the coast of East and West Antarctica, and in the Peninsula area, mostly on the islands around its tip.

The largest expanse of ice-free ground in Antarctica is the McMurdo Sound Oasis, in the mountains of South Victoria Land, on the western side of McMurdo Sound. Our estimate of the area of all ice-free ground, on Ross Island as well as on the continent, based on the ice-free ground shown on the USGS 1:250,000 map sheet — Ross Island and Vicinity — which covers the area between 76°30' and 78°45' S latitude and between 159° and 170° W longitude, is 6000 km², comparable with Clark's (1965) estimate of 4000 km² for the McMurdo dry valleys and related areas. In this region there are extensive areas of till and about 10% of the exposures are bare rock on steep ridges, etc. Other ice-free areas are scattered throughout the Transantarctic Mountains and the Ellsworth Mountains. None of these are as extensive as those of the McMurdo Sound region, but in total are significant in area. In these localities the proportion of bare rock appears to be higher. For purposes of this estimate of the carbon pool in soils we will assume that half the total ice-free areas in Antarctica are found in the Transantarctic Mountains and the Ellsworth Mountains, giving a total area of 23,000 km².

In the absence of any detailed information, we will assume that the area of ice-free ground around the coast of Antarctica is 11,500 km², with the remaining 11,500 km² found in the Peninsula.

Soils of the Transantarctic Mountains

Organic Carbon in Soils
The soils of the Transantarctic Mountains (Figure 4.3) are practically devoid of biological processes and have, with few exceptions, negligible contents of organic matter. They were described as ahumic by Tedrow and Ugolini (1966) and little evidence has been found during the extensive soil studies that have taken place in subsequent years to change this view (Campbell and Claridge, 1987).

Nevertheless, careful examination of the soils of this region show that habitable niches do exist and that organisms have colonized and radiated to fill most of them. The soils contain bacteria, yeasts, and algae in small amounts, while in favorable situations lichens and mosses grow. Organisms feeding on these: nematodes, amoebae, protozoa, and arthropods (mites, collembola) are also found. Even in situations seemingly hostile to life, such as bare rock outcrops, algae and lichens grow in a symbiotic relationship within the pores of some sandstones. In such places they exist in a kind of greenhouse, in a layer a few millimeters thick beneath the rock surface, where moisture is held between pores in the rock and temperatures can rise to well above freezing point for lengthy periods, due to high insolation during the summer when the sun is continually above the horizon (Friedmann, 1982).

The biomass and metabolic activity of the ahumic soils of the Transantarctic Mountains are considered the lowest of any soil ecosystem, and environmental constraints are at the limits of life for many groups of organisms (Cameron et al., 1970; Friedmann, 1993)

FIGURE 4.3 Ice-free area high in the Transantarctic Mountains with the East Antarctic Ice Sheet in the background. Biological activity is at a minimum here and the soil contains only traces of organic carbon.

Soils under the rare moss patches are shallow, commonly less than 10 cm deep, and stained slightly with organic matter. Moss patches are generally less than 0.25 m^2 in extent, even in the most favorable sites in the lee of boulders in old stream channels (Campbell and Claridge, 1987).

In places, algal peats have formed where shallow ponds had been, in which masses of cyanobacteria (blue-green algae) have grown and dried, leaving a layer of 5 to 6 cm of barely decomposed algae. These deposits are rare but are found in low-lying coastal situations (McCraw, 1967), or buried in palaeosols, or in reworked stream and lakeshore deposits.

Very few workers have measured the carbon content of soils of the Transantarctic Mountains, as they have been more interested in mineral weathering, salt accumulation, and staining as evidence of soil formation. However extensive studies by Cameron and co-workers (Cameron et al., 1970a,b; Cameron, 1974; etc.) and later work by Freckmann and Virginia (1997), have provided some data. Sheppard et al. (1994) analyzed soils from sites around Vanda Station in the Wright Valley, including some considered free from contamination. Carbon contents of some soils are shown in Table 4.1.

One site, the southernmost soil in the world, from Mt. Howe is significantly higher than the others and is excluded, as this soil may contain traces of coal, present in some of the sandstones from which the soil was formed. Other soils containing obvious traces of organic matter, such as the soils around Vanda Station, which on examination contained traces of dried algae blown from the nearby lake edge, or as in the case of one of the soils from Wheeler Valley containing obvious plant growth, are excluded. The mean value of 0.07 may be larger than the true value as the figure is dominated by the large number of analyses from Taylor Valley, which, being at low altitude with a relatively mild climate, may contain higher quantities of organic material than the majority of the soils of the Transantarctic Mountain region (Figures 4.4 and 4.5).

TABLE 4.1
Carbon Contents of Some Soils from the Transantarctic Mountains

Location	% C	Number	Ref.
Mt. Howe	0.35	1	Cameron, 1974
La Gorce Mts	0.02	1	Cameron, 1974
Taylor Valley	0.02	1	Cameron, 1974
Conrow Valley	0.03	1	Cameron, 1974
McKelvey Valley	0.03	1	Cameron, 1974
Victoria Valley	0.05	1	Cameron, 1974
Deception Island	0.02	1	Cameron, 1974
Taylor Valley	0.08	130	Freckmann and Virginia, 1997
Matterhorn	0.03	10	Cameron et al., 1970a
Wheeler Valley	0.03	13	Cameron et al., 1980b
Mean value	0.07	161	

The depth of soil to ice-cemented permafrost is between 10 and 50 cm (Campbell et al., 1997b). We will assume the maximum figure of 50 cm.

We will assume that the area of ice-free ground in the Transantarctic Mountains is 23,000 km^2, of which 10% is bare rock and 90%, or 21,000 km^2 is till on which soils have formed. Given an average carbon content of 0.07%, a soil density of 1.8 t m^{-3}, a figure close to that determined experimentally (Campbell et al., 1997c), the total organic carbon in soils of the Transantarctic Mountains is approximately 15 Mt.

Algae in Soils and Lakes

Soils containing more carbon, such as those under moss or containing algae from nearby lakes, are rare and probably make an insignificant contribution to the carbon pool. Algal peats although spectacular when found in this environment, cover an extremely small area and their contribution may also be ignored here. Although they were described by McCraw (1967), they were of insufficient extent to appear in his soil map of the Taylor Valley, the only detailed map of an extensive area of Antarctic soils to have been published.

Friedmann (1982) has estimated the total biomass of endolithic organisms in sandstones as 30 g m^{-2}, about half of which is carbon. Although not all rock outcrops are capable of supporting endolithic algae we will assume that 10% of the bare ground is rock, and all supports algae. Thus the carbon pool of endolithic algae in the Transantarctic Mountains is 0.345 Mt.

We estimate the total shoreline of the lakes in the McMurdo Oasis to be of the order of 200 km. Algae growing within the lakes dry out and blow around, where they may be found on the soil surface and under stones. At a maximum we estimate that a strip 10 m wide around the lake shore — narrower on steeply rising shores, wider in more gentle slopes, especially where there has been recent changes in lake level, carries soils with a carbon content of 0.5% in the upper 0.1 m according to values found by Sheppard et al. (1994). This would add a further 1.5×10^3 t to the carbon pool. If small ponds not marked on the maps, of which there are many within the McMurdo valleys, contribute an equal amount then the total contribution may be 3×10^3 t.

Soil Carbonates

Calcium carbonate deposits are found in some but not all Antarctic soils. In soils of the Hut Point Peninsula and the Taylor Valley, McMurdo Sound, calcite crusts form on the underside of surface stones, but these are rarely found elsewhere in Antarctica. Some of the soils of the Taylor Valley also contain free calcium carbonate. In some cases this is related to the presence of marble, which outcrops in this region, in the till on which the soil is formed (Campbell and Claridge, 1987).

FIGURE 4.4 On the edge of the Ross Sea, where temperatures are warmer than at inland sites, moss and lichens can grow in favorable situations and organic carbon may accumulate in the soils in some places.

Additionally, some calcium carbonate is currently being precipitated on the floors of lakes in the Taylor Valley as a result of biological activity, and some of the calcium carbonate deposits in the soils of the valley may have been precipitated in a similar manner in the larger lake which occupied the lower part of the valley during the last glacial maximum (Lawrence and Hendy, 1989).

Claridge (1965) measured calcium carbonate concentrations of between 0.7 and 4% in some soils of the Taylor Valley. We will assume that 10% of the ice-free area of the McMurdo Sound region, 500 km^2, carries calcareous soils with an average calcium carbonate concentration of 2%, corresponding to 0.24% C. This gives a contribution to the carbon pool of 50 Mt from this source.

Coastal Antarctica

Ice-free areas are scattered throughout the coastal areas of East Antarctica where the continental ice sheet has retreated from the coast. In most cases this retreat has taken place since the post-Wisconsin rise in sea level, but some areas may have been ice-free for longer periods. Temperatures

FIGURE 4.5 Moist soil surrounding the large boulder provides a suitable habitat for colonization by moss and lichen. In the Transantarctic Mountains such occurrences are restricted to those sites where water is present in the summer and seldom cover more than a few square meters.

are warmer than in the Transantarctic Mountains (Campbell and Claridge, 1987) and moisture is available for plant growth.

The ice-free areas are scattered and generally small in extent. One of the largest is the Vestfold Hills (Figure 4.1) where an area of over 400 km^2 is ice-free. The Vestfold Hills are low-lying and relatively level, but with prominent, if low, relief. Lakes occupy many of the basins and comprise about 20% of the area. The surface is largely bare rock, glacially striated, and till occurrences are patchy (Adamson and Pickard, 1986). The area offers a favorable habitat for plant growth, and wherever the surface is suitable, moses, lichens, and algae are found. Thus the soils may contain appreciable amounts of organic matter. Most soils are devoid of plant life, however, and it is probable that plant cover does not occupy more than 10 to 20% of the surface.

Conditions are similar in the Bunger Oasis (Wisniewski, 1983), which is about the same area as the Vestfold Hills, in the Schirmacher Oasis, where Molodezhnaya Station is situated (MacNamara, 1969), and in the Prince Charles Mountains, where large amounts of bare ground are also exposed. The most extensive and best-developed plant communities in Antarctica outside the Antarctic Peninsula are found in Wilkes Land, around Casey Station (Smith, 1990). Here, many hectares of windswept ice-free terrain support a relatively diverse cryptogramic flora, which forms dense stands of macrolichens, while in moister, more sheltered situations, bryophytes are abundant and locally form closed stands of 25 to 30 m^2, comprising a moss turf up to almost 30 cm deep.

The soils around Casey Station have been discussed by Blume and Bolter (1994) and Beyer et al. (1995). They mapped in detail an area of about 6 ha which may be representative of the Casey Station area, and by inference, of ice-free regions of coastal Antarctica. We will assume that the total ice-free area of coastal Antarctica is 11,500 km^2, which is probably an overestimate, and that the distribution of soils is similar in proportion to the small area mapped near Casey Station.

Tops of small hills are covered with a thin layer of fine-grained till, with a growth of mosses and lichens on the surface. There is only a very thin layer of organic-rich material on the surface, probably with about 20% of carbon in the upper 1 cm. These surfaces make up about 25% of the landscape and contribute 11.5 Mt to the carbon pool.

Sideslopes and footslopes are covered with thin till, 20 to 30 cm deep, on which soils, which Blume and Bolter (1994) describe as Dystri-gelic Leptosols and Lepti-gelic Podzols, occur under a largely lichen vegetative cover. The Leptosols are grayish-brown in color and contain about 1% of carbon throughout the profile. The Podzols have a thin grayish-brown surface horizon 1 cm thick underlain by a reddish-brown B horizon, and contain about 2.5% C throughout. These soils occupy about 30% of the mapped area and contribute 27.2 Mt C to the pool.

Deep tills in the valley floors are free of plant cover because of disturbance due to cryoturbation and occupy about 30% of the terrain. We will assume that they contain 0.1% C in the upper 100 cm, because of mixing of small amounts of organic matter throughout, and thus they contribute 8.0 Mt C to the pool.

In the lowest part of the topography, where moisture accumulates, Gelic Histosols, formed under a mixed cover of mosses and lichens, accumulate. These are stained a dark reddish-brown (5YR 2.5/1.5) to a depth of nearly 30 cm, and containing between 16 and 30% of organic carbon. Blume and Bolter (1994) also found an organic carbon content of between 16 and 27% in a soil under decomposing algae, which they described as a Fibri-gelic Histosol, and which would be very similar to the algal peats described by McCraw (1967), only much deeper. These occupy about 10% of the mapped area and contribute 60 Mt C to the pool.

The total contribution to the carbon pool from the soils of coastal Antarctica is thus 107 Mt.

Antarctic Peninsula

In the absence of any other data, we will assume that the ice-free areas of this region total 11,500 km². Much of the lower part of the Peninsula is mountainous, with a very heavy snowfall, but in places, bare ground occurs in relatively small patches. Even in the north, in the South Shetland Islands, exposures are small in extent. On Signy Island, for example, only 6.5 km² is ice-free, out of a total area of 18 km². Only 5% of Elephant Island is ice-free (O'Brien et al., 1979), about 25 km², while in the South Shetland Islands, the most studied island, King George Island, contains about 50 km² of ice-free ground, most of which is found on Fildes Peninsula, the site of a number of occupied bases. There are also extensive ice-free areas on the Trinity Peninsula, at the top of the Antarctic Peninsula, and throughout the Peninsula itself, mainly along the western coastline.

In the South Orkney Islands, the most northerly part of Antarctica to be considered in this discussion, the most studied area is Signy Island. About half of the island is ice-free, with a cool maritime climate. Temperatures range between –10 and +9°C, while precipitation averages about 400 mm annually. Winds are strong and frequent, while skies are mostly cloudy. Precipitation is frequent, although in small amounts — snow in the winter but frequently as light rain in the summer. Humidities are high and the soils are generally moist, and with regular freeze-thaw activity.

There is little vegetative cover on most slopes because of disturbance by cryoturbation and solifluction, but stable areas support widespread patches of mosses and lichens. Shallow peat, rarely exceeding 25 cm in depth, develops under moss carpets, but the slow rate of moss growth and the high microbial activity prevents significant organic matter accumulation. Under some moss species, peaty banks up to 2 to 3 m in depth can form, but with permafrost below an aerobic active layer 20 to 30 cm thick (Smith, 1979). Longton (1985) described the extent of the major vegetation types on Signy Island and elsewhere in the Peninsula region. The extent of plant cover and diversity of the vegetation decrease southward down the west coast of the Peninsula, while the eastern coast is even more sparsely vegetated.

There is some transient plant growth on soils on slopes and mineral debris, and these are richer in organic matter than the soils of coastal Antarctica. From the results quoted by Holdgate et al.

(1967) the mean carbon content of soils up to 140 cm deep is about 0.6%. Although carbon is added at the surface from decaying plant matter, the soils are mixed by cryoturbation. Approximately half of the ice-free area of the island is covered by these soils. Wherever vegetative growth is more extensive, especially in moist areas, there appears to be little decay of the organic debris and the upper part of the profile of wet soils where peat accumulates contains up to 46% carbon. Soils under moss and lichen may contain slightly less C, around 20 to 30%.

We will assume that half the remaining area is covered by soils containing 30% of carbon in the upper 25 cm, and 1% in the remaining material to 1 meter. The remaining half contains 45% carbon in the upper 25cm.

The soils of Elephant Island (O'Brien et al., 1979) appear to be similar to the soils on slopes of Signy Island, with carbon contents ranging between 0.3 and 2.8%. Here the topography seems too steep for stable sites to form and humus-containing layers to accumulate. On King George Island, plant cover is also not continuous, but under moss-lichen sod and algal crusts soil organic matter accumulates with organic matter contents of between 1 and 17% (Simonov, 1977; Zhu et al., 1991). The active layer above the permafrost is up to 1.5 m deep in favored sites. On the adjacent Trinity Peninsula (Everett, 1976) a similar distribution of soils is found. Soils on sloping topography have relatively low carbon contents while on level surfaces, peaty or organic soils with a surface layer approximately 25 cm deep of peaty or highly organic material with carbon contents up to 60% are found. It will be assumed that the same types of soil are found throughout the Peninsula area in about the same proportions.

Assuming a total ice-free area of 11,500 km^2 for the Antarctic Peninsula region, and that half of this is occupied by soils with a depth of 140 cm to permafrost and having an average carbon content of 0.6%, these soils will contribute a total of 70 Mt to the carbon pool. The remaining area, half, or 2875 km^2, carry soils under moss or lichen with an average carbon content in the upper 20 cm of 25% overlying a meter of material containing 0.6% C, will contribute 170 Mt. The peaty soils, with the upper 25 cm containing 45% C, will contribute 350 Mt, a total of approximately 600 Mt in the soils of the Peninsula region.

These calculations assume a bulk density of 1.5 m^{-3} for inorganic soils and of 1 t m^3 for peaty and organic soils.

Ornithogenic Soils

Soils inhabited by birds (ornithogenic soils) have a constant input of organic matter in the form of bird excreta, dead chicks, etc. resulting in a constant transfer of carbon from the oceans to the land. The most obvious of these are the rookeries occupied by Adelie penguins (*Pygoscelis adeliae*), where many pairs of individuals nest in close proximity to each other. Populations of these rookeries range from a few thousand up to half a million or more, depending of the availability of food sources in the sea. Individual rookeries may cover several hectares where topography and sea conditions are suitable. They are found all around the coastline of Antarctica, southernmost on Ross Island at 78° S, and as far north as the South Shetland Islands at 62° S.

Estimates of the total extent of penguin rookeries and the associated ornithogenic soils are hard to make, as the number of rookeries is still not known. However, Stonehouse (1985) estimated that the total population of Adelie penguins is of the order of 5 to 10 million, and that of chinstrap penguins (*Pygoscelis antarctica*) which nest on the northern part of the peninsula to be about 10 million. We will assume a total population of 20 million penguins nesting in Antarctica. If the nesting territory occupied by each pair of birds covers 2 m^2, then the maximum area of rookery and hence of ornithogenic soils is 20 km^2.

Ornithogenic soils are very similar throughout their range, from the southernmost rookeries on Ross Island (Heine and Speir, 1989; Speir and Cowling, 1984) to those on King George Island in the South Shetland Islands (Tatur and Myrcha, 1984). Adelie penguins make their nests from stones about 20 cm in diameter, so that ornithogenic soils occur in mounds consisting of stones and guano.

As the mounds build up, the guano decomposes and the proportion of stones increases. The carbon content of the soils range from 24% on Ross Island (Speir and Cowling, 1984) to 14% on King George Island (Tatur and Myrcha, 1984), falling to about 6% at depth. We will assume that the upper 20 cm of the profile contains 20% of C and the remaining 80 cm contains 10% C. This gives a total carbon pool of approximately 3 Mt.

Other predatory birds, such as skuas (*Catharacta maccormicki*) do not nest in large rookeries but in isolated sites along the coast. Thus they do not create large areas of ornithogenic soils but probably contribute the same amount of carbon to the soils. Estimates of the number of these birds nesting on land is between 1 and 10% of the number of penguins. We may assume that a further 0.3 Mt of carbon is added to the pool from this source, giving a total of 3.5 Mt.

Streams and Lakes

During the summer period in Antarctica, ground temperatures rise above freezing for considerable periods of time and lakes and streams become unfrozen. The longest river in Antarctica, the Onyx in Wright Valley flows for about 6 weeks each year, discharging an average of 3 million m^3 into Lake Vanda. Much of this water is lost by sublimation during the remainder of the year, although, in common with many other Antarctic lakes, the lake level is slowly rising (Chinn, 1993). The waters of these lakes and streams are relatively rich in dissolved inorganic nutrients and, because of the high insolation during the short summer period, are energy rich and biological productivity can be high.

The bottoms of many lakes are coated with thick, dark-green or purple mats of cyanobacteria which may be several centimeters thick. Algal mats may also form in streams, especially where the waters are shallow and velocities are low, such as in shallow reaches where the stream spreads out.

The freshwater lakes in the McMurdo Sound region cover an area of about 35 km^2, based on the latest 1:50,000 maps of the area. Some of these lakes are saline, warm, and anoxic at depth, others are fresh and remain frozen apart from a moat around the edge. In all of them, however, extensive algal and microbial mats form, and as a first assumption it will be assumed that the whole of the lake floors of all the lakes are covered with algal sediments varying in thickness from 2 mm near the lake shore to greater than 700 mm at depths below 10 m. Unpublished data (Hawes, I., personal communication) show that the carbon content of the top 1 cm of sediment is about 8 g m^{-2} and it will be assumed that this applies over the whole of the lake area. The contribution to the carbon pool from this source is thus 280 t, a very small figure in comparison with some of the other carbon pools.

Lakes are an extensive feature of the ice-free areas of coastal Antarctica, and assuming that the same conditions apply, and that they occupy about the same proportion of the total ice-free area, the contribution from lakes can be doubled, making it 560 t.

Downes et al. (1986) have studied dissolved and particulate organic carbon materials in streams and lakes of the McMurdo Sound region. In some streams the growths of cyanobacteria can be very luxuriant and fix considerable amounts of carbon during the short summer period when streams are flowing. In the McMurdo Sound region the total length of the major streams is approximately 70 km, with an average width of 2 m, of which 40 km contains appreciable growths of algae. Assuming a concentration of 10 g m^{-2} C for the algae coating the stream bed, the contribution from this source to the pool is 800 kg.

The lakes and stream waters themselves contain organic matter, both dissolved and particulate. According to Downes et al. (1986), the average dissolved organic carbon content of stream waters is 2.75 g m^{-2}, whereas McKnight et al. (1993) reported values averaging 0.5 g m^{-2}. Assuming that these differences represent seasonal variations in carbon production we will adopt a value of 1 g m^{-2}. Assuming a stream length of 70 km, measured from available maps, an average width of 2 m and an average depth of 20 cm, the total volume of these waters is 28×10^3 m^3, and the total dissolved carbon 28 kg, a negligible amount in contrast to other parts of the carbon pool.

If the lakes that receive these waters have the same composition as a minimum, then the carbon content of the lakes can be estimated. The largest and deepest lake, Lake Vanda, is approximately 6 km long and 1.5 km wide at its deepest point. It is 66 m deep. The total volume of the lake is approximately 270×10^6 m^3. The other lakes are generally shallower and somewhat smaller than Lake Vanda and probably contain in total the same amount of water, giving a volume of lake water of approximately 600×10^6 m^3. McKnight et al. (1993) estimate the total dissolved organic carbon in Lake Fryxell in the Taylor Valley, with a volume of 43×10^6 m^3, to be 330 t based on analyses of lake water and consideration of the annual input of dissolved organic carbon in the streams that feed the lake. Although the lakes differ markedly in chemistry, and in particular the changes in concentration of dissolved carbon with depth, we will assume that the same proportion of organic carbon to lake volume applies on average, and thus the total organic carbon contained within the lake waters is between 0.5 and 1 Mt.

The largest non-marine aquatic ecosystem in Antarctica is the McMurdo Ice Shelf, an ablation region of ice covered with thin till or marine sedimentary material, forming an interlinking system of lakes, pools, and streams occurring across more than 1500 km^2 (Howard-Williams et al., 1990). The base of the pools and lakes are coated with mats and films of cyanobacteria. Since the ice is ablating, the ponds and lakes are relatively mobile and the areas between the lakes have at some time been under water and contain dried algal matter.

According to Kaspar (personal communication) the sediments of the floors of the McMurdo Ice Shelf ponds contain on average about 3% organic matter or 1.2% C in a layer about 2 cm thick. The density of the sediment was about 1.8. Assuming that the distribution of organic matter is the same over the whole area, the McMurdo Ice Shelf contains about 0.6 Mt C.

The contribution to the organic carbon pool from streams and lakes from all these sources is then about 2 Mt.

Effects of Human Activity

Human activity in Antarctica, apart from the increasing amounts of carbon dioxide in the atmosphere which may accumulate in the ice cap, is limited to the small areas around bases and camps, mainly around the coast. Around such areas, especially in the earlier days of Antarctic exploration, rubbish, food scraps, spilt fuel, and other organic matter was spread around the area of the camps. Although procedures for waste disposal and fuel handling have been tightened up in recent years, the effects of earlier human activity remain. In one of the few studies of human influence that have been made, Sheppard et al. (1994) found that in the vicinity of Vanda Station, in the Wright Valley, total carbon values were much higher than for the pristine soils sampled well away from the base. Where wastewater had been disposed of for long periods, for example, carbon contents of up to 2.5% were encountered in the surface layer. Over the whole station area, 500 m^2, the average organic carbon content to a depth of about 1 m was 0.3%.

Vanda Station is small compared with the McMurdo Station-Scott Base complex on Ross Island, occupying about 1.5 km^2. There are about 30 occupied or abandoned bases on ice-free ground around the coasts of Antarctica, most of them smaller than the McMurdo-Scott Base complex. Assuming that on average each occupies or influences 1 km^2, the total area occupied by bases may be 30 km^2. Given a soil depth of 1 m, and a carbon content of 0.3%, similar to that found for Vanda Station, the contribution to the carbon pool arising from human contamination of soils is 0.09 Mt.

DISCUSSION

The contribution to the total carbon pool from the various sources discussed here is shown in Table 4.2. As explained in the text, most of the figures quoted here are highly speculative, as assumptions have been made in many directions in order to arrive at a final figure. However, we feel that in the absence of further information these figures at least give an indication of the order of magnitude of the contribution of the various sources to the carbon pool.

TABLE 4.2
Contribution to the Total Carbon Pool of
Antarctica from Various Sources (Rounded).
The Last Column is a Measure of the
Assumptions Made in the Calculation

Type of pool	Contribution	Accuracy
Ice shelves	1,555 Mt	Speculative
Ice sheet organic carbon	250 Mt	Very speculative
Dissolved gases	400 Mt	Fair
Transantarctic Mountains	15 Mt	Fair
Soil carbonates	50 Mt	Fair
Coastal soils	107 Mt	Speculative
Peninsula soils	600 Mt	Speculative
Lakes and rivers	2 Mt	Fair
Ornithogenic soils	3.5 Mt	Fair
Human activity	0.09 Mt	Speculative
Total	3,050 Mt	Speculative

The major contribution appears to be the organic-rich layer on the underside of floating ice. This may not strictly be part of the fixed pool since it is in exchange with the oceans, but can be regarded as fixed. Should the climate become warmer, and the ice shelves eventually disintegrate, which the popular press has inferred to be happening at the present time, then this carbon will be released.

The figure for organic carbon in the Antarctic ice sheet is very speculative, as it is based on very few data, but indicates that this vast mass of ice can effectively isolate external material that it contains. The amount of CO_2 and other gases trapped in the ice cap is even greater and this could be released to the atmosphere if the ice cap melts. It may, however, be dissolved in the increased volume of sea water that would eventuate from the melting of this ice.

Considerable amounts of carbon are stored in the organic-rich layers in coastal soils and those of the Peninsula. This comes about because the slow rate of decomposition of organic matter allows very high amounts of organic matter to accumulate near the soil surface, which, in some cases are higher than would be found in peats formed in temperate climates.

We consider that the figure for organic carbon for the soils of the Transantarctic Mountains is reasonable because the area is well defined and the soils are nearly all of low carbon content. However, the figure could be refined if more analyses were available.

For the soils of the coastal regions and of the Antarctic Peninsula area there is some doubt as to the areal extent of the ice-free areas, and this figure could be better defined. What needs to be evaluated, however, is the extent of the areas without plant growth, and of those with plant cover. Soil maps are desirable, but few are available, most workers being concerned with describing soil-plant associations rather than mapping. Thus the figures given for carbon content are very speculative, and could change markedly as better information becomes available.

In the event of an amelioration in the Antarctic climate due to climatic warming, the vegetation pattern found in the coastal regions of Antarctica and on the Antarctic Peninsula should spread further south. Unless the temperature warms to a very great extent, it can be expected that moss and lichen patches, at present of very limited extent in the Transantarctic Mountains, will increase in area, and the soils under them to thicken and organic matter content to increase. Thus a climatic warming can be expected to increase the amount of carbon stored in the soils. Plant growth may also become more abundant in the northern regions, but it is possible that most likely sites are already occupied by plants, and that the area covered by them would not increase markedly with increasing temperatures.

Other contributions, such as the carbon sources in lakes and rivers, and that of ornithogenic soils, are minor in relation to the other contributors to the pool.

The effect of human activity is small in comparison with most of the other pools of carbon in the Antarctic, but measurable and probably increasing.

CONCLUSIONS

These first estimates of the amounts and spatial distribution of the carbon stored in the ice and soils of Antarctica may be compared with estimates made for other parts of the world. For example Tate et al. (1997) have derived a total carbon stock of 6680 Mt for the soils and vegetation of New Zealand. The figure we have arrived at is about half that for New Zealand, but only about 950 Mt of this are stored in the soils. Climatic warming should permit this figure to increase markedly.

REFERENCES

Adamson, D.A. and Pickard, J., Physiography and geomorphology of the Vestfold Hills, in *Antarctic Oasis: Terrestrial Environments and History of the Vestfold Hills*, Academic Press, Sydney, 1986, p. 99.

Archer, S.D., Leakey, R.J.G., Burkill, P.H., Sleigh, M.A., and Appleby, C.J., Microbial ecology of sea ice at a coastal Antarctic site: community composition, biomass, and temporal change, *Mar. Ecol. Prog. Ser.*, 135, 179, 1996.

Balks, M.R., Campbell, D.I., Campbell, I.B., and Claridge, G.G.C., Interim results of 1993/94 soil climate, active layer and permafrost investigation at Scott Base, Vanda, and Beacon Heights. Special Report #1, Department of Earth Sciences, University of Waikato, Hamilton, NZ, 1995.

Bardin, V.I. and Suvetova, Y.A., Basic mathematic characteristics for Antarctica and budget of the Antarctic ice cover. Scientific reports of the Japanese Antarctic Research Expedition, Special Issue I, 92, 1967.

Barnola, J.M., Pimienta, P., Raynaud, D., and Korotkevich, Y.S., CO_2-climate relationship as deduced from the Vosok ice core: a reexamination based on new measurements and a reevaluation of air dating, *Tellus*, 43B, 83, 1991.

Barnola, J.M., Anklin, M., Porcheron, J., Raynaud, D., Schwander, J., and Stauffer, B., CO^2 evolution during the last millennium as recorded by Antarctic and Greenland ice, *Tellus*, 47B, 264, 1995.

Beyer, C., Sorge, C., Blume, H.P., and Schulten, H.R., Soil organic matter composition and transformation in gelic histosols of coastal continental Antarctica, *Soil Biol. Biochem.*, 10, 1279, 1995.

Blume, H.P. and Bolter, H.P., Soils of Casey Station, Wilkes Land, Antarctica, Proc. First Int. Conf. Cryopedol., Russian Academy of Sciences, *Puschino*, 2, 96, 1994.

Bouma, W.J., Pearman, G.I., and Manning, M.R., Eds., *Greenhouse, Coping with Climate Change*, CSIRO, Melbourne, 1996.

Bull, C., Snow accumulation in Antarctica, in *Research in the Antarctic*, Publ. # 93, Quam, L.O., Ed., American Association for the Advancement of Science, Washington, D.C., 1971, pp. 367-421.

Burdecki, F., Climate in the Graham Lang region, in Meteorology of the Antarctic, Van Rooy, M.P., Ed., Department of Transport, Capetown, South Africa, 1967, pp. 153-171.

Burton, H.R. and Campbell, P.J., The climate of Vestfold Hills, Davis Station, Antarctica, with a note on its effects on the hydrology of a deep saline lake, ANARE Sci. Rept. Ser. D, Meteorological Publ. No. 129, Australian Government Publishing Service, Canberra, 1980.

Cameron, R.E. and Benoit, R.E., Microbiology and ecological investigation of recent cinder cones, Deception Island, Antarctica, *Ecology*, 51, 802, 1970.

Cameron, R.E., King, J., and David, C.N., Microbiology, ecology, and microclimatology of soil sites in dry valleys of Southern Victoria Land, Antarctica, in *Antarctic Ecology*, Vol. 2, Holdgate, M.W., Ed., Academic Press, London, 1970a.

Cameron, R.E., King, J., and David, C.N., Soil microbiological ecology of Wheeler Valley, Antarctica, *Soil Sci.*, 109, 110, 1970b.

Cameron, R.E., Application of low-latitude microbial ecology to high-latitude deserts, in *Polar Deserts and Modern Man*, Smiley, T.H. and Zumberge, J.H., Eds., University of Arizona Press, Tucson, 1974.

Campbell, I.B. and Claridge, G.G.C., *Antarctica: Soils, Weathering Processes and Environment*, Elsevier, Amsterdam, 1987.

Campbell, I.B., Claridge, G.G.C., and Balks, M.R., Moisture content in the soils of the McMurdo Sound and Dry Valley region, Antarctica, in *Ecosystem Processes in Antarctic Ice-Free Landscapes,* Lyons, W.B., Howard-Williams, C., and Hawes, I., Eds., A.A. Balkema, Rotterdam, 1997a.

Campbell, I.B., Claridge, G.G.C., Campbell, D.I., and Balks, M.R., The soil environment of the McMurdo Dry Valleys, Antarctica, in *The McMurdo Dry Valleys, Antarctica: A Cold Desert Ecosystem,* Priscu, J., Ed., Antarctic Res. Ser., American Geophysical Union, Washington, D.C., 1997b. p. 297.

Campbell, D.I., MacCulloch, R.J.L., and Campbell, I.B., Thermal regimes of some soils in the McMurdo Sound region, Antarctica, in *Ecosystem Processes in Antarctic Ice-Free Landscapes,* Lyons, W.B., Howard-Williams, C., and Hawes, I., Eds., A.A. Balkema, Rotterdam, 1997c.

Carmack, E.C., Large scale physiography of polar oceans, in *Polar Oceans: Part A. Physical Science,* Smith, W.O., Ed., Academic Press, San Diego, 1990, p. 171.

Cheng, Z., Zhijiu, C., and Hei-gang, X., Characteristic of the active layers on Fildes Peninsula of King George Island, Antarctica, *Antarct. Res.,* 2, 24, 1991.

Chinn, T.J., Physical hydrology of the Dry Valley lakes, in *Physical and Biochemical Processes in Antarctic Lakes,* Antarctic Res. Ser. No. 59, American Geophysical Union, Washington, D.C., 1993, pp. 1-51.

Claridge, G.G.C., The clay mineralogy and chemistry of some soils from the Ross Dependency, Antarctica, *N.Z. J. Geol. Geophys.,* 8, 186, 1965.

Claridge, G.G.C. and Campbell, I.B., Origin of nitrate deposits, *Nature,* 217, 428, 1968.

Clark, R.H., The oases in the ice, in *Antarctica,* Hatherton, T., Ed., Reed, Wellington, New Zealand, 1965.

Downes, M.T., Howard-Williams, C., and Vincent, W.F., Sources of organic nitrogen, phosphorus and carbon in Antarctic streams, *Hydrobiologica,* 134, 215, 1986.

Everett, K.R., A survey of soils in the region of the South Shetland Islands and adjacent parts of the Antarctic Peninsula, Insitute of Polar Studies Report 58, Ohio State University, Columbus, OH, 1976.

Fogg, G.E., Observations on the snow algae of the South Orkney Islands, *Philos. Trans. R. Soc. London,* B252, 279, 1967.

Fox, A.J. and Cooper, P.R., Measured properties of the Antarctic Ice Sheet derived from the SCAR digital database, *Polar Record,* 30, 201, 1994.

Freckmann, D.W. and Virginia, R.A., Low-diversity Antarctic soil nematode communities: distribution and response to disturbance, *Ecology,* 78, 363, 1997.

Friedmann, I.E., Endolithic microorganisms in the Antarctic cold desert, *Science,* 215, 1045, 1982.

Friedmann, I.E., *Antarctic Microbiology,* John Wiley & Sons, New York, 1993.

Harrison, W.D., Permafrost response to surface temperature change and its implications for the 40,000-year-old history of Prudhoe Bay, *J. Geophys. Res.,* 96, 683, 1991.

Heatwode, H., Swinger, P., Spain, P., Kerry, E., and Donaldson, J., Biologic and chemical characteristics of some soils from Wilkes Land, Antarctica, *Antarct. Sci.,* 1, 225, 1989.

Heine, J.C. and Spier, T.W., Ornithogenic soils of the Cape Bird Adelie penguin rookery, Antarctica, *Polar Biol.,* 2, 199, 1989.

Holdgate, M.W., Terrestrial ecosystems in the Antarctic, *Philos. Trans. R. Soc. London,* B279, 5, 1970.

Holdgate, M.W., Allen, S.E., and Chambers, M.J.G., a preliminary investigation of the soils of Signy Island, South Orkney Islands, *Br. Antarct. Surv. Bull.,* 12, 53, 1967.

Holland, H.D., *The Chemistry of the Atmosphere and Oceans,* John Wiley & Sons, New York, 1978.

Houghton, J.T., Meira-Filho, L.G., Callendar, B.A., Harris, N., Kattenberg, A., and Maskell, K., Climate Change 1995, *The Science of Climate Changes,* Cambridge University Press, Cambridge, U.K., 1996.

Howard-Williams, C., Pridmore, R.D., Broady, P.A., and Vincent, W.F., Environmental and biological variability in the McMurdo Ice Shelf ecosystem, in *Antarctic Ecosystems: Ecological Change and Conservation,* Kerry, K.R. and Hempel, G., Eds., Springer-Verlag, Berlin, 1990.

Knox, G.A., Primary production and consumpyion in McMurdo Sound, Antarctica, in *Antarctic Ecosystems: Ecological Change and Conservation,* Kerry, K.R. and Hempel, G., Eds., Springer-Verlag, Berlin, 1990, pp. 115-128.

Koster, E.A., Nieuwenhuizen, M.E., and Judge, A.G., Permafrost and climate change: an annotated bibliography. Glaciological Data Report 27, University of Colorado, Boulder, 1994.

Lawrence, M.J. and Hendy, C.H., Carbonate deposition and Ross Sea ice advances, Fryxell Basin, Taylor Valley, Antarctica, *N.Z. J. Geol Geophys.,* 32, 267, 1989.

Longton, R.E., Terrestrial habitats — vegetation, *Key Environments — Antarctica,* Bonner, W.N. and Walton, D.H., Eds., Pergamon Press, Oxford, U.K., 1985.

Lorius, C., Bouzel, J., Ritz, C., Merlivat, L., Barkov, N.I., Korotkevich, Y.S., and Kotlyakov, V.M., A 150,00-year climatic record from Antarctic ice, *Nature*, 316, 591, 1985.

Lucas, M. and Lindesay, J.A., Global climate change: environmental and climatic links between Antarctica and South Africa, *S. Afr. J. Antarct. Res.*, 21, 193, 1991.

MacCulloch, R., The Microclimatology of Antarctic Soils, M.Sc Thesis, University of Waikato, Hamilton, N.Z., 1996.

MacNamara, E.E., Soils and geomorphic surfaces in Antarctica, *Biul. Peryglacjalny*, 20, 299, 1969.

McCraw, J.D., Soils of Taylor Dry Valley, Victoria Land, Antarctica, with notes on soils from other localities in Victoria Land, *N.Z. J. Geol. Geophys.*, 10, 498, 1967.

McKnight, D.M., Aiken, G.R., Andrews, E.D., Bowles, E.C., and Harnish, R.A., Dissolved organic material in Dry Valley lakes: a comparison of Lake Fryxell, Lake Hoare, and Lake Vanda, in *Physical and Biogeographical Processes in Antarctic Lakes*, Antarctic Res. Ser. No. 59, American Geophysical Union, Washington, D.C., 1993, p. 119.

Nakazawa, T., Machida, T., Esumi, K., Tanaka, M., Fujii, Y., Aoki. S., and Watanbe, O., Measurements of CO_2 and CH_4 concentrations in air in a polar ice core, *J. Glaciol.*, 39, 209, 1993.

O'Brien, R.M.G., Romans, J.C.C., and Robertson, L., Three soil profiles from Elephant Island, South Shetland Islands, *Br. Antarct. Surv. Bull.*, 47, 1, 1979.

Pecherzewski, K., Organic carbon (DOC and POC) in waters of Admiralty Bay, King George Island, South Shetland Islands, *Pol. Polar Res.*, 1, 67, 1980.

Pickard, J., The Vestfold Hills: a window on Antarctica, in *Antarctic Oases: Terrestrial Environments and History of the Vestfold Hills*, Pickard, J., Ed., Academic Press, Sydney, 1986, pp. 334-354.

Post, W., Emmanuel, W.R., Zinke, P.J., and Stangenberger, G., Soil carbon pools and world life zones, *Nature*, 298, 156, 1982.

Raynaud, D., Jouzel, J., Barnola, J.M., Chappelaz, J., Delmas, R.J., and Lorius, C., The ice record of greenhouse gases, *Science*, 259, 926, 1993.

Sheppard, D.S., Campbell, I.B., Claridge, G.G.C., and Deely, J.M., Contamination of soils around Vanda Station, Antarctica, Sci. Rep. 94/20, Institute of Geological and Nuclear Sciences, Lower Hutt, New Zealand, 1994.

Shumskiy, P.A., The Antarctic Ice Sheet, in Results of Research in Antarctica during 10 years, National Committee for Antarctic Research, Academy of Sciences of the U.S.S.R., Moscow, 1967, pp. 27-75.

Simonov, I.M., Physio-geographic description of the Fildes Peninsula, South Shetland Islands, *Polar Geogr.*, 1, 223, 1977.

Smith, R.I.L., Peat-forming vegetation in the Antarctic, in *Classification of Peat and Peatlands*, Kivinen, E., Heikurainen, L., and Pakarinen, P., Eds., International Peat Society, Helsinki, 1979, p. 58.

Smith, R.I.L., Plant Community Dynamics in Wilkes Land, Antarctica, *Proc. N.I.P.R. Symp. Polar Biol.*, 3, 229, 1990.

Speir, T.W. and Cowling, J.C., Ornithogenic soils of the Cape Bird Adelie Penguin Rookeries, Antarctica. I. Chemical Properties, *Polar Biol.*, 2, 199, 1984.

Stonehouse, B.C., Birds and Mammals — Penguins, in *Key Environments — Antarctica*, Bonner, W.N. and Walton, D.H., Eds., Pergamon Press, Oxford, U.K., 1985, pp. 266-292.

Tarnocai, C. and Lacelle, B., Soil organic carbon map of Canada, Map Eastern Cereals and Oilseed Research Centre, Agriculture and Agrifood Research Branch, Ottawa, Canada, 1996.

Tate, K.R., Giltrap, D.J., Claydon, J.J., Newsome, P.F., Atkinson, I.E.A., Taylor, M.D., and Lee, R., Organic carbon stocks in New Zealand's terrestrial ecosystems, *J. R. Soc. N.Z.*, 27, 315, 1997.

Tatur, S.A. and Myrcha, A., Ornithographic soils on King George Island, South Shetland Islands (Maritime Antarctic Zone), *Pol. Polar Res.*, 5, 31, 1984.

Tedrow, J.C.F. and Ugolini, F.C., Antarctic soils, in *Antarctic Soils and Soil-Forming Processes*, Tedrow, J.C.F., Ed., Antarctic Res. Ser. No. 8, American Geophysical Union, Washington, D.C., 1966, p. 161.

Tingey, R.J., The geological evolution of the Prince Charles Mountains, an Antarctic Archaen Cratonic Block, in *Antarctic Geoscience*, Craddock, C., Ed., University of Wisconsin Press, Madison, WI, 1982, p. 455.

Van der Wal, R.S.W., Van Roijen, J.J., Raynaud, D., Van der Berg, K., De Jong, A.F.M., Oerlemans, J., Lipenkov, V., and Heubrechts, P., From $^{14}C/^{12}C$ measurements towards radiocarbon dating of ice, *Tellus*, 46B, 94, 1994.

Wilson, A.T., Surface of the ocean as a source of airborne, nitrogenous material and other plant materials, *Nature*, 184, 99, 1959.

Wisniewski, E., Bunger Oasis: the largest ice-free area in the Antarctic, *Terra*, 95, 178, 1983.

5 Carbon and Nitrogen Storage in Upland Boreal Forests

J.S. Bhatti and M.J. Apps

CONTENTS

INTRODUCTION

In upland forest ecosystems, the main carbon (C) pools are associated with living biomass, detritus (including coarse woody debris) on the forest floor, and in the underlying mineral soil. Forest floor and mineral soils in global boreal forests contain approximately 200 Pg C (excluding peat), three times more than the estimated biomass pool (Apps et al., 1993). The dynamics of the forest floor, mineral soil, and biomass C pools are highly connected — the organic C content of the forest floor and the mineral soil is the net result of the processes relating to primary production, decomposition, and the factors controlling these processes. At the same time, potential biomass productivity is constrained by attributes of the forest floor and mineral soil, especially available N, soil drainage, and soil temperature (Vitousek et al., 1997; Bonan and Van Cleve, 1992; Vitousek and Howarth, 1991). Understanding the interacting processes, and the quantitative relationships between them are keys to improved projections of C budget responses to climate change and to potential management interventions (IGBP, 1998).

A number of attempts have been made to estimate the soil organic C (SOC) pool in Canadian soil using both models and various data sets. The soil C module of the Carbon Budget Model of Canadian Forest Sector (CBM-CFS2) (Kurz et al., 1992) estimated the different SOC pools in boreal forest ecosystems using a dynamic model. Tarnocai (1998) compiled data for SOC pools at the soil landscape level and estimated soil C at the soil order and ecoclimatic provinces level. Using data sets to calculate the C content at the ecoclimatic province scale, the presence of organic soil — which could be up to 10 m thick and averaging 200 to 400 cm (Tarnocai, 1984) — must be considered, especially in the arctic, subarctic, and boreal regions. Nevertheless many soil profiles in the boreal region are less than 10 cm thick (Siltanan et al., 1997). Failure to recognize these distinctions in the analysis of the boreal regions can easily lead to estimates of average C contents which are overly influenced by organic soil profiles atypical of the upland boreal forest. Here, we estimate the soil C content of upland forests in the western boreal region of Canada. Pedon data, presented in this study, are used to examine the processes and relationship that influence the C cycle in the boreal forest.

The objectives of this study are to: (i) estimate the amount of C in aboveground biomass and its relationship with drainage class, soil clay content, and total N; (ii) estimate the pools of C and N in forest floor and mineral soil; (iii) establish the relationship between these pools; and (iv) validate the C to N relationship against independent field data. Relating aboveground C to site conditions and estimating the different pools of C and N in forest floor and mineral soil are essential for the formulation and evaluation of the process models that are needed to assess the effect of climate change on soil C pools (Peng et al., 1998; Melillo, 1996; Rastetter et al., 1992).

METHODS

DESCRIPTION OF THE DATABASE

Data from two different sources were obtained which include the Boreal Forest Transect Case Study (BFTCS) part of the BOReal Ecosystem-Atmospheric Study (BOREAS) (Halliwell et al., 1997a,b) and soil profile and organic C database for Canadian forest and tundra mineral soils compiled by Siltanen et al. (1997). The BFTCS transect is oriented along an ecoclimatic gradient, ranging from agricultural grasslands in southern Saskatchewan through the boreal forest to the tundra in northern Manitoba (Figure 5.1). In total, there were 92 sites with 19 sites having no mineral horizons. In the Siltanen et al. (1997) report, there were a total of 374 boreal west sites with the majority of sites in Alberta, Saskatchewan, and Manitoba, from which 82 sites were excluded due to missing data (Figure 5.1). The dominant plant species on these sites included *Populus tremuloides* (aspen), *Picea mariana* (black spruce), *Pinus banksiana* (jack pine), and *Picea glauca* (white spruce).

Along the BFTCS, 97 sites at 84 distinct geographical locations, were sampled (Halliwell et al., 1997a). At each sampling point, a minimum number of 15 trees were sampled, with the size of the plots ranging from 25 to 100 m². For each tree, the data recorded included diameter at breast height (DBH), species, and height along with other variables. Stand density was calculated from the overstory data. Aboveground biomass (excluding foliage) was calculated for each live tree in the sample plot and was multiplied by the number of stems per hectare and the values were then summed. The biomass of each tree was estimated using a linearized equation (Singh, 1982). In this equation, biomass is a function of DBH and height. The database complied by Siltanen et al. (1997) does not have aboveground biomass information.

For BFTCS, soil samples were collected from a soil pit at each site. Each soil pit was excavated approximately 1 m or to bed rock and the profiles described using Canadian Soil Classification System (Agriculture Canada Expert Committee on Soil Survey, 1987). Samples from each horizon were analyzed for bulk density, C, N, and other chemical properties. The procedure for chemical analysis is described in detail by Halliwell and Apps (1997b). Only sites with a mineral soil horizon were included in the analysis; organic soils were not included as the processes affecting the site productivity and C accumulation are considerably different in such soils.

The pedons included in the BFTCS database have information on soil drainage class, forest floor thickness, bulk density, mineral soil horizon thickness, bulk density, % clay content, % silt content, total N content, and C content. We compiled the pedon information and calculated total C and N of the forest floor and of the mineral soil horizon to the depth of 100 cm, but for mineral soil with lithic contact (shallow soils over bedrock) the pools are calculated for the depth to bedrock contact. In the Siltanen et al. (1997) database, the average sampling depth of a mineral profile in

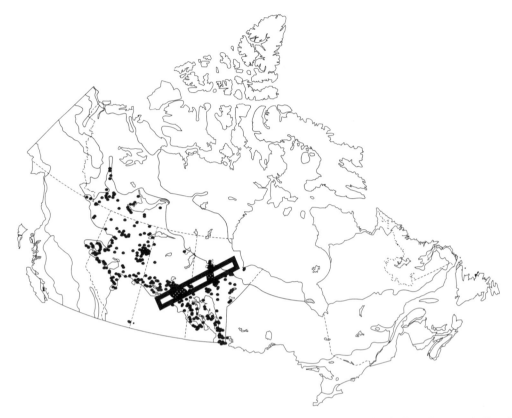

FIGURE 5.1 Map showing the location of sites in the soil profile database in the boreal west ecoclimatic province forests; the rectangle represents the BFTCS study area.

the western boreal region was 50 cm. For each horizon, a total quantity of soil C was calculated by multiplying organic C, bulk density, and horizon thickness. Similar calculations were done for total N content in each horizon. Loss on ignition (LOI) was converted to organic C by dividing by 1.724 (Kalra and Maynard, 1991). Horizons with missing data were assigned values from an adjoining, genetically similar horizons in the pedon, or assigned the average of values from the adjoining genetically similar horizon above and below it. The C values thus calculated for each layer were then combined to obtained the total C and N content in each pedon. Missing bulk density information was calculated using the empirical relationship between bulk density and soil organic C content (Grigal et al., 1989). Six classes of soil drainage were recognized and coded according to the following classes: (1) rapid, (2) well, (3) moderately well, (4) imperfect, (5) poorly, (6) very poorly drained.

STATISTICAL ANALYSIS

Using the BFTCS database, regression analyses were used to ascertain the relationship between aboveground biomass with clay content and total N. For aboveground biomass, the data were divided into two subsets on the basis of soil drainage class. The first subset included sites with very rapid, rapid, well, and moderately well drained. The second set included sites with imperfect, poor, and very poor soil drainage classes. Regression analyses were also performed to find the best predictive equation with forest floor soil organic C content, mineral soil organic C content as a dependent variable, and total N content as an independent variable. The best regression equation was chosen based on the adjusted r^2.

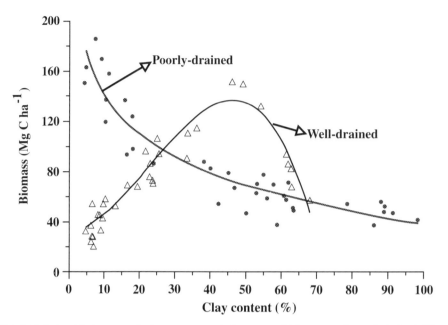

FIGURE 5.2 Relationship between % clay content and aboveground biomass under well-drained (Y = 3.19 + 0.20*X + 0.14*X² – 0.0021*X³; r² = 0.89; n = 34) and poorly-drained (Y = 240 - 4.4*ln(X); r² = 0.88; n = 35) conditions in the boreal forests.

RESULTS AND DISCUSSION

ABOVEGROUND BIOMASS

Aboveground biomass in the BFTCS database varied between 22 and 154 Mg ha⁻¹ with an average of 76 Mg ha⁻¹ for well-drained and 40 to 187 Mg ha⁻¹ with an average of 86 Mg ha⁻¹ under poorly drained sites. The higher biomass under poorly drained conditions is due to the presence of many mature black spruce stands in the BFTCS transect. In both the categories, there were some low-productivity stands. These biomass values were within the range of biomass values for the boreal forest in Alberta (70 Mg ha⁻¹), Saskatchewan (70 Mg ha⁻¹), and Manitoba (30 Mg ha⁻¹) calculated from independent inventory data (Bonnor, 1985). Conversion of biomass to C indicates that C storage in these boreal forests ranges from 11 to 77 Mg C ha⁻¹ for well-drained sites and 20 to 93 Mg C ha⁻¹ for poorly drained sites. These values are lower than the 50 to 150 Mg C ha⁻¹ reported for the Lake States forest (Grigal and Ohmann, 1992). The aboveground biomass was related to the soil clay content both under well-drained and poorly drained conditions (Figure 5.2). The significance of the relationships between soil clay content and aboveground biomass suggests that soil clay content plays a major role both in providing soil nutrients and the physical environment for plant growth. Under well-drained conditions, aboveground biomass increases with increasing clay content over an initial range, but then decreases rapidly as clay content increases above 45%. Clay content above a threshold of about 45% appears to limit biomass growth and this may be due to a restriction of soil rooting volume and poor aeration. Wang and Klinka (1996) also reported that fine-textured soil may introduce anoxic conditions which can adversely affect vegetation growth.

Under poorly drained conditions, there is an exponential decrease in aboveground biomass with increases in clay content (Figure 5.2). On coarser soil parent material, podzolization can create an iron (Fe) pan in the B horizon. This structural development can restrict drainage and lead to the formation of poorly drained soil in the boreal forest (Ugolini and Mann, 1979). Poor drainage

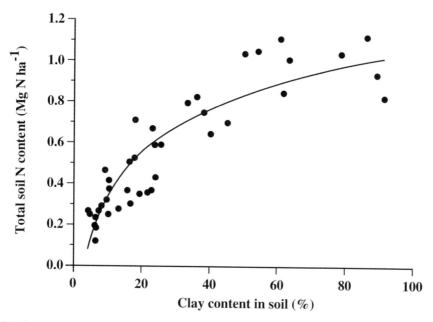

FIGURE 5.3 Relationship between soil clay content (%) and total N content ($Y = 0.31*\ln(X) - 0.37$; $r^2 = 0.82$; $n = 42$) in upland boreal forests.

adversely influences the available rooting volume, soil aeration, and soil temperature. Under poorly drained conditions, the less favorable conditions for root growth are directly expressed in aboveground productivity (Gale and Grigal, 1987). Lower aboveground biomass of spruce-fir forests in Maine (Williams et al., 1991), and willows (Labrecque et al., 1994), has been reported for poorly drained sites as compared to well-drained sites. Under waterlogged conditions, Jeglum (1974) concluded that with increases in clay content, black spruce growth was retarded due to anoxic or hypoxic conditions and reduced nutrient availability. Labrecque et al. (1994) also observed that trees on poorly drained sites took up less nutrient relative to those on well-drained sites. In boreal forests, drainage conditions worsen with increasing clay content, and the total N contents increase at a lower rate (Figure 5.3). Furthermore, with lower soil temperatures and reduced aeration, poorly drained conditions result in reduced N mineralization (Kimmins, 1996) and ultimately lower N availability (Klinka et al., 1994).

FOREST FLOOR ORGANIC CARBON CONTENT

The forest floor soil organic matter content (LFH + LH) was observed to vary about an average of 37 Mg ha[-1] with values between 16 Mg ha[-1] and 50 Mg ha[-1]. Nitrogen concentrations ranged from 0.24 to 2.34% with an average of 1.14% in the boreal forest. The average C:N ratio is 45 for the boreal forest. The observed variability in C and N content could be due to variability in three variables used in calculations namely, the concentration of C or N, bulk density, and the depth of the forest floor. The mean values of forest floor organic C are close to 33 Mg C ha[-1] reported by Vogt et al. (1986) for the boreal forest. The results from this study differ significantly from the estimated surface soil organic C contents of 118 Mg ha[-1] reported by Tarnocai (1998). We suggest, however, that the higher C results reported by Tarnocai (1998) were due to the inclusion of organic soils, agricultural soils, and other soil types along with the upland forested soils in his estimation of C in the boreal zone. Our values are much lower than the average of 83 Mg C ha[-1] in southeast Alaska (Alexander et al., 1989) but higher than the 20 Mg C ha[-1] for the mountainous region of Oregon (Homann et al., 1995), and the 17 Mg C ha[-1] in Minnesota, Wisconsin, and Michigan forests (Grigal and Ohmann, 1992).

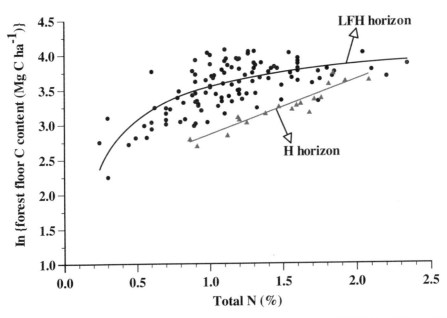

FIGURE 5.4 Relationship between soil C and total N in the forest floor horizons (LFH) (Y = 4.75 - 1.20 [$X^{0.5}$]$^{-1}$; $r^2 = 0.62$) and H horizon (Y = 2.07 + 0.79*X; $r^2 = 0.89$) in boreal forest soils.

The carbon in forest floor soil organic matter represents a very active portion of the forest ecosystem. The litter flux into this pool is the sum of all organic material transferred to the soil compartments. As expected, forest floor organic C contents were significantly related to the N concentration. However, this relationship is not linear: as N concentration increases, there does not appear to be a direct increase in LFH + LH (Figure 5.4) organic C content. These relationships indicate that soil organic C saturates at N concentration above 1.3% in LFH + LH. It appears that there is a threshold level, or a limiting process, above which no accumulation of forest floor organic C takes place. This would suggest that at N concentrations above 1.3%, the decomposition of forest floor organic C is not limited by the N supply and there will be net mineralization. Van Cleve et al. (1986) studied a range of forest site types in interior Alaska and concluded that forest floor N concentration had strong influence on soil N mineralization rates.

In the humus layer (H), the average organic C content was 27.5 Mg ha^{-1} (17 to 39 Mg ha^{-1}) with a N concentration of 1.51% (0.86 to 2.08%) with C:N ratio of 32. Since the H layer is an advanced stage of decomposition due mainly to the action of microorganisms (Jenny, 1980), the average N concentration was higher with lower C content than that of LFH+LH layers. A linear relationship was evident between H layer organic C content and N concentration (Figure 5.4) indicating that as the concentration of N increases in the H layer, there was corresponding increases in the organic C content. Total N is a useful index of available soil N at a regional scale (Yin, 1992; Binkley and Hart, 1989). Therefore, as available N increases, site productivity is also expected to increase.

MINERAL SOIL ORGANIC CARBON CONTENT

The mineral soils showed more variability in the soil organic C content than did the organic litter layers. For the mineral soil, the average C content was 55 Mg C ha^{-1}, with a range from 9 to 231 Mg C ha^{-1}. The average N content was 1210 kg ha^{-1} with a range from 18 kg ha^{-1} to 5860 kg ha^{-1} in the western boreal zone. The average depth of sampling for the mineral horizon in this data set was only 50 cm and data were not extrapolated to a depth of 100 cm.

There were considerable variations in the variables used to calculate the soil organic C and total N. These results indicate that upland boreal forests retain significant amounts of N along with

TABLE 5.1
Mineral Soil Organic C Content (Mg C ha⁻¹) in Relation to Drainage Classes for Upland Boreal Forest.

Drainage class	Mineral soil organic C (Mg C ha⁻¹)					
	Mean	Maximum	Minimum	Median	Std. dev.	nᵃ
Rapid	18	61	4	12	14	26
Well	36	142	8	31	29	87
Moderately well	44	189	14	35	41	97
Imperfect	32	134	7	24	31	34
Poor	21	55	4	19	12	24

ᵃ Number of observations.

C. Melillo (1996) made a similar observation that the soil was a major sink of N in forest ecosystems of the northeastern U.S. The amount of C measured in this study is comparable to the organic C in mineral soils (up to 100 cm) of 111 to 190 Mg C ha⁻¹ reported by Post et al. (1982) for global boreal forests, but much lower than the 505 Mg C ha⁻¹ calculated by Tarnocai (1998). As noted previously, however, Tarnocai's (1998) calculation includes organic soils (peat) which were excluded in our calculations. The values we report are comparable to those reported for organic C in the mineral soils (to 100 cm) of other regions. The average for 169 forest sites in Minnesota, Wisconsin, and Michigan was 105 Mg C ha⁻¹ (Grigal and Ohmann, 1992) while in north-central U.S., the average for mineral soils was 107 Mg C ha⁻¹ (Franzmeier et al., 1985). The average for 149 forest profiles in southeast Alaska was 185 Mg C ha⁻¹, with some profiles extending to a depth of 150 cm (Alexander et al., 1989). Differences in all soil-forming factors, including climate, vegetation, parent material, topography, and especially moisture regime (Goulden et al., 1998) may contribute to the differences between soil organic C of the western boreal zone and other regions. The world average for all soils is 117 Mg C ha⁻¹ to a 100-cm depth (Eswaran et al., 1993).

The soil drainage regime had a strong influence on the soil organic C content (Table 5.1). As drainage classes go from rapid to moderately well-drained, soil C increases, and under imperfect and poor drainage classes, soil C content decreased. Lower organic C may be due to higher productivity under well-drained conditions than under poorly drained conditions (Labrecque et al., 1994). Total N has a linear relationship with mineral soil organic C (Figure 5.5). This indicates that as the soil N content increases there were linear increases in soil C. The N in soil is influenced by the same factors as the organic C contents — namely, the soil-forming factors of climate, topography, vegetation, parent material, and age. Relations between soil organic C and edaphic conditions (Jenny, 1980) have been studied extensively at both local and regional levels. Soil clay content explained 42% variability in mineral soil C and 54% in total N content. In a literature review, Martin and Haider (1986) concluded that mineral soil C increases with an increase in the amount of clay in soil. The climatic factors that control the plant decomposition process in the short term, however, have no significant influence on mineral soil organic C content. The clay content is related to the formation of organo-mineral complexes and soil aggregates (Oades, 1988) thus physically protecting the soil C from mineral degradation.

OVERALL RELATIONSHIP

The relationship developed between forest floor (LFH + LH) organic C and mineral soil C with N concentration and N content, respectively, was tested using BFTCS data. The predicted C pools were compared with the measured mean forest floor and mineral soil C content. Based on the paired *t*-test, there was no difference between the predicted and measured forest floor organic C contents. The calculated forest floor organic C contents were higher then the measured values and

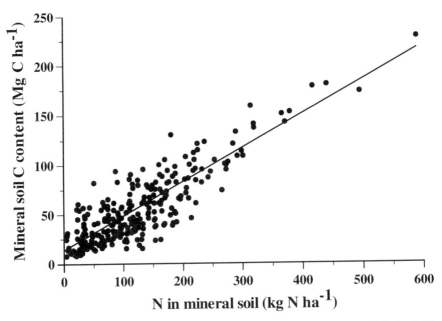

FIGURE 5.5 Relationship between soil C and total N in the mineral soil ($Y = 0.35*X + 13.1$; $r^2 = 0.75$; n = 284) in boreal forest soils.

were within 10% of the 1:1 line (Figure 5.6a). The relationship between mineral soil C and N contents was also tested using BFTCS data set and it was found that for some sites, the model underpredicts the C content (Figure 5.6b). This could be due to the fact that measured C content in the BFTCS data included both well-drained and poorly drained sites but the boreal west data set used in model development included only the upland forest sites. This is further support for the generality of the relationships.

IMPLICATIONS FOR CLIMATE CHANGE

The LFH and H layer organic C is part of the labile organic matter (LOM) pool in soil and respond quickly to changing external conditions. This C pool has a fast turnover rate. Since the C:N ratio of this material is high (Post et al., 1985), a large amount of CO_2 will be released through decomposition of C in the pool, with an accompanying small release of N. Mineral soil C, however, is mainly comprised of compounds having long turnover times and contains low C:N material. It therefore decays with relatively small amounts of CO_2 release and large net releases of N per unit of mineral soil C. If this is the case, with climate change and declines in the soil organic C pool, there will be corresponding increases in available N which can contribute to increased plant productivity (Schimel et al., 1994). The realization of the potential increase in plant productivity would be determined by other factors such as changes in species and competitive interactions, water availability, and the effect of temperature increase on photosynthesis and respiration. In the long term, as C approaches equilibrium any increase in plant growth may ultimately result in higher C storage in mineral soil.

CONCLUSIONS

Total N is known to be an important variable for determining present regional, continental, and circumpolar boreal forest C stores and for projecting their future change. Aboveground biomass varied between 22 and 154 Mg ha[-1] (mean = 76 Mg ha[-1]) for well-drained and 40 to 187 Mg ha[-1] (mean = 86 Mg ha[-1]) for poorly drained sites. Soil clay content affected the aboveground biomass

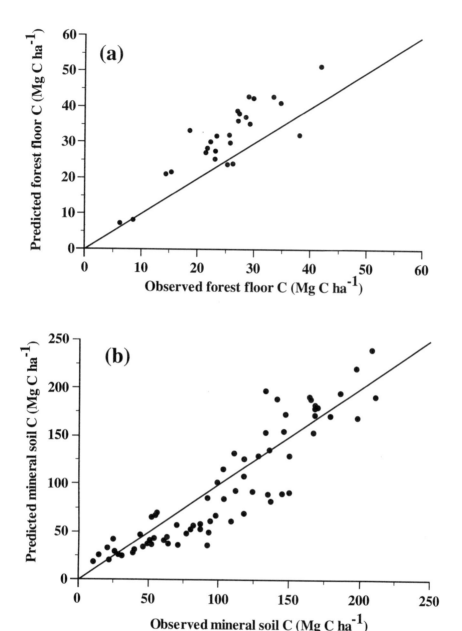

FIGURE 5.6 Observed vs. predicted soil C in forest floor (a) and mineral soil (b) organic C content for BFTCS sites.

by influencing the soil nutrient regime, physical environment, and moisture regime. Observed values of boreal forest soil organic carbon (SOC) on the forest floor range from 16 to 50 Mg C ha^{-1} (mean = 37 Mg C ha^{-1}) and in mineral soil, from 9 to 231 Mg C ha^{-1} (mean = 55 Mg C ha^{-1}). Soil drainage was also found to strongly influence both aboveground biomass and mineral SOC. Variability in concentration, bulk density, and depth of soil all contributed to the variation in forest floor and mineral soil C and N contents. Regression analyses indicated that SOC is related to the total N content, but on the forest floor C content is strongly dependent on the N content. Site properties such as soil texture and drainage have significant influence on the C sequestration capacity of boreal forest. With higher N availability, boreal forest productivity will increase under a changed climate, resulting ultimately in their higher C sequestration capacity. Such gains may, however, not be realized during a transient change in climate.

ACKNOWLEDGMENTS

Special thanks to D.H. Halliwell and R.M. Siltanen for valuable discussion, information, and assistance. Funding for this study was contributed by the Sustainable Forest Management Network of Centres of Excellence (NCE), and the Energy from the Forest (ENFOR) program of the Federal Panel on Energy Research and Development (PERD).

REFERENCES

Agriculture Canada, Expert Committee on Soil Survey, The Canadian System of Soil Classification, 2nd ed., Pub. 1646, Research Branch, Agriculture Canada, Ottawa, 1987.

Alexander, E.B., Kissinger, E., Huecker, R.H., and Cullen, P., Soils of southeast Alaska as sinks for organic carbon fixed from atmospheric carbon dioxide, in Proceedings of Watershed '89, a Conference on the Stewardship of Soil, Air, and Water Resources, Alexander, E.B., Ed., U.S. Department of Agriculture Forest Service, Juneau, AK, Mar 21-23, 1989.

Apps, M.J., Kurz, W.A., Luxmoore, R.J., Nilsson, L.O., Sedjo, R.J., Schmidt, R., Simpson, L.G., and Vinson, T., The changing role of circumpolar Boreal forests and tundra in global C cycle, *Water, Air, Soil Pollut.,* 70, 39, 1993.

Binkley, D., and Hart, S.C., The components of nitrogen availability in forest soils, *Adv. Soil Sci.,* 10, 57, 1989.

Bonnor, G.M., Inventory of forest biomass in Canada, Inf. Rep. Fo42-80/1985, Canadian Forestry Service, Ottawa, Canada, 1985.

Bonan, G.B. and Van Cleve, K., Soil temperature, nitrogen mineralization, and carbon source-sink relationships in boreal forests, *Can. J. For. Res.,* 22, 629, 1992.

Eswaren, H., ven den Berg, E., and Reoch, P., Organic carbon in soils of the world, *Soil Sci. Soc. Am. J.,* 57. 192, 1993.

Franzmeier, D.P., Lemme, G.D., and Miles, R.J., Organic carbon on soils of north central U.S., *Soil Sci. Soc. Am. J.,* 49, 702, 1985.

Gale, M.R. and Grigal, D.F., Vertical root distributions of northern tree species in relation to successional status, *Can. J. For. Res.,* 17, 829, 1987.

Goulden, M.L., Wofsy, S.C., Harden, J.W., Trumbore, S.E., Crill, P.M., Gower, S.T., Fries, T., Daube, B.C., Fan, S.M., Sutton, D.J., Bazzaz, A., and Munger, J.W., Sensitivity of boreal forest carbon balance to soil thaw, *Science,* 279, 214, 1998.

Grigal, D.F. and Ohmann, L.F., Carbon storage in upland forests of the Lake States, *Soil Sci. Soc. Am. J.,* 56, 935, 1992.

Grigal, D.F., Brovold, S.L., Nord, W.S., and Ohmann, L.F., Bulk density of surface soils and peat in north central U.S., *Can. J. Soil Sci.,* 69, 895, 1989.

Halliwell, D.H. and Apps, M.J., BOReal Ecosystem-Atmosphere Study (BOREAS) biometry and auxiliary sites: over-story and under-story data. Inf. Rep. Fo42-266/2-1997E, Northern Forest Research Center, Canadian Forest Services, National Resources Canada, Edmonton, Alberta, 1997a.

Halliwell, D.H. and Apps, M.J., BOReal Ecosystem-Atmosphere Study (BOREAS) biometry and auxiliary sites: soils and detritus data. Inf. Rep. Fo42-266/3-1997E, Northern Forest Center, Canadian Forest Services, National Resources Canada, Edmonton, Alberta, 1997b.

Homann, P.S., Sollins, P., Chappell, H.N., and Stangenberger, A.G., Soil organic carbon in a mountainous forested region: relation to site characteristics, *Soil Sci. Soc. Am. J.,* 59, 1468, 1995.

IGBP Terrestrial Carbon Working Group, The terrestrial carbon cycle: implications for the Kyoto protocol, *Science,* 280, 1393, 1998.

Jeglum, J.K., Relative influence of moisture-aeration and nutrients on vegetation and black spruce growth in northern Ontario, *Can. J. For. Res.,* 4, 114, 1974.

Jenny, H., *The Soil Resources,* Springer-Verlag, New York, 1980.

Kalra, Y.P. and Maynard, D.G., Methods Manual for Forest Soil and Plant Analysis, Inf. Rep. NOR-X-319, Northern Forest Research Center, Natural Resources Canada, Edmonton, Alberta, 1991.

Kimmins, J.P., Importance of soil and role of ecosystem disturbance for sustained productivity of cool temperate and boreal forests, *Soil Sci. Soc. Am. J.,* 60, 1643, 1996.

Klinka, K., Wang, Q., and Kayahra, G.J., Quantitative characterization of nutrient regime in some boreal forest soils, *Can. J. Soil Sci.,* 74, 29, 1994.

Kurz, W.A., Apps, M.J., Webb, T.M., and McNamee, P.J., The Carbon Budget of the Canadian Forest Sector: Phase 1, Inf. Rep. NOR-X-326, Northern Forest Research Center, National Resources Canada, Edmonton Alberta, 1992.

Labrecque, M., Teodorescu, T.I., Babeux, P., Cogliastro, A., and Daigle, S., Impact of herbaceous competition and drainage conditions on the early productivity of willows under short-rotation intensive culture, *Can. J. For. Res.,* 24, 493, 1994.

Martin, J.P. and Haider, K., Influence of mineral colloids on turnover rates of soil organic carbon, in *Interactions of Soil Minerals with Natural Organics and Microbes,* Huang, P.M. and Schnitzer, M., Eds., SSSA Spec. Publ. 17, Soil Science Society of America, Madison, WI, 1986, p. 284.

Melillo, J.M., Carbon and nitrogen interactions in the terrestrial biosphere, in *Global Change and Terrestrial Ecosystems,* Walker, B. and Steffen, W., Eds., Cambridge University Press, Cambridge, U.K., 1996, p. 431.

Oades, J.M., The retention of organic matter in soils, *Biogeochem.,* 5, 35, 1988.

Peng, C.H., Apps, M.J., Price, D.T., Nalder, I.A., and Halliwell, D.H., Simulating carbon dynamics along the Boreal Forest Transect Case Study (BFTCS) in central Canada. I. Model testing, *Global Biogeochem. Cycles,* 12, 381, 1998.

Post, W.M., Emanuel, W.R., Zinke, P.J., and Stangenberger, A.G., Soil carbon pools and world life zones, *Nature,* 208, 156, 1982.

Post, W.M., Pastor, J., Zinke, P.J., and Stangenberger, A.G., Global pattern of soil nitrogen storage, *Nature,* 317, 613, 1985,

Rastetter, E.B., McKane, R.B., Shaver, G.R., and Mellilo, J.M., Changes in C storage by terrestrial ecosystems: how C, N interaction restrict responses to CO_2 and temperature, *Water, Air, Soil, Pollut.,* 64, 327, 1992.

Schimel, D.S., Braswell, B.H., Holland, E. A., McKeown, R., Ojima, D.S., Painter, T.H., Parton, W.J., and Townsend, A.R., Climate, edaphic, and biotic controls over-storage and turnover of carbon in soils, *Global Biogeochem. Cycles,* 8, 279, 1994.

Siltanen, R.M., Apps, M.J., Zoltai, S.C., Mair, R.M., and Strong, W.L., A Soil Profile and Organic Carbon Database for Canadian Forest and Tundra Mineral Soils, Inf. Rep. Fo42-271/1997E, Northern Forest Research Center, National Resources Canada, Edmonton Alberta, 1997.

Singh, T., Biomass equations for ten major tree species of the prairie provinces, Inf. Rep. NOR-X-242, Northern Forest Research Center, National Resources Canada, Edmonton, Alberta, 1982.

Tarnocai, C., The amount of organic carbon in various soil orders and ecoprovinces in Canada, in *Soils and Global Change,* Lal, R., Kimble, J.M., Follett, R.F., and Stewart, B.A., Eds., CRC Press, Boca Raton, FL, 1998, p. 81.

Tarnocai, C., Peat Resources in Canada, Division of Energy, Peat Energy Program NRCC 24140, National Research Council Canada, Ottawa, 1984.

Ugolini, F.C. and Mann, D.H., Biopedolgial origin of peatlands in southern Alaska, *Nature,* 281, 366, 1979.

Van Cleve, K., Heal, O.W., and Roberts, D., Bioassay of forest floor nitrogen supply to plant growth, *Can. J. For. Res.,* 16, 1320, 1986.

Vitousek, P.M., Aber, J.D., Howarth, R.W., Likens, G.E., Matson, P.A., Schindler, D.W., Schlesinger, W.H., and Tilman, D.G., Human alteration of the global nitrogen cycle: sources and consequences, *Ecol. Appl.,* 7, 737, 1997.

Vitousek, P.M. and Howarth, R.W., Nitorgen limitation on land and sea: how can it occur?, *Biogeochem.,* 13, 87, 1991.

Vogt, K.A., Grier, C.C., and Vogt, G.J., Production, turnover, and nutrient dynamics of above- and belowground detritus of world forests, *Adv. Ecol. Res.,* 15, 303, 1986.

Wang, G.G. and Klinka, K., Classification of moisture and aeration regimes in subboreal forest soils, *Environ. Monit. Assess.,* 39, 451, 1996.

Williams, R.A., Hoffman, B.F., and Seymour, R.S., Comparison of site index and biomass production of spruce/fir stands by soil drainage class in Maine, *For. Ecol. Manage.,* 41, 279, 1991.

Yin, X., Empirical relationships between temperature and nitrogen availability across North American forests, *Can. J. For. Res.,* 22, 707, 1992.

6 Carbon Pools in Soils of the Arctic, Subarctic, and Boreal Regions of Canada

C. Tarnocai

CONTENTS

INTRODUCTION

Northern soils, especially those affected by permafrost, contain large amounts of organic carbon. Post et al. (1982) estimated that 27% of the world's soil carbon occurs in tundra and boreal forest ecosystems. Since a large part of Canada lies within these regions, a significant portion of the world's soil carbon occurs in Canadian territory.

    The Arctic, Subarctic, and Boreal ecoclimatic provinces (Ecoregions Working Group, 1989) cover approximately 75% of the area of Canada. Some information concerning amounts of soil carbon in the various ecoclimatic provinces is available from other researchers, including values for various global life zones (Post et al., 1982), and for the ecological provinces in Canada (Kurz et al., 1992), which also includes an estimate of 76.4 Gt (gigatonne; 1 Gt = 10^9 tonnes = 10^{12} kilograms = 10^{15} grams) for the mass of organic carbon in Canadian forest soils. Most of these estimates use a relatively small number of pedons to represent large areas. For example, estimates for the Arctic and Boreal areas of the entire world (Post et al., 1982) use a total of 308 pedons, with only 48 pedons being from Arctic areas and 260 from Boreal areas. More recently, Tarnocai (1997), using the Canadian Soil Organic Carbon Database, provided soil organic carbon estimates for all Canadian ecoclimatic provinces and subprovinces. These estimates indicate that the Arctic, Subarctic, and Boreal regions contain approximately 87.9% (231.7 Gt) of the total amount of soil organic carbon occurring in all Canadian soils.

In this chapter, the surface and total soil organic carbon masses and contents in the Arctic, Subarctic, Boreal, and portions of the Cordilleran ecoclimatic provinces and regions will be given both by region and on the basis of the soil orders occurring in these areas. In addition, some carbon pools will be identified and discussed, including the carbon pools in deep soil layers (>1 m depth).

STUDY AREA

The study area includes all of the Arctic, Subarctic, and Boreal ecoclimatic provinces and the subarctic and boreal portions of the Cordilleran and Interior Cordilleran ecoclimatic provinces (Figure 6.1). These ecoclimatic provinces, which are described in detail by the Ecoregions Working Group (1989), are hereinafter referred to as the Arctic, Subarctic, Boreal and Cordillera. Short descriptions of these ecoclimatic provinces are given in the following sections.

ARCTIC

The Arctic, which covers approximately 2375×10^3 km^2 (land area; soil area 1509×10^3 km^2), has a triangular shape. Its southern border is the Arctic tree line, the northern limit of trees. This border extends eastward from the Yukon–Alaska border (at approximately lat. 69.5° N, long. 142° W) across the middle of Hudson Bay, to the Atlantic coast (at approximately lat. 57.5° N, long. 62° W). Its northernmost point is the northern tip of Ellesmere Island. The Arctic is characterized by a nearly continuous shrub–tundra vegetation in the south, grading to a sparse cover of dwarf shrub, herb, moss, and lichen vegetation in the north. Permafrost is continuous in this region and the dominant soils are Cryosols (Soil Classification Working Group, 1998). The Arctic is subdivided into High, Mid- and Low Arctic regions (Figure 6.1), hereinafter referred to as the High Arctic, Mid-Arctic, and Low Arctic. Detailed descriptions of these regions are found in Ecoregions Working Group (1989).

SUBARCTIC

The Subarctic, which lies south of the arctic tree line, covers approximately 1712×10^3 km^2 (land area; soil area 1500×10^3 km^2). Its southern boundary extends eastward from the Yukon–Alaska border (at approximately lat. 65° N, long. 142° W) to just south of the tip of James Bay, and then to the Atlantic coast (at approximately lat. 58° N, long. 54° W). The Subarctic is characterized by an open-canopy coniferous forest having a ground cover of ericaceous shrubs and lichens. Permafrost ranges from continuous in the north to widespread discontinuous in the south. The dominant soils are Cryosols, Brunisols, and Podzols. The Subarctic is subdivided into High, Mid- and Low Subarctic regions (Figure 6.1). These regions, which are described in detail in Ecoregions Working Group (1989), are hereinafter referred to as the High Subarctic, Mid-Subarctic, and Low Subarctic.

BOREAL

The Boreal, which covers approximately 2521×10^3 km^2 (land area; soil area 2218×10^3 km^2), lies south of the Subarctic. Its southern boundary extends eastward from near the Yukon–Alaska border (at approximately lat. 57° N, long. 130° W) across the northern shores of Lake Superior, to the northeastern part of Nova Scotia. The Boreal is characterized by a closed-canopy coniferous forest or a mixed coniferous forest with aspen and birch. Open-canopy coniferous forests occur on

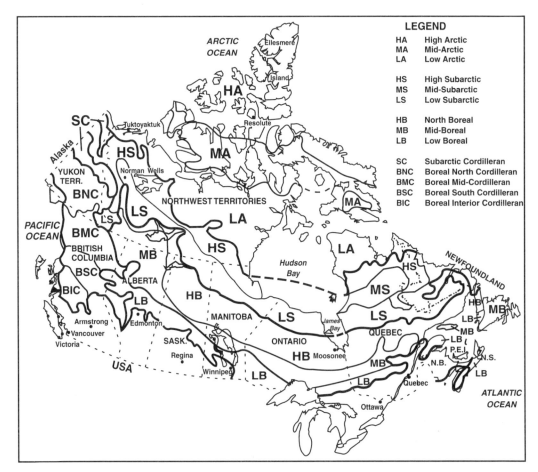

FIGURE 6.1 Arctic, Subarctic, Boreal, and portions of the Cordilleran ecoclimatic provinces and regions of Canada.

poorly drained sites, which are primarily peatlands. Although permafrost is discontinuous or spo-radic in the north, no permafrost occurs in the south. The dominant soils are organic soils, Podzols, Brunisols, and Luvisols. The Boreal is subdivided into High, Mid- and Low Boreal regions (Figure 6.1). These regions, which are described in detail in Ecoregions Working Group (1989), are hereinafter referred to as the High Boreal, Mid-Boreal, and Low Boreal.

Cordillera

The Cordillera, which, for purposes of this chapter, consists of the subarctic and boreal portions of the Cordilleran and Interior Cordilleran ecoclimatic provinces, covers approximately 919×10^3 km^2 (land area; soil area 713×10^3 km^2). The Cordillera is composed of mountainous areas in British Columbia, Alberta, the Yukon, and along the western boundary of the Northwest Territories. Below the tree line the Cordillera is characterized by open-canopy coniferous forests in the Subarctic areas and closed-canopy mixed coniferous and deciduous forests in the Boreal areas. Permafrost is continuous in the Subarctic portion of the Cordillera and discontinuous in northern portions of the Boreal, with occasional alpine permafrost in southern portions. Brunisols are the dominant soils, with Cryosols also occurring in the north and Luvisols occurring in the south. Organic soils occur very sporadically throughout the area. The Cordillera is subdivided into Subarctic, Boreal North, Boreal Mid-, Boreal South, and Boreal Interior Cordilleran regions (Figure 6.1). These regions are described in detail in Ecoregions Working Group (1989).

METHODS

Data Analysis

Soil organic carbon values were calculated using information stored in the Soil Organic Carbon Database, which provides data for all soil areas throughout Canada. A detailed description of the database, including the structure, all of the attributes, and the methods of calculating carbon values, is given in Lacelle (1998). For purposes of this chapter, information relating to the soil orders occurring in northern areas of Canada (the Arctic, Subarctic, Boreal, Cordilleran, and Interior Cordilleran ecoclimatic provinces and regions) was extracted from the Soil Organic Carbon Database and analyzed.

The soil organic carbon contents, which were corrected for coarse fragments (Tarnocai, 1997) were calculated on the basis of soil areas, not land areas. These carbon contents and the soil organic carbon masses of both the surface soil (0–30 cm) and the total soil were calculated for each of the soil orders in each region. It should be noted here that, in the Canadian system, the surface organic horizons of mineral soils (designated LFH in the Canadian system and O in the American system) are considered to be part of the surface layer. For mineral soils, in most cases, the total soil values are based on a depth of one meter, but, for mineral soils with lithic contact within less than one meter (shallow soils over bedrock), they are based on the depth to the contact. For organic soils these values are based on the full depth of the peat deposit and also include the organic-rich mineral layer underlying the peat.

The data for organic carbon contents and masses in deep soil layers were obtained by using core samples taken from a small study area (6577 km²) in the Mackenzie River valley. The locations of these six sites (Table 6.1) were entered in the Soil Organic Carbon Database in order to determine the polygons to which these sites belong. The carbon layer file was then updated for the five polygons that composed the study area and the organic carbon contents and masses for the 0–30, 0–100, 0–200, and 0–300 cm depths were calculated.

Sampling Method for Deep Soil Layers

The samples used to determine the amount of organic carbon in deep soil layers were collected in the Mackenzie River valley (Table 6.1). Sampling was carried out in March, 1994 using a modified CRREL power auger (Tarnocai, 1993). This sampling tool, which can core frozen materials, provides a 7.6 cm diameter core. The frozen cores were slit in the field and samples were collected for laboratory analysis.

Laboratory Methods

The organic carbon percentages given in the database were determined using either a modified Walkley and Black method to measure organic carbon by wet oxidation, or the Leco induction furnace method, subtracting carbonate carbon to obtain organic carbon (Sheldrick, 1984). The carbon concentrations shown in Figure 6.2 were determined using a Leco auto-analyzer.

RESULTS AND DISCUSSION

Soil Carbon in Ecoclimatic Provinces

The four ecoclimatic provinces and the associated regions that compose the study area contain 243 Gt soil organic carbon, which is 92.2% of the total soil organic carbon occurring in all Canadian soils (Table 6.2). The following sections discuss the amounts of soil organic carbon in the various ecoclimatic provinces in more detail.

TABLE 6.1
Locations of Sites and Site Parameters

Parameter	Site I-2[a]	Site G-1[a]	Site PS-1[a]	Site 3A-1[a]	Site 7A-1[a]	Site 7B-1[a]
Latitude (N)	68°06'39"	65°45'43"	65°17'23"	64°54'45"	63°36'43"	63°36'34"
Longitude (W)	133°28'29"	127°54'44"	126°52'58"	125°34'53"	123°38'31"	123°37'52"
Elevation (m)	30	245	70	60	240	250
Parent material	Till	Till	Lacustrine	Alluvial	Till	Till
Texture	Silty clay	Clay	Silt loam	Silt loam	Silty clay loam	Silt loam
Drainage	Imperfect	Imperfect	Imperfect	Imperfect	Imperfect	Imperfect
Patterned ground	Earth hummocks	Earth hummocks	Earth hummocks	Earth hummocks	Earth hummocks	Earth hummocks
Vegetation[b]	b-l-ds-m	b-l-ds-m	b-m-l-s	b-m-l-s	b-m-l-s	b-m-s-l
Soil classification (Canada)[c]	OETC	OETC	RTC	HETC	HETC	TMOC
Soil classification (U.S.)[d]	TP	TP	TA	TP	TP	TA
Coring depth (cm)	220	244	279	275	280	275
Carbon content (%) at maximum depth	1.7	0.9	2.3	3.6	2.1	2.2

[a] Site: These sites, which are listed by number, all occur in the Northwest Territories. Site I-2 is located along the Dempster Highway in the Inuvik area and site G-1 is located at Gibson Cap. The remaining four sites are located along the Norman Wells pipeline: site PS-1 at Pump Station 1 (in the Norman Wells area), 3A-1 at Great Bear River, and 7A-1 and 7B-1 at Table Mountain.

[b] Vegetation: b-black spruce, l-lichen, m-moss, s-shrub, ds-dwarf shrub.

[c] Soil classification (Canada): OETC-Orthic Eutric Turbic Cryosol, RTC-Regosolic Turbic Cryosol, HETC-Histic Eutric Turbic Cryosol, TMOC-Terric Mesic Organic Cryosol (Soil Classification Working Group, 1998).

[d] Soil class (U.S.): TP-Typic Psammiturbel, TA-Typic Aquaturbel (USDA Soil Survey Staff, 1998).

Arctic

The total soil organic carbon mass in the Arctic is 43 Gt, while the surface soil organic carbon mass is 16 Gt (Table 6.3). Almost all (99.4%) of the soil organic carbon mass in this region is contained in Cryosols (Figure 6.3). Other soils (Brunisols, organic soils, and Regosols) contain only 0.6% of the soil organic carbon mass.

The Low Arctic contains the largest masses of total and surface soil organic carbon in the Arctic, while the High Arctic contains the smallest (Table 6.2).

In the Low and Mid-Arctic both mineral and organic Cryosols are major soil organic carbon pools. In mineral Cryosols this carbon commonly occurs as a surface organic layer or an organic-rich mineral layer. Considerable amounts of organic carbon are also stored in cryoturbated mineral horizons and cryoturbated subsurface organic horizons. In these regions, organic Cryosols are associated primarily with low- and high-centered lowland polygons. The organic material is generally 2–2.5 m thick, especially in high-centered lowland polygons, but can be as thick as 10 m.

Although the surface organic horizons become thinner and very patchy in the High Arctic, a high amount of carbon is still stored in both the surface and subsurface horizons.

Subarctic

The total soil organic carbon mass in the Subarctic is 76 Gt, while the surface soil organic carbon mass is 17 Gt (Table 6.4). The largest mass of organic carbon is contained in Cryosols (58.1%), followed by organic soils (30.2%) and Podzols (6.7%) (Figure 6.3).

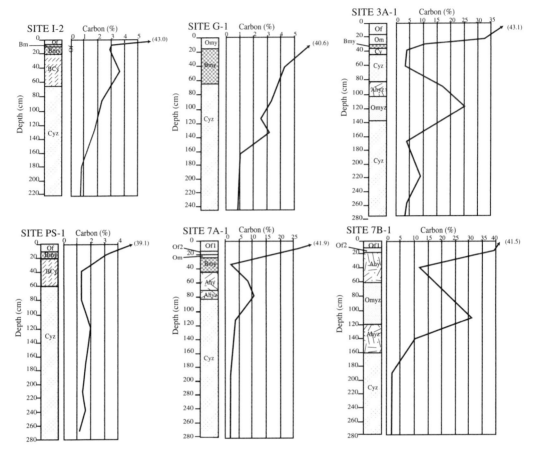

FIGURE 6.2 Carbon contents (%) in both surface and deep layers of six Cryosols from the Mackenzie River valley.

The largest masses of total and surface soil organic carbon are found in the Low Subarctic, while the Mid-Subarctic, which has the smallest area, has the smallest soil organic carbon masses (Table 6.2).

Mineral soils occurring in the Subarctic usually have thick surface organic horizons and cryoturbated, organic-rich subsurface horizons. Organic soils associated with peatlands are a common feature of the soil landscape. The combination of organic-rich mineral soils and the widespread occurrence of organic soils (peatlands) makes the Subarctic an important soil carbon pool for northern regions. It should be noted that, although organic soils cover a relatively small proportion of the Subarctic (10.5%), they have the highest total soil organic carbon content (144.8 kg m^{-2}) of all soils in these northern regions.

Boreal

The total soil organic carbon mass in the Boreal is 112 Gt, while the surface soil organic carbon mass is 26 Gt (Table 6.5). Organic soils contain the largest mass of soil organic carbon (68.3%), followed by Cryosols (11.4%), Podzols (9.6%), and Brunisols (3.4%) (Figure 6.3).

The largest masses of total and surface soil organic carbon are found in the High Boreal, while the smallest are found in the Low Boreal (Table 6.2).

TABLE 6.2
Amount of Soil Organic Carbon in the Arctic, Subarctic, Boreal, and Portions of the Cordilleran Ecoclimatic Provinces and Region

Ecoclimatic provinces and regions	Soil carbon content (kg m^{-2})		Soil carbon mass (Gt)		Area (10^3 km^2)	
	Surface	Total	Surface	Total	Land[a]	Soil
Arctic	10.78	28.52	16.27	43.04	2375	1509
High Arctic	7.00	19.66	2.20	6.19	681	315
Mid-Arctic	14.43	34.31	5.75	13.68	476	399
Low Arctic	10.46	29.14	8.31	23.17	1218	795
Subarctic	11.05	50.43	16.57	75.65	1712	1500
High Subarctic	10.58	47.58	6.49	29.19	725	614
Mid-Subarctic	9.15	20.69	0.85	1.93	126	93
Low Subarctic	11.63	56.13	9.23	44.52	861	793
Boreal	11.88	50.55	26.36	112.14	2521	2218
High Boreal	12.36	59.18	11.89	56.92	1085	962
Mid-Boreal	12.82	52.25	10.63	43.34	933	829
Low Boreal	8.99	27.81	3.84	11.88	503	427
Cordilleran	5.65	14.60	4.66	12.05	919	825
Subarctic Cordilleran	5.42	9.91	0.61	1.11	137	112
Boreal North Cordilleran	5.98	15.43	1.50	3.88	274	251
Boreal Mid-Cordilleran	5.69	14.34	1.11	2.79	225	194
Boreal South Cordilleran	5.72	17.30	1.14	3.44	212	199
Boreal Interior Cordilleran	4.45	12.11	0.30	0.83	70	68
Total	10.55	40.13	63.86	242.88	7527	6053
All of Canada			72.8	263.5	8878	7233

[a] Includes nonsoil areas such as rockland, glacier ice, and urban land.

TABLE 6.3
Amount of Organic Carbon in Various Soil Orders in the Arctic

Soil orders		Soil carbon content (kg m^{-2})		Soil carbon mass (Gt)		Soil area (km^2)
Canada	U.S.	Surface	Total	Surface	Total	
Brunisol	Inceptisol	6.34	18.01	0.09	0.26	14,687
Cryosol	Gelisol	10.83	28.63	16.17	42.76	1,493,483
Organic	Histosol	9.52	42.46	0.00[a]	0.01	207
Regosol	Entisol	4.28	5.17	0.00[b]	0.00[c]	327
All soils		10.78	28.52	16.27	43.04	1,508,704

[a] Actual value is 0.00197 Gt = 1.97 × 10^9 kg.
[b] Actual value is 0.00140 Gt = 1.40 × 10^9 kg.
[c] Actual value is 0.00169 Gt = 1.69 × 10^9 kg.

The Boreal is the largest soil carbon pool in all of Canada because of the prevalence of organic soils (peatlands) in this area. The Mid- and High Boreal, in particular, provide optimal climatic conditions for peat development. The major soil carbon pool is thus derived from these peatlands, which not only have the highest total soil organic carbon contents (133.9 kg m^{-2} for organic soils, 77.6 kg m^{-2} for Cryosols) in the Boreal, but also cover huge areas.

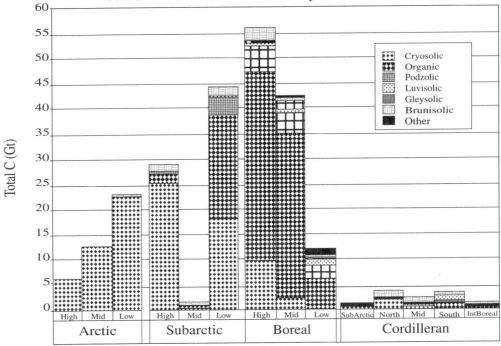

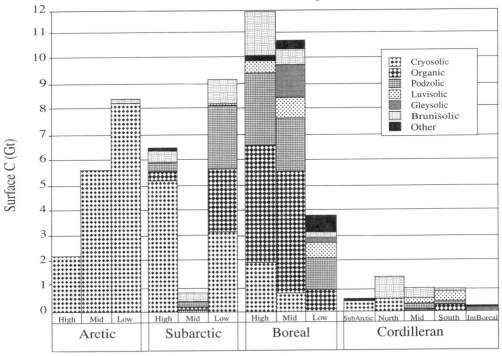

FIGURE 6.3 Total (upper) and surface (lower) soil organic carbon masses for each soil order in the various ecoclimatic regions.

TABLE 6.4
Amount of Organic Carbon in Various Soil Orders in the Subarctic

Soil orders		Soil carbon content (kg m⁻²)		Soil carbon mass (Gt)		Soil area (km²)
Canada	U.S.	Surface	Total	Surface	Total	
Brunisol	Inceptisol	5.33	9.81	1.82	3.35	341,085
Cryosol	Gelisol	12.03	62.57	8.45	43.95	702,523
Gleysol	Aqu suborders	10.09	30.67	0.04	0.12	3,801
Luvisol	Boralf and Udalf	6.57	15.40	0.06	0.15	9,513
Organic	Histosol	19.33	144.82	3.05	22.88	157,975
Podzol	Spodosol	11.29	18.63	3.07	5.07	272,108
Regosol	Entisol	5.88	10.55	0.08	0.14	13,184
All soils		11.05	50.43	16.57	75.65	1,500,189

TABLE 6.5
Amount of Organic Carbon in Various Soil Orders in the Boreal

Soil orders		Soil carbon content (kg m⁻²)		Soil carbon mass (Gt)		Soil area (km²)
Canada	U.S.	Surface	Total	Surface	Total	
Brunisol	Inceptisol	5.82	9.07	2.42	3.77	415,607
Chernozem	Boroll	7.93	13.56	0.54	0.93	68,317
Cryosol	Gelisol	17.14	77.59	2.81	12.74	164,133
Gleysol	Aqu suborders	14.19	22.75	1.85	2.97	130,511
Luvisol	Boralf and Udalf	5.81	10.49	1.83	3.31	315,225
Organic	Histosol	18.26	133.88	10.44	76.55	571,762
Podzol	Spodosol	12.33	22.42	5.91	10.75	479,316
Regosol	Entisol	7.49	15.44	0.48	0.99	64,178
Solonetz	Mollisol and Alfisol	7.44	14.89	0.07	0.14	9,404
All soils		11.88	50.55	26.36	112.14	2,218,453

Cordillera

The total soil organic carbon mass in the Cordillera is 12 Gt, while the surface soil organic carbon mass is 5 Gt (Table 6.6). Cryosols contain the largest mass of soil organic carbon (26.5%), followed by Organic soils (22.2%) and Brunisols (20.8%).

The largest masses of total and surface soil organic carbon are found in the Boreal North Cordilleran, while the smallest are found in the Boreal Interior Cordilleran, which also has the smallest area (Table 6.2).

Although the Cordillera is included in this chapter, it is not considered to be a significant soil carbon pool. It is given here only because it falls within the broad geographic regions encompassed by the Subarctic and Boreal.

CARBON IN DEEP SOIL LAYERS

The total soil organic carbon masses given in Tables 6.2 through 6.6 refer to the carbon found in the upper (0–1 m) section of the mineral soils. Data indicate, however, that considerable amounts

TABLE 6.6
Amount of Organic Carbon in Various Soil Orders in the Cordillera

Soil orders		Soil carbon content (kg m^{-2})		Soil carbon mass (Gt)		Soil area (km^2)
Canada	U.S.	Surface	Total	Surface	Total	
Brunisol	Inceptisol	4.49	9.29	1.22	2.51	270,585
Chernozem	Boroll	5.96	10.92	0.03	0.05	4,604
Cryosol	Gelisol	7.24	18.84	1.23	3.19	169,417
Gleysol	Aqu suborders	12.08	25.45	0.21	0.45	17,608
Luvisol	Boralf and Udalf	3.96	8.29	0.72	1.51	181,764
Organic	Histosol	19.11	102.33	0.50	2.67	26,117
Podzol	Spodosol	5.11	11.08	0.68	1.47	133,035
Regosol	Entisol	3.38	8.28	0.07	0.18	21,740
Solonetz	Mollisol and Alfisol	11.42	21.71	0.01	0.01	480
All soils		5.65	14.60	4.66	12.05	825,350

TABLE 6.7
Soil Organic Carbon Contents and Masses for Various Depth Ranges on a 6577-km^2 Study Area in the Mackenzie River Valley

Depth of layer (cm)	Soil carbon content (kg m^{-2})	Soil carbon mass (Gt)	% of total soil carbon mass in layer
0 to 30	15.20	0.10	18.9
30 to 100	28.89	0.19	35.8
100 to 200	27.37	0.18	34.0
200 to 300	9.12	0.06	11.3
Total: 0 to 300	80.58	0.53	100

of carbon also exist below the 1 m depth in these soils (Table 6.7). In order to determine how significant these deep carbon pools might be, soils were sampled below the 1 m depth on six sites (in five polygons) composing the small study area (6577 km^2) in the Subarctic part of the Mackenzie River valley (Table 6.1). Soil organic carbon values for these five Turbic Cryosols (mineral soils) and one Organic Cryosol were calculated not only for the surface layer (0–30 cm) and the 30–100 cm depth, but also for depths of 100–200 cm, 200–300 cm, and 0–300 cm (Table 6.7). All of these soils have a high organic carbon content in the surface organic horizon. At lower depths there is usually a second peak in carbon content at depths of about 60 to 140 cm (Figure 6.2). This depth coincides approximately with the depth of the fluctuating permafrost table, which is associated with organic-rich mineral or organic horizons. The permafrost table is very dynamic and fluctuates according to climatic and environmental conditions. The build-up of organic matter along the permafrost table has been attributed to cryoturbation (Bockheim and Tarnocai, 1998).

The data suggest that a large soil carbon pool exists below the 1 m depth in these soils, since the 0–100 cm depth contains only 54.7% of the total soil organic carbon mass present to the 300 cm depth. Below the permafrost table, therefore, the soil still contains nearly as much carbon as occurs in the active layer (a layer approximately 1 m deep). In some cases, such as at site 3A–1, the soil still contains 8.0% carbon at the 220 cm depth and 3.6% carbon at the 275 cm depth (Figure 6.2; Table 6.1).

Since the samples used for calculation of the deep soil carbon represent the dominant soils and parent materials of the Mackenzie River valley, this indicates that the soil organic carbon mass of the area is underestimated if the total carbon mass is given for only the 0–100 cm depth. When the total soil organic carbon mass stored in these Cryosols is calculated to a depth of 3 m it becomes obvious not only that a large amount of organic carbon in these high-latitude soils may not have been accounted for in previous studies, but also that the values given for Cryosols in Tables 6.3 through 6.6 provide information about only a portion of the actual carbon pool in these soils. A significant amount of carbon may still exist in the deep layers of these permafrost soils.

Since all of this deep-soil carbon is in the permafrost layer, the origin of this carbon must be determined. For soils developed on alluvial material (e.g., site 3A–1), the deep organic matter most likely resulted from deposition. For soils developed on other parent materials, however, alternate explanations must be found. One possibility is that it could be due to the dynamic nature of the permafrost and the permafrost landscape. Climatic and environmental changes can cause the permafrost table to fluctuate. In the past, when the active layer may have been much deeper than it is now, cryoturbation could have caused the deeper soil layers to be enriched with translocated organic matter. This deepening of the active layer can also lead to instability in the landscape, causing soil materials to be redistributed by solifluction, flow slides, and other erosional processes. Another possibility for the origin of the carbon in the deep soil layers is the migration of water from warmer surface soil layers to the permafrost layer. This mechanism causes soluble organic matter to be translocated from the active layer to the perennially frozen subsoil. A similar phenomenon was suggested by Tarnocai (1972) to explain the increase of Ca in the frozen layer of Organic Cryosols.

MAJOR SOIL CARBON POOLS

Soils

Two soil orders, Cryosols and organic soils, are major carbon pools, containing 84.3% (205 Gt) of the total soil organic carbon mass in the study area and 77.7% of that in all Canadian soils. In addition, they are the dominant soils in these regions, covering large areas. Their surface and total soil organic carbon contents are also generally the highest of all the soils in the area. The high surface soil organic carbon contents suggest that large amounts of carbon could be directly affected if environmental change occurs.

Cryosols occur throughout the study area and contain 103 Gt of carbon, or 42.3% of the total soil organic carbon in these areas (Figure 6.3). They contain 99.4% (43 Gt) of the soil organic carbon in the Arctic (17.6% of that in the study area), while in the Subarctic, where they are still the dominant soils, they contain 58.1% (44 Gt) of the carbon (18.1% of that in the study area). Cryosols contain 11.4% (13 Gt) of the soil organic carbon in the Boreal (5.2% of that in the study area).

Organic soils, which occur primarily in the Subarctic, Boreal, and Cordillera, contain 102 Gt of total soil organic carbon, or 42.0% of the total soil organic carbon in the study area (Figure 6.3). These soils occur only rarely in the Arctic, but contain 30.2% (23 Gt) of the total soil organic carbon in the Subarctic (9.4% of that in the study area) and become the dominant soils in the Boreal, where they contain 68.3% (77 Gt) of the carbon (31.5% of that in the study area).

When the carbon masses of organic soils and Cryosols are combined, their value as soil carbon pools in these regions is obvious. They contain 99.4% of the soil organic carbon in the Arctic, 88.3% in the Subarctic, 79.6% in the Boreal, and 48.6% in the Cordillera. Further, in the Boreal they contain 36.8% of the total soil organic carbon in the study area and in the Subarctic they contain 27.5% of the carbon.

Regions

Although the Arctic, Subarctic, and Boreal are all soil organic carbon pools, containing 231 Gt carbon (Table 6.2), the richest soil carbon pool lies in the region from the Mid-Boreal, through the High Boreal, to the Low Subarctic (Figure 6.3). These areas, which contain 145 Gt soil organic carbon, have 54.9% of the total Canadian soil carbon (in 35.8% of the total soil area of Canada)

TABLE 6.8
Amount of Organic Carbon in Various Soil Orders in the Hudson Bay Lowland

Soil orders		Soil carbon content (kg m⁻²)		Soil carbon mass (Gt)		Soil area (km²)
Canada	U.S.	Surface	Total	Surface	Total	
Brunisol	Inceptisol	5.15	7.01	0.08	0.11	15,689
Cryosol	Gelisol	13.33	92.53	1.13	7.83	84,600
Gleysol	Aqu suborders	7.54	12.57	0.05	0.08	6,710
Luvisol	Boralf and Udalf	5.99	7.91	0.01	0.02	2,105
Organic	Histosol	12.49	112.44	2.73	24.60	218,809
Podzol	Spodosol	12.71	15.63	0.02	0.03	1,654
Regosol	Entisol	6.47	8.93	0.05	0.07	7,460
All soils		12.09	97.13	4.07	32.73	337,027

and 59.6% of the soil organic carbon in the study area (in 42.7% of the soil area of these regions). In addition, the soil organic carbon contents of the surface layer and the total soil are 12.3 kg m⁻² and 56.0 kg m⁻², respectively, compared to values of 10.6 kg m⁻² and 40.1 kg m⁻² for the whole study area. In contrast, the area lying to the north of these three regions is similar in size, but contains only half the mass of soil carbon (74 Gt) (Table 6.2) and has surface and total soil carbon contents of 10.7 kg m⁻² and 33.5 kg m⁻², respectively.

Physiographic Areas

Significant carbon pools may exist in a number of physiographic areas, such as the Mackenzie Valley, the Yukon Coastal Plain, the Hudson Bay Lowland, and the Manitoba Plain (Bostock, 1970, a, b). The most significant of these areas is probably the Hudson Bay Lowland, which is one of the world's largest peatlands. This area, which lies in a band along the southwest side of Hudson Bay, covers approximately 339,974 km² (land area; soil area 337,027 km²). It is dominated by organic soils and Cryosols, which together cover 90.0% of the soil area. The total soil organic carbon mass in this area is 33 Gt, while the surface soil organic carbon mass is 4 Gt (Table 6.8). The total organic carbon mass contained in the Hudson Bay Lowland is 12.4% of the total soil organic carbon mass of Canada and 13.5% of that of the four ecoclimatic provinces presented here. Most of this area lies within the Low Subarctic and High Boreal regions, which together have a total soil carbon mass of 101 Gt and a surface mass of 21 Gt. The Hudson Bay Lowland thus contains 32.3% of the total carbon mass of these regions and 19.3% of the surface carbon mass in 19.2% of the soil area.

Organic soils, which cover the largest area, also contain the largest mass (75.2%) of soil organic carbon followed by Cryosols (23.9%). All other soils in the area contain less than 1% of the total organic carbon mass.

CONCLUSIONS

1. Two soil orders, organic soils and Cryosols, are major carbon pools, containing 205 Gt carbon, which is 84.3% of the total soil organic carbon mass in the study area and 77.7% of that in all Canadian soils.
2. The Arctic, Subarctic, Boreal, and subarctic and boreal portions of the Cordilleran ecoclimatic provinces contain 243 Gt carbon, which is 92.2% of the total soil organic carbon mass in all Canadian soils.

3. Within this area, the Low Subarctic and Mid- and High Boreal regions are major soil organic carbon pools, containing 145 Gt carbon, which is 59.6% of the total soil organic carbon mass of the study area and 54.9% of the total Canadian soil carbon mass.

4. The Hudson Bay Lowland, most of which lies within the Low Subarctic and High Boreal regions, is one of the world's largest peatlands. It contains 33 Gt total soil organic carbon mass, 99.1% of which occurs in organic soils and Cryosols. The total soil organic carbon mass of this area is 32.3% of the total soil organic carbon mass of the combined Low Subarctic and High Boreal regions, 13.5% of that in the study area, and 12.4% of the organic carbon mass in all Canadian soils.

5. Approximately half of the total soil organic carbon mass in some Cryosols may occur below the 1 m depth (between 1 and 3 m). The large amount of carbon found in these deep pools indicates that the values of total soil organic carbon given in this chapter may be underestimated.

REFERENCES

Bockheim, J.G. and Tarnocai, C., Recognition of cryoturbation for classifying permafrost-affected soils, *Geoderma*, 81, 281, 1998.

Bostock, H.S., Physiography of Canada, Map # 1254A, scale 1:5,000,000, Geological Survey of Canada, Ottawa, 1970a.

Bostock, H.S., Physiographic subdivisions of Canada, in *Geology and Economic Minerals of Canada*, Douglas, R.J.W., Ed., Dept. of Energy, Mines and Resources Canada, Geological Survey of Canada, Economic Geology Report No. 1, 1970b, p. 9.

Ecoregions Working Group, Ecoclimatic Regions of Canada. First Approximation, Ecoregions Working Group of the Canada Committee on Ecological Land Classification, Ser. 23, including map, Sustainable Development Branch, Canadian Wildlife Service, Conservation and Protection, Environment Canada, Ottawa, 1989.

Kurz, W.A., Apps, M.J., Webb, T.M., and McNamee, P.J., *The Carbon Budget of the Canadian Forest Sector: Phase 1*, Inf. Rep. NOR-X-326, Northwest Region, Northern Forestry Center, Edmonton, Alberta, 1992.

Lacelle, B., Canada's soil organic carbon database, in *Soil Processes and the Carbon Cycle*, Advances in Soil Science Series, Lal, R., Kimble, J.M., Follett, R.L.F., and Stewart, B.A., Eds., CRC Press, Boca Raton, FL, 1998, p. 93.

Post, W.M., Emanuel, W.R., Zinke, P.J., and Stangenberger, G., Soil carbon pools and world life zones, *Nature*, 298, 156, 1982.

Sheldrick, B.H., Ed., *Analytical Methods Manual*, Land Resource Research Institute, Agriculture Canada, Ottawa, 1984,

Soil Classification Working Group, *The Canadian System of Soil Classification*, 3rd. ed., Publ. 1646, NRC Research Press, Ottawa, 1998.

Tarnocai, C., Some characteristics of cryic organic soils of northern Manitoba, *Can. J. Soil Sci.*, 52, 485, 1972.

Tarnocai, C., Sampling frozen soils, in *Soil Sampling and Methods of Analysis*, sponsored by Canadian Society of Soil Science, Lewis Publishers, Boca Raton, FL, 1993, chap. 71, p. 755.

Tarnocai, C., The amount of organic carbon in various soil orders and ecological provinces in Canada, in *Soil Processes and the Carbon Cycle*, Advances in Soil Science Series, Lal, R., Kimble, J.M., Follett, R.L.F., and Stewart, B.A., Eds., CRC Press, Boca Raton, FL, 1997, p. 81.

USDA Soil Survey, Keys to Soil Taxonomy, U.S. Department of Agriculture, Washington, D.C., 1998.

Section II

Natural and Anthropogenic Disturbance

7 Simulated Carbon Dynamics in the Boreal Forest of Central Canada under Uniform and Random Disturbance Regimes

M.J. Apps, J.S. Bhatti, D.H. Halliwell, H. Jiang, and C.H. Peng

CONTENTS

INTRODUCTION

Forest biomass, detritus, and soil are the three major pools of carbon of forest ecosystems. Understanding the factors controlling these pools and the exchange of C amongst them and with the atmosphere is critical for estimating the role of forests in the global C cycle. Changes in forest ecosystem C pools are mainly driven by the dynamics of the living biomass. Accumulations of organic C in litter and soil change significantly as forest stands are subjected to various disturbances, such as fire, insects, and harvesting. Disturbances transfer biomass C to detritus and soil C pools where it decomposes at various rates in years following the disturbance.

    Most forest gap and population dynamic models assume that disturbances occur at regular intervals. Natural agents of disturbance, however, have a highly stochastic component and are often regarded as unpredictable. While the spatially averaged return interval may be relatively uniform over time, this does not necessarily require that a given location be subjected to repeated disturbances at regular intervals. Fires and insect outbreaks — the two dominant natural disturbance agents of the Canadian boreal forests — do not naturally occur at uniform intervals (Kurz et al., 1995b).

Climatic shifts on scales of decades to centuries have produced substantially different fire disturbance regimes (Clark, 1989). Changes in fire disturbance regimes over scales of decades and centuries are important explanatory factors of structural, compositional, and functional changes in existing forests (Steijlen and Zackrisson, 1988; Kurz et al., 1995a). They may also be an important determinant for future forests under a changing climate (Kurz and Apps, 1995).

Kurz and Apps (1995, 1996) have shown that much of boreal forest C dynamics can be explained by changes in the disturbance frequency. With increasing disturbance frequency, a greater proportion of the forest is found in younger age-classes having lower C reservoirs. Consequently, such changes tend to result in C losses from the forest ecosystem if other conditions are unchanged (Apps and Price, 1996). Similarly, a decreasing disturbance regime tends to result in an increase in forest ecosystem C.

The range of time scales governing the C dynamics in biomass C, detritus C, and soil C introduce more temporal complexity. There is a significant lag in the release of C through decay and decomposition from forest floor and mineral soil pools relative to the uptake of C by regrowing vegetation biomass. Subsequent regrowth of biomass and reestablishment of litter input to these pools are similarly delayed and have different temporal characteristics. At a given point, the C in the main forest pools thus depends on the history of that site, and in particular, the temporal scales associated with these different processes in relation to the disturbance return interval. As already observed, there is no *a priori* reason to assume that this return interval is constant (except in the case of plantation harvesting) yet this is the assumption commonly made. One of the primary objectives of this chapter is to explore the implications of this assumption for the simulation of forest carbon budgets of the boreal forests of central Canada.

In this chapter, two simulation models that have previously been calibrated and tested for Saskatchewan forests are used to examine the relationships between biomass C, litterfall C, and soil C pools under uniform and random disturbance regimes. CENTURY 4.0, a point model, is used to estimate changes in forest C pools for two site near Prince Albert at the southern boarder of central Canada's boreal forest. The Carbon Budget Model of the Canadian Forest Sector (CBM-CFS2), a spatially distributed model, is similarly used to examine the forest C dynamics at the scale of the entire boreal forest of Saskatchewan. Comparisons of these models with each other and with observed data are used to provide a validation of the soil module of CBM-CFS2.

MATERIALS AND METHODS

UNIFORM AND RANDOM DISTURBANCE REGIME SEQUENCES

The uniform disturbance regime was represented by a sequence of fires at precise intervals of 100 years over 7000 years. To represent random disturbance regimes with the same average return interval, a simple algorithm was used to generate 70 random events with annual probability of disturbance equal to 0.01. Only sequences of 70 events within 1% of the 7000 years interval were kept (70 fires over 7000 ± 1% years); 15 different such random sequences were generated and used for CBM-CFS2 simulations and 6 were used for CENTURY simulations.

THE CENTURY MODEL AND ITS PARAMETERIZATION

CENTURY 4.0, a process-based point level biogeochemistry model, simulates the long-term (100-10,000 years) dynamics of carbon (C), nitrogen (N), phosphorus (P), and sulfur (S) for different plant-soil ecosystems. The model has been described in detail by Parton et al. (1987, 1993) and Metherell at al. (1993) while Peng et al. (1998) and Peng and Apps (1998) report its verification and application in the boreal forests of central Canada. CENTURY's forest production module partitions biomass into several compartments: foliage, fine and coarse roots, fine branches, and large wood. Carbon (C) and nitrogen (N) are allocated to the different plant parts using a fixed

TABLE 7.1
Site Parameters Used in the CENTURY Model for Two Locations:
PA1 and PA2 Near Prince Albert

Parameters	PA1	PA2
Latitude	53.37 N	53.10 N
Longitude	101.08 W	105.08 W
Mean monthly minimum (°C)	−25.25	−25.25
Mean monthly maximum (°C)	24.45	24.45
Annual precipitation (mm)	398	398
Dominant vegetation	*Pinus banksiana*	*Populus termuloides*
Soil type	Developing Spodosols	Developing Spodosols
Soil texture		
Clay (%)	13	11
Sand (%)	73	60

allocation scheme (Peng et al., 1998). Gross primary productivity (GPP) is calculated as a function of maximum gross productivity, moisture, soil temperature, and live leaf-area-index (LAI).

The model operates on a monthly time step. The major input variables for the model include both biotic and abiotic site factors including monthly mean maximum and minimum air temperature, monthly precipitation, soil texture, atmospheric and soil N inputs, plant lignin content, and initial values for soil C, N, P, and S. If the monthly temperature and precipitation data and soil texture are known, other input variables can be estimated internally by the model. Site-specific parameters and initial conditions, such as soil texture (clay, silt, and sand content), bulk density, soil pH, soil C content for the 0-20 cm layer, soil C content for the 0-100 cm layer, and drainage characteristics of soil were obtained from field data (Siltanen et al., 1997) (Table 7.1). Mean maximum and minimum monthly temperature and monthly precipitation were calculated by CENTURY 4.0 using the Canadian Atmospheric Service (AES) climate station 30-year normals (1950-1980) (AES, 1983).

CENTURY 4.0 simulations were run for about 7000 years using an average fire return interval of 100 years to simulate changes in boreal forest ecosystem C pools at two sites along the southern limits of the boreal forest near Prince Albert, Saskatchewan. Simulations were performed using both a uniform disturbance regime (100-year) and 6 different random disturbance sequences whose temporal average was 100 years over the period.

CBM-CFS2 MODEL

CBM-CFS2 is a spatially distributed simulation model which accounts for C pools and fluxes in forest ecosystems (Kurz and Apps, 1998). In the model, the simulation area is divided into spatial units having broadly similar vegetation characteristics (Apps and Kurz, 1993). Within each of these, the model simulates the dynamics of groups of stands (State Variable Objects, or SVOs) having similar species, productivity, stocking, and age-class characteristics. Biomass growth curves, derived from forest inventory data (Kurz et al., 1992) are associated with each SVO and used to simulate changes in both the above- and belowground biomass. Litter fall and mortality are derived from the growth curves and used with a soil decomposition model (Apps and Kurz, 1993; Kurz and Apps, 1998) to account for changes in litter and soil pools between disturbances. The number of these SVOs, and the area associated with each, changes during the simulation as disturbances are applied to the region. During a disturbance event, transfers of biomass C to litter (including coarse woody debris) and to the forest product sector (in the case of harvesting) and from both biomass and litter to the atmosphere (in the case of fire) are specified as proportions of the SVO totals.

In each SVO forest vegetation is represented by 12 pools consisting of 4 aboveground (foliage, submerchantable, merchantable, and other) and 2 belowground (coarse and fine roots) biomass pools for both a hardwood and a softwood group of species. Soil and litter dynamics are represented by four soil/detritus C pools (designated as very fast, fast, medium, and slow pool) having different decomposition rates influenced by pool type, mean annual temperature, and stand conditions that are associated with the SVO. In this way, the amount of C in biomass and soil pools simulated in the model is an explicit function of ecosystem type and past disturbance history. In this chapter, only two C pools, fast and slow, are discussed as they are more easily compared to the existing experimental data and to the pools represented in CENTURY. In addition, the size of the other two CBM-CFS2 pools (very fast and medium) were small and play relatively minor roles in the overall dynamics. The fast pool receives input from other small-sized biomass (branches and tree tops of trees of merchantable size, all biomass of trees of submerchantable size, and all coarse roots). The slow soil C pool represents humified organic matter and receives C from the three other pools (very fast, fast, and medium).

In the Saskatchewan boreal forest, the major tree species are jack pine (*Pinus banksiana*), white spruce (*Picea glauca* (Moench) Voss), black spruce (*Picea mariana* (Mill.) B.S.P.) and trembling aspen (*Populus tremuloides* Michx.). Of the 457 growth curves developed from the national forest inventory by Kurz and Apps (1998) the major proportion of the Saskatchewan boreal forest area (24.5 M ha) is associated with 6 different growth equations (Figure 7.1) under present conditions. Growth curve 4 is not shown in the figure, as the area presently associated with it is less than 0.005% of the total area and it is almost identical to growth curve 3. For this study, a simplified version of CBM-CFS2 was used with six SVOs representing the six growth equations. This simplified model does not use the spatial database from the forest inventory of Saskatchewan boreal forest but does represent all the possible conditions for the region. At each time step, CBM-CFS2 calculates and reports C pools, C fluxes, and areas associated with each SVO. The total Saskatchewan boreal forest biomass C, litterfall C, and soil C pools are then estimated using an area-weighted sum over all SVOs.

To avoid initialization artifacts, simulation results were examined for periods between 2001 and 7000. With the random sequences, the number of actual disturbances during the analysis period (2001-7000) varies from 41 to 54.

RESULTS AND DISCUSSION

RANDOM VS. UNIFORM DISTURBANCE REGIMES USING CENTURY

Simulation results with uniform and random disturbance regimes for total ecosystem C, biomass C, litter pool C, active soil pool C, and slow soil pool C are presented in Table 7.2. The results show that higher biomass is simulated for both aspen (7–35%) and for jack pine (14-32%) under the random disturbance regime than for the uniform case. The biomass C simulated by CENTURY appears to overestimate biomass with this disturbance interval: even with the uniform regime, the results were about 35% higher than estimated from the forest inventory data by Bonnor (1985). One reason for the bias could be due to the climate data used for the model. As there were no climate station data for the actual sites from which biomass measurements were obtained, climatic data from the Prince Albert airport, located several kilometers south of the sites, were used. Another reason could be an artifact of CENTURY's failure to represent regeneration delay (Figure 7.1). In CENTURY, the growth curve is assumed to start its exponentially limited growth with a substantial initial biomass value (2.2 kg m^{-2}) following disturbance. This results in higher C accumulation rates in the early stages of recovery from disturbance than are generally found in natural boreal ecosystems. The CENTURY model lacks detailed representation of species and stand dynamics (Parton et al., 1993), so Table 7.2 shows the same simulated biomass for jack pine and aspen stands.

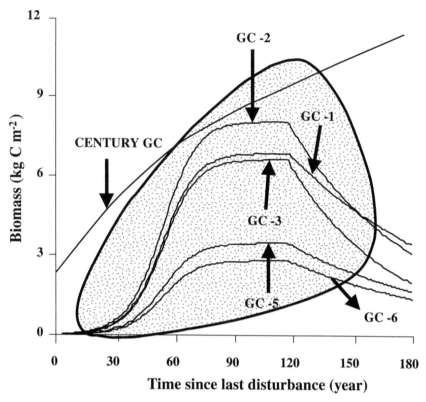

FIGURE 7.1 A typical CENTURY growth curve and five growth curves used in the CBM-CFS2 model to simulate forest growth. The shaded area represents the range of observed biomass in Saskatchewan's boreal forest (Halliwell et al., 1995).

TABLE 7.2
C in the Total Ecosystem, Aboveground Biomass, Litter Pool, and the Active and Slow Soil Pools of the Upper 20 cm of Soil as Simulated by CENTURY 4.0 with Uniform and Random Disturbance Regimes (the Range of Results is Shown for the Latter)

	Uniform			Random	
C pool	PA1	PA2		PA1	PA2
			kg m^{-2}		
Ecosystem	21.1	20.7		22.7 to 24.6	21.4 to 24.8
Biomass	6.8	6.8		7.8 to 9.0	7.3 to 9.2
Litter pool	3.9	3.9		3.9 to 4.1	4.0 to 4.1
Slow soil	5.2	4.9		5.2 to 5.3	4.9 to 5.2
			gm^{-2}		
Active soil	60	61		63 to 67	63 to 70

The litter pool C simulated with CENTURY was between 3.9 and 4.2 kg C m^{-2}. This compares very well with the value of 3.7 kg C m^{-2} for boreal forest litter pool reported by Matthews (1997). Interestingly, there was only a small difference (less than 5%) in the values of the slow soil C pool under the uniform and random disturbance regimes. The simulated soil C content of 4.9 kg m^{-2} (0–20 cm) was within 5% of the observed values reported by Siltanen et al. (1997). The approximately 6% difference in simulated values for slow soil pool C between aspen and jack pine may

TABLE 7.3
Biomass C, Litterfall C, Fast Soil Pool C, and Slow Soil Pool C
Simulated with CBM-CFS2 for Saskatchewan Boreal Forests
with Uniform and Random Disturbance Regimes.

C pool	Uniform	Random
Biomass (kg m^{-2})	2.40	2.67 to 2.70
Litterfall (kg m^{-2} year^{-1})	0.31	0.35 to 0.38
Fast soil (kg m^{-2})	1.42	1.75 to 1.83
Slow soil (kg m^{-2})	18.3	18.8 to 20.1

be related to differences in soil mineralogy and texture (Peng et al., 1998) for the two sites. Schimel et al. (1994) showed that the interaction of soil organic matter with Fe and Al oxides and clay may promote stabilization of organic C in soils.

RANDOM VS. UNIFORM DISTURBANCE REGIMES USING CBM-CFS2

Under random disturbance regimes, simulated values for C in biomass, litterfall, fast, and slow soil pools were, respectively, 11-12%, 13-22%, 23-29%, and 3-10% higher than under the uniform disturbance regime (Table 7.3). The average biomass C values of 2.40 kg C m^{-2} and 2.67-2.70 kg C m^{-2}, simulated by CBM-CFS2 under uniform and random disturbance regimes, respectively, were close to the biomass C measurements for boreal forest of 2.41 kg C m^{-2} reported by Simpson et al. (1993) through direct measurements.

Litterfall rates over the 5000-year period (2001–7000) under the uniform disturbance regime were 0.31 kg C m^{-2} yr^{-1}, and increased to 0.35–0.38 kg C m^{-2} yr^{-1} under the random regimes. These rates are within the range of litterfall rates of ≤0.27 kg C m^{-2} yr^{-1} reported by Matthews (1997), 0.46 kg C m^{-2} yr^{-1} for jack pine, and 0.26 kg C m^{-2} yr^{-1} for black spruce stands in Ontario observed by N.W. Foster (personal communication, 1998). The average soil C estimated by CBM-CFS2 was 18.3 kg C m^{-2} under uniform disturbances and 18.8–20.1 kg C m^{-2} under the random regimes. These simulated soil C values were close to the soil C content of 13.5–19.5 kg C m^{-2} to the depth of 100 cm reported by Pastor and Post (1988) for northern boreal forest in North America. The simulated soil C contents with CBM-CFS2, however, were 60% higher than those reported by Bhatti and Apps (1998) for the upland boreal forest from field data reported by Siltanen et al. (1997) but much lower than the 50 kg C m^{-2} estimated from interpreted soil carbon maps by Tarnocai (1998).

There are several explanations for the higher CBM-CFS2 estimates relative to the Bhatti and Apps (1998) estimates based on field measurement data of Siltanen et al. (1997), which we believe to be the most accurate. First, the CBM-CFS2 soil module contains a relatively simple representation of the processes governing soil organic C dynamics, including a crude parameterization of the partitioning of litter decomposition products between SOM and the atmosphere (Kurz et al., 1992). Second, both the nature of the historical disturbances (fire) and the average frequency (100 years) is clearly a simplification of a much more complex history. While not a true validation of the model for that reason, the fact that the simulation results are in such good agreement with the observed data gives added confidence in the representation of soil and litter in CBM-CFS2.

As expected, the simulated litterfall rate is strongly dependent upon the biomass C (Figure 7.2); increasing biomass C increases the litterfall rate. With the lower biomass C that is found under the uniform disturbance regime, a lower litterfall rate is observed in comparison to the random regime. This difference in the litterfall C rate between the two disturbance regimes may be related to the difference in age-class structure (Figure 7.3). Under the uniform disturbance regime, the forest develops as an even-age "normal" forest (MacLaren, 1996) and transfers the same amount of C

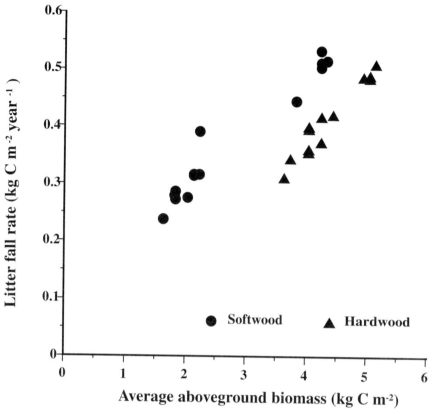

FIGURE 7.2 Relationship between average litter fall C rate and total biomass C (Y = 0.097*X + 0.113 for softwood; Y = 0.117*X − 0.086 for hardwood) in CBM-CFS2, for Saskatchewan boreal forests.

during each disturbance cycle (Figure 7.3a). In contrast, under the random disturbance regime, each stand develops as an even-aged cohort, but with a declining exponential (inverse J-curve) distribution of stand ages throughout the forest as a whole (Van Wagner, 1978) (Figure 7.3b). As a consequence, the litter production varies considerably amongst the stands making up the forest: stands in the younger age classes have lower litterfall rates compared to those in the older age classes. Moreover, the physical size of individual litter components differs across the age distribution. In the model, litterfall material is estimated separately for aboveground foliage, dead wood, bark, branches, as well as, belowground fine and coarse roots. Both these factors (litterfall rates and size of litter components), as well as others not discussed here, can influence both the quantity and the quality (decomposability) of litter input to the detritus and soil pools. The model simulation tracks one stand over time rather than a distribution of stands in a forest, but the same age class effect appears in the time average of the simulation.

Averaging the results of 5000 years of CBM-CFS2 simulations, it is found that deciduous species have lower C litterfall rate when compared to coniferous species at the same biomass (Figure 7.2). In CBM-CFS2, the foliage litterfall rate was three time higher for the deciduous growth curves than for the coniferous growth curves. Root, branch, bark, and dead stem components of stand biomass, however, contribute up to 80% of the litterfall for the coniferous growth curves as compared with 60% for the deciduous growth curves. This appears to be consistent with the existing, although sparse, information in the literature. Deciduous foliage litterfall C (sugar maple *Acer saccharum* Marsh.) in Ontario was observed to be lower than that from mature jack pine (N.W. Foster, personal communication; Morrison, 1991). Foster et al. (1995) also reported that in mature jack pine stands, foliage contributes as little as 20% of the total litterfall C every year and that dead trees and

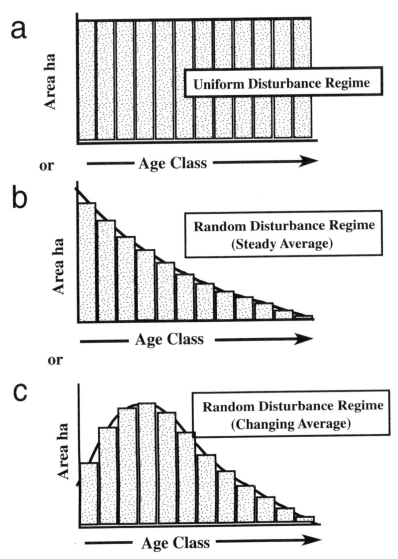

FIGURE 7.3 Boreal forest age-class structure under uniform (a), and random disturbance regimes (b and c). With random disturbances, the middle figure (b) represents a steady average disturbance frequency, and the bottom figure (c) under an increasing average disturbance frequency.

belowground turnover contribute a large portion of the total litterfall C. Therefore, coniferous species may indeed generate higher litterfall rates than deciduous species (Vogt et al., 1986), as simulated in CBM-CFS2.

The simulated soil C content in the boreal forest is directly related to the simulated litterfall C (Figure 7.4). As the litterfall rate increases, there is a steady increase in the soil C content. The size of the slow soil C pool is much higher than the fast soil C pool (Table 7.2). Decomposition of the slow soil C pool is also slower than the fast soil C pool because the edaphic environment of the soil causes slow pool soil C to become stabilized with mineral colloids that protect them from microbial attack (Anderson, 1991). In addition, much of the total litterfall (particularly in older SVOs) is transferred to an intermediate pool on the basis of size (Kurz et al., 1992). In a comparative decomposition study of forest leaves and roots, the latter always decayed at a signif-icantly slower rate because of the higher proportion of polyphenolics present in the root tissue

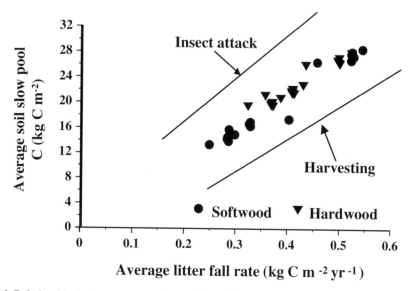

FIGURE 7.4 Relationship between average slow soil pool C and litterfall C rate (Y = 55.06*X + 0.29) in Saskatchewan boreal forest simulations; the solid line represents national trends for influence of disturbance by insect attacks and harvesting.

(Bloomfield et al., 1993). This slower rate of decomposition of roots, branches, and coarse woody debris is inherent in the parameters used by Kurz et al. (1992) to define the decomposition of the litter and contributes to sequestration of C in the soil.

The proportions of biomass C transferred to soil and detritus pools are also significantly influenced by the type of disturbance (Kurz et al., 1995). It can be assumed that biomass C litter transfer would be greatest for insect-induced stand mortality, intermediate for fire (as some C will be released directly to the atmosphere as combustion products), and smallest for harvesting (where most of the boles are transferred to the forest product sector) (Figure 7.4). The amount of C transferred to the soil will influence future rates of C and nutrient cycling (Bhatti et al., 1998) as well as other ecosystem functions.

The results of the simulations emphasize that the assumption of uniform disturbance regimes affect predictions of the C storage in boreal forest. With simulated random disturbance regimes, the C storage in all forest ecosystem compartments (biomass C, detritus C, and soil C) was higher than that obtained from a uniform regime of the same average return frequency.

INFLUENCE OF DISTURBANCE FREQUENCY ON SOIL C DYNAMICS

Carbon accumulating in soil is stable for long times when the difference between litter input and decomposition release is changing only slowly. The soil carbon thus may not fluctuate significantly until a disturbance event occurs at the site. The effect of such disturbance is both to provide a pulse of litter carbon (particularly coarse woody debris) and to change the site microclimate and decomposition variables (Apps and Kurz, 1993). Simulated soil C dynamics under random disturbance regimes were different than that under the uniform regime (Figure 7.5).

With random disturbance regimes, soil C was strongly affected by the previous disturbance pattern, which influences the present forest age-class structure. During a period of apparent low disturbance frequency, the soil C stocks were higher. With higher forest biomass C contributing to a higher litterfall rate, the balance between decomposition losses and litter inputs is achieved at a higher soil carbon value. Therefore, soil appears to act as a C sink during a transition to a lower disturbance frequency, a result previously noted by Kurz and Apps (1995). In contrast, during a transition to an apparently higher disturbance frequency, the soil C pool decreases and the soil

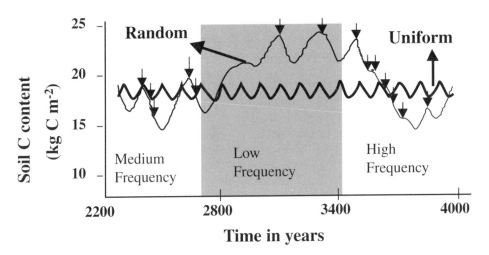

FIGURE 7.5 Soil C content in Saskatchewan boreal forests under random and uniform disturbance regimes over 1500 years; the three intervals represent periods of apparent medium, low, and high disturbance frequency, although over the 1500 years of the entire simulation the number of fire events is the same as the uniform regime; small arrows represent the disturbance events in random disturbance sequence.

appears to act as a source of C. This observed phenomenon arises for three reasons; (1) at the higher disturbance frequency, there is a higher proportion of younger age stands which generate lower litter transfers (as described previously), and for the same reason; (2) there is decreased input of coarse woody debris due to the decrease in proportions of older-age stands (Harmon et al., 1990), and (3) an increased rate of decomposition of detritus and soil C pool due to changed microclimate and higher exposure associated with the younger stands.

Although not presently simulated within CBM-CFS2, it is possible that under very high disturbance frequencies, regeneration may fail or be significantly delayed if the time between disturbances is shorter than the time required by boreal species to reach reproductive age. For instance, in Saskatchewan forests burned areas could be 10^6 ha or more, resulting in large areas of even-aged jack pine, whose serotinous cones require high temperatures to open. Therefore increased disturbance frequency could also increase the regeneration delay and dramatically change the C sequestration capacity of boreal forests. More importantly, as Figure 7.5 dramatically shows, the soil of any given site may act as a sink or a source of C depending on the actual disturbance history, even though the average return period for that point (or for the region) is constant. At any particular time, if a site is in a period of apparent low disturbance frequency, the soil C content would be higher than it would have in a period of apparent high disturbance frequency. It is not possible to account for these difference in soil C content under an assumed uniform disturbance regime.

COMPARING CENTURY AND CBM-CFS2 GROWTH CURVES

The growth curves used in the two models are quite different from each other (Figure 7.1). In CENTURY 4.0, it is assumed that at the time of disturbance due to fire, all aboveground biomass C is lost from the system to the atmosphere. Biomass begins its exponential regrowth initialized from the portion of biomass assumed to survive the disturbance. In contrast CBM-CFS2 transfers only a portion of biomass and litter soil C to the atmosphere. Instead most of the biomass C is transferred to litter C pools for subsequent decay into the atmosphere and slow soil pools.

CENTURY's exponential biomass growth curves assume that tree growth is controlled by soil moisture, soil temperature, shading, and a limiting nutrient that further constrains plant growth (Parton et al., 1993). As long as nutrient and water supply are not limiting factors, trees will keep on growing at a given site. In reality this likely is not true for boreal forest species. CENTURY also ignores the delay in regeneration following disturbance that is commonly observed in boreal forest stands. As shown in Figure 7.1, CENTURY starts exponential tree growth immediately following disturbance. In reality, there is typically a delay in the reestablishment of an overstory which depends upon many factors, including microclimatic and other site conditions for seed establishment, the presence of seed sources, and the prevalence of competitors such as grass (Kurz and Apps, 1998).

In CBM-CFS2, vegetation growth goes through four distinct stages of growth (Figure 7.1) (Kurz and Apps, 1994): regeneration, immature, mature, and over-mature. During the regeneration phase, it is assumed that there is only a small, approximately linear increase in biomass C associated with grasses and shrubs. Detritus C and soil C pool sizes decrease (Figure 7.5) as the decomposition losses, elevated due to higher soil temperature (Keenan and Kimmins, 1993), exceed the reduced input from biomass C. During this phase, there is a net loss of C from the forest stand (SVO) which acts as a source of atmospheric C. During the immature phase (also called the stem-exclusion phase), a logistic growth curve is followed, significant litterfall occurs and the canopy closes, changing the site microclimate. In the mature phase, there is little or no further increase in live biomass, and depending on the decomposition rates there may be a slight decrease in the total soil C pool during this growth phase (Figure 7.5). The biomass C in the over-mature phase is often less than that in the mature phase as indicated by inventory data of Canadian forests (Bonnor, 1985) and recent growth and yield curves for Russian forests (Venevsky and Shvidenko, 1997). This is quite different from the situation normally assumed for forests undergoing gap-phase replacement. During the over-mature phase, there is a net decline in the biomass C pool resulting in significant increases of large-size litterfall to the detritus and soil C pools. With no change in decomposition rate during this over-mature phase, there is a net gain of C by the detritus and soil C pools. Whether the entire stand (SVO) is a net source or sink depends on the ecosystem ecology — specifically it depends on the rate of stand breakup relative to the litter and soil decomposition rates. If no disturbance intervenes, new cohorts eventually emerge into the opening canopy and regrowth begins (Kurz et al., 1992). CBM-CFS2 is believed to represent the mature and over-mature phases of stand C dynamics and related processes more realistically than the CENTURY model.

IMPLICATIONS FOR CLIMATE CHANGE RESPONSE

Simulations with CBM-CFS2 suggest that total ecosystem C (biomass, detritus, and soil) are higher under random disturbance regimes than with uniform ones having the same average return interval. If true, this suggests that most models of boreal forest dynamics may either underpredict the amount of ecosystem C or have compensating and unintended errors. For the Saskatchewan boreal forest simulations reported here, the ecosystem C was 0.62 Gt C (16%) higher using the random regime than with a uniform regime. The magnitude of the difference depends on the relative ratios of ecosystem time constants and the disturbance return frequency. Moreover, these errors which pertain to a system in long-term equilibrium, are probably exacerbated when the disturbance regime is changing over time (Figure 7.5).

The differences in C dynamics between the two regimes can be understood in terms of the differences in the age class structure of the forest (Apps, 1997). Variations in litter input rates cause most of the variation in the soil carbon pools (Figure 7.4). Fire disturbance regimes are affected by climatic conditions through seasonal episodes of drought and elevated temperatures that affect fuel loads, and ridging high pressure systems that change the probability of lightning (Weber and Flannigan, 1997). Numerous studies have reported higher disturbance frequencies in the last 20 years (Blais, 1983; Clark, 1990; Kurz et al., 1995a). This period has also been a period of elevated temperatures in central and western Canada (Gullett and Skinner, 1992). These two factors have influenced the contemporary rate of net forest C storage in at least three ways: (1) detritus and soil C pool decomposition rates will have increased both due to regional warming and the reduced canopy shading associated with the increase in younger age-stands; (2) decomposable litter and soil C pools associated with disturbance regimes will have increased; and (3) C uptake will have decreased, at least initially, due to the increase in younger age stands with low initial growth rates and regeneration delay.

Using CBM-CFS2, Kurz and Apps (1996) showed that the factors just listed have likely resulted in a net C release from the boreal forests of Canada as a whole. In a companion paper Kurz and Apps (1995) showed that this source period is likely to continue for at least a decade, even if the disturbance regime returns to the averages of previous decades through increased fire protection efforts, for example. This hysteresis arises because of the decade-scale lag times in the dynamics of the biomass, litter, and soil pools. It appears that under changing conditions, some parts of the boreal forest may act as a sink of atmospheric C while others act as a source. Further, this spatial complexity can prevail for considerable lengths of time.

IMPLICATIONS FOR THE MITIGATION OF CLIMATE CHANGE

The factors discussed in the previous section highlight the importance of maintaining a strong fire protection and prevention program in Canada. Barney and Stocks (1983) reported a reduction in the area burned in the southern part of Canada in the 1970s and a number of reports have demonstrated the efficacy of protection measures in central and western Canada. These protection measures would have contributed to preservation of older age-classes (Blais, 1983) relative to what would have occurred without them. Protected areas, therefore, experience an apparently lower disturbance frequency and both the biomass and the soils likely act as a C sink (see middle portion of Figure 7.5) helping to offset some of the losses in the other less-protected areas. (The national scale analyses of Kurz and Apps [1998] used recorded disturbance statistics that include the results of suppression and hence explicitly incorporate these offsets.)

The question arises: How effective must fire suppression be in order to have a major influence on the C budget of these forests? Boreal forest ecosystems are adapted to a relatively short (<100 year) natural disturbance cycle (Van Wagner, 1978). Reducing the fire disturbance frequency will increase the average forest age (skewing the age-class distribution to the right) which may increase the forest susceptibility to insect attack. It may also increase the risk of much larger fires than those that were suppressed (Romme and Despain, 1989). Insect-induced disturbances, however, are associated with much higher litterfall, resulting in pulsed inputs of C to detritus and soil pools. They may also preferentially attack select specific age classes and thereby have rather different impacts on the ecosystem C storage than fire. It is beyond the scope of this chapter to examine these effects in detail; it suffices to note that protection measures can perhaps delay but not prevent eventual C release from the ecosystem (Kurz et al., 1995a). It should also be noted that a protected forest acts as a C sink (i.e., removes more C from the atmosphere than is returned) only during the period of transition from one disturbance regime to another. If the new regime can be maintained, the sink phenomenon saturates as C uptake through photosynthesis becomes balanced by releases by respiration, decomposition, and disturbance. Finally, if the protection is not maintained, or the risk exceeds the protection measures, and the disturbance regime increases, the forest will again become a source. This can be comprehended from random simulations of soil C content presented in the transition from low to high disturbance frequency (Figure 7.5).

A number of management options exist to increase the C sequestration through management of the boreal forest. Price et al. (1997) carried out a comprehensive assessment of carbon pools and fluxes in a boreal-cordilleran forest region in Canada and reported that forest management practices might contribute to an increase in total carbon storage in the region. Binkley et al. (1998) provided a useful summary of the main ways that a "natural" forest may be managed for increased carbon sequestration, together with an economic analyses. As this chapter focuses on the role of disturbance, only options that are related to influencing the disturbance regime will be discussed. There are three main options: protection, reducing regeneration delay, and harvesting practices.

Many boreal forest stands are in the regeneration phase for 10 to 30 years, showing little net biomass accumulation. Through seeding and planting of areas that have undergone recent disturbance, it may be possible to eliminate this delay and thereby increase its C sequestration (Price et al., 1997).

Harvesting may be regarded as a form of disturbance. It differs from natural disturbances in its impact on the on-site C stores (through the removal of boles), its influence on age-class structure, and the development of an off-site C store in the form of forest products (Apps et al., 1998). By combining protection against natural disturbance with harvesting, with an appropriate rotation cycle, it is possible to realize increased net C storage in both the forest ecosystem and forest products (Price et al., 1997; Apps et al., 1998).

CONCLUSIONS

CBM-CFS2 was compared against observed data for biomass C, litterfall rate, and soil C content in central Canada. The simulated biomass C and litterfall rates were consistent with the observed data. On average, CBM-CFS2 overestimated the soil C content for Saskatchewan boreal forests by 60% as compared to field measurements. The estimates, however, are quite within the range of soil C content reported for the boreal zone.

Simulation of C dynamics in CENTURY and CBM-CFS2 are quite different. CENTURY makes the assumption that at the time of disturbance, all stand biomass C is lost from the system to the atmosphere. In CBM-CFS2, only the specified portion of biomass C is transferred to the atmosphere and, instead, most of the biomass C is transferred to litter C pools where it subsequently decays into the atmosphere and slow soil pools. In CENTURY, biomass begins its exponential regrowth initialized from a sizable portion of biomass that is assumed to survive the disturbance, while in CBM-CFS2 the forest stand first goes through a regeneration phase which typically shows only a small linear growth in biomass C.

Simulated ecosystem C, biomass C, litter pool C, fast, and slow soil pool C were higher under a random disturbance regime than under a uniform regime having the same average return frequency. These differences arise from differences in the vegetation age-class structure associated with the two regimes. While the influence of age-class structure on forest-scale vegetation biomass (and hence phytomass C) is well known, the influence of this structure on litter and soil C pools has generally been ignored. This influence is, in part, due to the functional relationship between litterfall (of all types) and forest stand development (stand age, or time since disturbance). This relationship begins with the disturbance itself: during or shortly after most disturbances (including fire), there are important pulses of detritus transferred onto the forest floor. Litterfall (above and belowground) continues to vary with the stage of stand development until a subsequent disturbance repeats the cycle of regrowth.

The relationship also reflects species differences. Higher amounts of C accumulate in detritus and soil C pools under coniferous forest stands, suggesting that coniferous forest soils may sequester C for longer periods of time compared with deciduous forests. The variations in climate over the last 20 years appear to have resulted in a higher disturbance frequency cycle. With higher disturbance rates, boreal forests have younger age-class forests and greater releases of C from detritus and soil C pools. Therefore, central Canada's boreal forests appear to be acting as a source of C as a result of such climate variations (Kurz and Apps, 1996, 1998). This situation is expected to worsen as climatic change proceeds with a projected shortening of fire return intervals (Weber and Flannigan, 1997). With efficient fire prevention, it may be possible to reduce the loss of C, or even increase the amount of C sequestration by the boreal forest sector, but such opportunities have limited scope and feasibility if not combined with a larger suite of forest resource practices. These include reducing regeneration delay, substitution of natural disturbance by harvesting with an appropriate rotation length, and the judicious use of forest products. In addition, such practices must take into account the response of both the forest ecosystems and the disturbance regime to future changes in climate.

ACKNOWLEDGMENTS

Special thanks to W. A. Kurz (ESSA Technologies Ltd., Vancouver) and Ralph Mair for valuable discussion, information, and assistance. Funding for this study was contributed by the Sustainable Forest Management Network of Centres of Excellence (NCE), and the Energy from the Forest (ENFOR) program of the Federal Panel on Energy Research and Development (PERD).

REFERENCES

Anderson, J.M., The effect of climate change on decomposition processes in grassland and coniferous forests, *Ecol. Appl.,* 1, 326, 1991.

Apps, M.J., Biomass burning: accounting for fire in the IPCC guidelines. Unpublished paper for IPCC Workshop on Biomass Burning, Land-Use Change, and Forestry, Rochhampton, Australia, 1997.

Apps, M.K., Kurz, W.A., and Bhatti, J.S., Energy, bioenergy, and the carbon budget of the Canadian forest product sector, *J. Environ. Sci. Policy,* 2, 25-41, 1999.

Apps, M.K. and Price, D.T., Eds., Introduction, in *Forest Ecosystems, Forest Management and the Global Carbon Cycle,* Vol. 40, NATO ASI Series. I. Global Environmental Change, Springer-Verlag, Heidelberg, 1996.

Apps, M.K., and Kurz, W.A., The role of Canadian forests in the global carbon balance, in Carbon Balance of World's Forested Ecosystems: Toward a Global Assessment, Publ. Acad. No. 3/1993, Helsinki, Finland, 1993.

AES (Atmosphere Environment Services), Canadian Climate normals 1951-1980, temperature, precipitation: Prairie Provinces, Environmental Canada, Downsview, Ontario, 1983.

Barney, R.J., and Stocks, B.J., Fire frequencies during the suppression period, *The Role of Fire in Northern Circumpolar Ecosystems,* Wein, R.W. and Maclean, D.A., Eds., SCOPE 18, John Wiley & Sons, Chichester, U.K., 1983.

Bhatti, J.S. and Apps, M.J., Carbon and nitrogen storage in upland boreal forests, in *Global Climate Change and Cold Ecosystems,* Lal, R., Kimble, J., Eswaran, H., and Stewart, B.A., Eds., CRC Press, Boca Raton, FL, 2000.

Bhatti, J.S., Foster, N.W., Oja, T., Moayeri, M.H., and Arp, P.A., Modeling potential sustainable biomass productivity in jack pine forest stands, *Can. J. Soil Sci.,* 78, 105, 1998.

Blais, J.R., Trends in the frequency, extent, and severity of spruce budworm outbreaks in eastern Canada, *Can. J. For. Res.,* 13, 539, 1983.

Binkley, C.S., Apps, M.J., Dixon, R.K., Kauppi, P., and Nilsson, L.O., Sequestering carbon in natural forests, *Crit. Rev. Environ. Sci. Technol.,* 27, S23, 1998.

Bloomfield, J., Vogt, K.A., and Vogt, D.J., Decay rate and substrate quality of fine roots and foliage of two tropical tree species in the Luquillo experimental forests, Puerto Rico, *Plant Soil,* 150, 233, 1993.

Bonnor, G.M., Inventory of Forest Biomass in Canada, Canadian Forest Service, Petawawa National Forestry Institute, Chalk River, Ontario, 1985.

Clark, J.S., Ecological disturbances as a renewal process: theory and application to fire history, *OIKOS,* 56, 17, 1989.

Clark, J.S., Fire and climate change during the last 750 years in northwest Minnesota, *Ecol. Monogr.,* 60, 135, 1990.

Foster, N.W., Morrison, I.K., Hazlett, P.W., Hogan, G.D., and Salerno, M.I., Carbon and nitrogen cycling within mid- and late-rotation jack pine, in *Carbon Forms and Functions in Forest Soils,* McFee, W.W. and Kelly, J.M., Eds., Soil Science Society, Madison, WI, 1995.

Gullett, D.W. and Skinner, W.R., The State of Canada's Climate, Temperature Change in Canada 1895-1991, SOE Rep. 92-2, Atmospheric Environmental Service, Environment Canada, Ottawa, 1992.

Halliwell, D.H., Apps, M.J., and Price, D.T., A survey of the forest site characteristics in a transect through the central Canadian boreal forest, *Water, Air, Soil Pollut.,* 82, 257, 1995.

Harmon, M.E., Ferrell, W.K., and Franklin, J.F., Effects of carbon storage on conversion of old-growth forests to young forests, *Science,* 247, 699, 1990.

Keenan, R.J. and Kimmins, J.P., The ecological effects of clear-cutting, *Environ. Rev.,* 1, 121, 1993.

Kurz, W.A. and Apps, M.J., A 70-year retrospective analysis of carbon fluxes in the Canadian forest sector, *Ecol. Appl.,* 9, 526-547, 1999.

Kurz, W.A. and Apps, M.J., Retrospective assessment of carbon flows in Canadian boreal forests, in *Forest Ecosystems, Forest Management and Global Carbon Cycle*, Apps, M.J. and Price, D.T., Eds., Springer-Verlag, Berlin, 1996, p. 173.

Kurz, W.A. and Apps, M.J., An analysis of future carbon budgets of Canadian boreal forests, *Water, Air, Soil Pollut.*, 82, 321, 1995.

Kurz, W.A. and Apps, M.J., The carbon budget of Canadian forests: a sensitivity analysis of changes in disturbance regimes, growth rates, and decomposition rates, *Environ. Pollut.*, 83, 55, 1994.

Kurz, W.A., Apps, M.J., Stocks, B.J., and Volney, W.J.A., Global climate change: disturbance regimes and biospheric feedbacks of temperate and boreal forests, in *Biotic Feedbacks in the Global Climate System*, Woodwell, G.M. and Mackenzie, F.T., Eds., Oxford University Press, New York, 1995a, p. 119.

Kurz, W.A., Apps, M.J., Beukema, S.J., and Lekstrum, T., Twentieth century carbon budget of Canadian forests, *Tellus*, 47B, 170, 1995b.

Kurz, W.A., Apps, M.J., Webb, T.M., and McNemee, P.J., The Carbon Budget of the Canadian Forest Sector: Phase I, Inf. Rep. NOR-X-326, Northern Forestry Center, Forestry Canada, Edmonton, Alberta, 1992.

MacLaren, J.P., Plantation forestry 3/4 and its role as a carbon sink: conclusions from calculations based on New Zealand's planted forest estate, in *Forest Ecosystems, Forest Management and the Global Carbon Cycle*, Apps, M.J. and Price, D.T., Eds., Springer-Verlag, Berlin, 1996, p. 257.

Matthews, E., Global litter production, pools, and turnover times: estimates from measurement data and regression models, *J. Geophys. Res.*, 102, 18771, 1997.

Metherell, A.K., Harding, L.A., Cole, C.V., and Parton, W.J., CENTURY soil organic matter model environ-ment, Technical documentation Agroecosystem Version 4.0, Tech. Rep. No, 4, Great Plains System Research Unit, USDA-ARS, U.S. Department of Agriculture, Fort Collins, CO, 1993.

Morrison, I.K., Addition of organic matter and elements to the forest floor of an old-growth *Acer saccharum* forest in the annual litter fall, *Can. J. For. Res.*, 21, 462, 1991.

Parton, W.J., Schimel, D.S., Cole, C.V., and Ojima, D.S., Analysis of factors controlling soil organic levels in Great Plains grasslands, *Soil Sci. Soc. Am. J.*, 51, 1173, 1987.

Parton, W.J., Scurlock, J.M.O., Ojima, D.S., Gilmanov, T.G., Scholes, R.J., Schimel, D.S., Kirchner, T., Menaut, J.C., Seastedt, T., Garcia Moya, E., Kamnalrut, A., and Kinyamarid, J.I., Observations and modeling of biomass and soil organic matter dynamics for the grassland biome worldwide, *Global Biogeochem. Cycles*, 7, 785, 1993.

Pastor, J. and Post, W.M., Response of northern forests to CO_2-induced climate change, *Nature*, 334, 55, 1988.

Peng, C.H. and Apps, M.J., Simulating carbon dynamics: the Boreal Forest Transect Case Study (BFTCS) in central Canada. II. Sensitivity to climate change, *Global Biogeochem. Cycles*, 12, 393, 1998.

Peng, C.H., Apps, M.J., Price, D.T., Nalder, I.A., and Halliwell, D.H., Simulating carbon dynamics along the Boreal Forest Transect Case Study (BFTCS) in central Canada. I. Model testing, *Global Biogeochem. Cycles*, 12, 381, 1998.

Price, D.T., Halliwell, D.H., Apps, M.J., Kurz, W.A., and Curry S.R., Comprehensive assessment of carbon stocks and fluxes in a boreal-cordilleran forest management unit, *Can. J. For. Res.*, 27, 2005, 1997.

Romme, W.H. and Despain, D.G., The Yellowstone fires, *Sci. Am.*, 261, 37, 1989.

Schimel, D.S., Braswell, B.H., Holland, E.A., Mckeown, R., Ojima, D.S., Painter, T.H., Parton, W.J., and Townsend, A.R., Climatic, edaphic, and biotic controls over storage and turnover of carbon in soils, *Global Biogeochem. Cycles*, 8, 279, 1994.

Siltanen, R.M., Apps, M.J., Zoltai, S.C., Mair. R.M., and Strong, W.L., A Soil Profile and Organic Carbon Database for Canadian Forest Tundra Mineral Soils, Inf. Rep. Fo42-271/1997E, Canadian Forest Service, Northern Forest Center, Edmonton, Alberta, 1997.

Simpson, L.G., Botkin, D.B., and Nisbet, R.A.,The potential aboveground carbon storage of North American forests, *Water, Air, Soil Pollut.*, 70, 197, 1993.

Steijlen, I. and Zackrisson, O., Long-term regeneration dynamics and successional trends in northern Swedish coniferous forest stand, *Can. J., Bot.*, 65, 839, 1987.

Tarnocai, C., The amount of organic carbon in various soil order and ecological provinces in Canada, in *Soil Processes and the Carbon Cycle*, Lal, R., Kimble, J.M., Follett, R.F., and Stewart, B.A., Eds., CRC Press, Boca Raton, FL, 1998, p. 81.

Venevsky, S. and Shvidenko, A., Modeling of Stand Growth Dynamics with a Destructive Stage, Proc. 7th Annu. Conf. Int. Boreal For. Res. Assoc., St. Petersburg, Russia, 1997, p. 28.

Van Wagner, C.E., Age-class distribution and the forest fire cycle, *Can. J. For. Res.*, 8. 220. 1978.

Vogt, K.A., Grier, C.C., and Vogt, D.J., Production, turnover, and nutrient dynamics of above- and belowground detritus of world forests, *Adv. Ecol. Res.*, 15, 303, 1986.

Weber, M.G. and Flanningan, M.D., Canadian boreal forest ecosystem structure and function in a changing climate: impact of fire regimes, *Environ. Rev.*, 5, 145, 1997.

8 Soil and Crop Management Impact on SOC and Physical Properties of Soils in Northern Sweden

L. Ericson and L. Mattsson

CONTENTS

INTRODUCTION

Agricultural soils in northern Sweden are generally rather rich in soil organic carbon (SOC) (Andersson et al., 1997). This is partly due to low temperatures and high precipitation. More important is the fact that crop production has been, and still to a great extent is, dominated by perennial grasses and clover for hay or silage. In 1996 almost 70% of the arable land in the four northernmost counties in Sweden was ley, used as fodder in dairy and beef production (SCB, 1997).

There are plenty of studies that confirm the importance of SOC for soil structure and soil physical properties (Angers and Girroux, 1996; Oades, 1984; Tisdall and Oades, 1980, 1982; and Schnitzer, 1992). Soils in the main agricultural areas in northern Sweden are mainly silty, especially in the coastal areas and the big river valleys. They are structurally weak and hence susceptible to slaking and crust formation. At most farms where dairy production dominates, problems with bad soil structure caused by decreasing amounts of SOC are small. At specialized farms with a high proportion of, e.g., cereals, potatoes, or other horticultural crops, loss of SOC can be a problem.

There are only a few studies on the impact of crop and soil management on SOC in northern Sweden. In this chapter data on SOC from seven long-term field experiments from three different studies, are reported. The emphasis is on a study in three field experiments in northern Sweden, made in the mid 1980s. Some results from this study have been published earlier (Ericson, 1994).

SOC IN SEVEN LONG-TERM EXPERIMENTS IN NORTHERN SWEDEN

EXPERIMENT SITES

Four sites in northern Sweden were used for the field experiments in the different studies reported below (Figure 8.1). Data for these sites are listed in Table 8.1.

STUDY 1 — INCORPORATION OF STRAW

Since 1980 we have conducted a field experiment at Röbäcksdalen, Umeå including four levels of nitrogen combined with removal and incorporation of straw (Table 8.1). The experiment has been cropped with a barley (*Hordeum vulgare*) and oats (*Avena sativa*). The amount of SOC was sampled three times during the 16 years of cropping (Table 8.2). Because there is only one sample for each nitrogen level, we have chosen to use N-levels as subsamples for the treatment of straw. This approach can be questionable, but the risk to overestimate differences between the two straw treatments is small, as fertilizer levels will add more variation to data.

After 16 years of cropping, levels of SOC have diverged, although there has been a reduction in SOC in both treatments. We calculated the amount of SOC that has been lost. In both treatments SOC has decreased. Incorporation of straw has reduced loss of SOC with 2.4 Mg ha^{-1}. That corresponds to approximately 17% of the total amount of straw incorporated during 16 years of cropping, which fits well with other findings in Sweden (Johansson, 1994). There is no significant difference in yield between treatments.

STUDY 2 — MANURE AND COMMERCIAL FERTILIZERS

Three long-term field experiments with different crop rotations, two levels of fertilizers, and one level of manure were started in 1965. The designs of these experiments are shown in Table 8.3. The experiments were placed on three different sites (Table 8.1). SOC was determined on samples taken from one occasion in 1988, after 23 years of cropping (Table 8.4). There is only one sample from each crop and fertilizer level. There are significant differences in SOC between treatments only at Öjebyn. One reason for this can be that the levels of SOC are much lower at this site. It is likely that cropping at the other sites will lead to a lowering of SOC in all crops and at all levels. This can overrun the effect of manure. One must also bear in mind that the data set is rather small, which makes it more difficult to reveal differences between treatments.

FIGURE 8.1 Map showing the location of the experiment sites used in the studies reported.

TABLE 8.1
Latitude, Soil Data, Mean Temperature (°C) of the Few Sites and Studies in Which the Experimental Sites Were Used

Site	Latitude	Soil type	Clay content (%)	Mean temperature Year	Mean temperature July	Study
Offer	63°08'	Silty clay loam	20 to 25	2.8	16.0	2
Ås	63°15'	Gravely loam	20	2.9	14.5	2 and 3
Röbäcksdalen	63°49'	Clayey silt loam	10 to 15	3.1	16.1	1, 2, and 3
Öjebyn	65°19'	Silt loam	5 to 10	2.0	16.7	3

TABLE 8.2
Changes in SOC During 16 years of Cropping

Treatment	1981 SOC	1988 SOC	1996 SOC	Added straw (43% C)	Differences in soil C	Remaining C (% added in straw)
				Mg ha^{-1}		
Straw removed	74.1	73.1	62.1	0	12	—
Straw incorporated	76.2	75.4	66.6	32 (13.8)	9.6	17

TABLE 8.3
Treatments in the Field Experiments in Study 2

Main plots

Treatments
1. Low level of nitrogen, phosphorus, and potassium in commercial fertilizers
2. Twice the amount of commercial fertilizers compared with treatment 1
3. Manure, 10 Mg ha^{-1} per year, replenishment with commercial fertilizers up to the same level as in treatment 1

Subplots

Crop rotations
A	Barley undersown with ley	1st ley	2nd ley	Fodder rape	Potatoes	Italian ryegrass
B	Barley undersown with ley					
C	Fodder rape all years					
D	Italian ryegrass all years					
E	Potatoes all years					
F	Barley all years					
G	Barley	Barley	Oats	Barley	Barley	Oats
H	Barley	Barley	Potatoes	Barley	Barley	Potatoes
I	Barley undersown with ley all years					

Note: Barley (*Hordeum vulgare*); fodder rape (*Brassoca napus*); potatoes (*Solanum tuberosum*); Italian ryegrass (*Lolium multiflorum*); ley was a mixture of red clover (*Trifolium pretense*), timothy (*Phleum pretense*), and meadow fescue (*Festuca* sp.).

TABLE 8.4
SOC in a Long-term Experiment with Different Crop Rotations and Three Levels of Fertilizers; Means for 9 Samples at Each Treatment; Figures Followed by the Same Letter are Not Significantly Different ($p = 05$, Fisher's LSD Test)

Treatment	SOC %, 1988		
	Röbäcksdalen	Ås	Öjebyn
Low fertilizer	3.40a	2.40a	1.36a
High fertilizer	3.44a	2.57a	1.45a
Manure	3.30a	2.62a	1.63b

TABLE 8.5
Treatments — Four Six-Year Crop Rotations; Study 3

A	B	C	D
Barley undersown with ley	Barley undersown with ley	Barley undersown with ley	Barley undersown with ley
1st ley	1st ley	1st ley	Ley as green manure
2nd ley	2nd ley	2nd ley	Winter rye
3rd ley	3rd ley	Winter rye	Peas
4th ley	Green fodder (oats/peas)	Green fodder (oats/peas)	Potatoes
5th ley	Fodder rape	Potatoes	Carrots/Swedes

Note: Barley (*Hordeum vulgare*); fodder rape (*Brassica napus*); potatoes (*Solanum tuberosum*); Oats (*Avena sativa*); winter rye (*Secale cereale*); peas (*Pisum sativum*); carrots (*Daucus carota*); Swedes (*Brassica napus var. napobrassica*); ley was a mixture of red clover (*Trifolium pretense*), timothy (*Pheleum pretense*), and meadow fescue (*Festuca* sp.).

STUDY 3 — SOIL PHYSICAL PROPERTIES AND SOC IN THREE LONG-TERM EXPERIMENTS

MATERIAL AND METHODS

In the mid-1950s three long-term crop rotation experiments were initiated in the northern part of Sweden. The aim was to compare forage yields in three different cropping systems. A fourth system without any forage crops was also added to the plan. Manure was added to the crop rotations with forage crops according to the number of animals that could be fed from the different rotations. Treatments are listed in Table 8.5. In 1987 different soil physical parameters were examined.

EXPERIMENTAL DESIGN

All crops in the rotations were grown in two replicated plots, giving 48 plots in each field; 6 years (cycle) $\times$ 4 (rotations) $\times$ 2 (replicated plots). The design differs among the different sites. At Ås and Röbäcksdalen a split-plot design with rotations as main plots and phases as subplots were used. At Offer phases formed the main plots and crops subplots. Two plots in each rotation were chosen for sampling and measurements, which was done in the summer of 1987.

Saturated Hydraulic Conductivity

Saturated hydraulic conductivity (K_{sat}) was measured in the laboratory on soil cores 5 cm long and 7.2 cm in diameter taken from the plough pan in one plot in each rotation. A constant head apparatus was used with a hydraulic gradient of 2 (hydraulic head = 10 cm) (Andersson, 1955).

Bulk Density in the Topsoil

Bulk density in the topsoil was determined using a method developed by Håkansson (1990). A steel frame 0.3 m in height with side lengths of 0.707 m is driven into the soil. By measuring the volume of the soil in its natural state, weighing the soil of the layer investigated, and determining the actual water content one can determine bulk density. Four measurements were made in two plots in each rotation.

Aggregate Distribution

Aggregate distribution was determined by dry sieving the seedbed after sowing in spring. A quadratic steel frame with 0.4 m side length was pressed into the soil by hand. A two-sided steel angle, with one side measuring 0.25 m and the other 0.4 m, was attached to the frame. That results in an area of 0.25 times 0.4 m surrounded with steel at three sides. From this area the loose material in the seedbed was put on sieves to separate aggregates in three sizes, less than 2 mm, 2–5 mm, and larger than 5 mm, respectively. This was done in two plots for each rotation. Six measurements were made in two plots in each rotation. The loose soil inside the bigger frame was used to determine sowing depth by measuring the volume. A more extensive description of the method is found in Kritz (1983).

Root Depth

Root depth was examined in a profile pit in one plot in each rotation. All plots examined were cropped with barley. Roots in a sector of 10 cm in width were counted at every 10 cm of depth until no roots could be found.

Soil Organic Carbon and Nitrogen

SOC in each plot was analyzed at several occasions during the run of the experiments. The number of samples varied among the different sites. Analysis of SOC was made by dry combustion (Ströhlein apparatus) and emitted CO_2 was measured with an IR-detector. Also, total nitrogen in the soil was analyzed by the Kjeldahl method.

Trends in Barley Yields

Barley undersown with ley was the only common crop in all rotations. In this experiment, however, it is difficult to make a direct comparison of barley yields because of differences in fertilizer levels and differences in preceding crops in the different treatments. Instead, trends in barley yields were calculated for the different rotations. The period 1963-1987 was chosen for all sites and was split into four six-year rotation periods. All plots of barley during, e.g., 1963-1968, are considered to be barley in rotation period 1, and so on. If rotation and rotation period interacted, one further studies orthogonal contrasts, indicating which of the trends, the linear, quadratic, or cubic, are significant. If the linear component was significant the slope for barley yield for each plot was calculated according to Fisher and Yates (1963). The mean slope for each rotation was determined and an analysis of variance and a multiple comparison of the means (Fisher's LSD) was performed. A more comprehensive description of the model is found in Ericson (1994).

TABLE 8.6
Saturated Hydraulic Conductivity (m × 10⁻³ day⁻¹) in the Lower Topsoil (Approximately 20 to 25 cm Deep); Figures Followed by the Same Letter Are Not Significantly Different (p = 05, Fisher's LSD Test)

	Experiment site	
	Offer	Röbäcksdalen
Rotation	K_{sat} (mm day⁻¹)	K_{sat} (mm day⁻¹)
A	6.15a	0.35a
B	3.44a	3.83a
C	0.37a	0.60a
D	0.13a	0.17a

TABLE 8.7
Bulk Density in the Topsoil (0 to 30 cm) at the Different Sites; Figures Followed by the Same Letter Are Not Significantly Different (p = 05, Fisher's LSD Test)

	Experiment site		
	Offer	Ås	Röbäcksdalen
Rotation	Bulk density	Bulk density	Bulk density
		mg m⁻³	
A	1.05a	1.33a	1.18a
B	1.16b	1.43ab	1.19a
C	1.23b	1.49b	1.39b
D	1.25b	1.53b	1.50b

RESULTS AND DISCUSSION

SATURATED HYDRAULIC CONDUCTIVITY

Saturated hydraulic conductivity (K_{sat}) at the different sites is shown in Table 8.6. As can be seen there is a vast variation between replicates; there are no significant differences between treatments at the 5% level. However, there is a trend at Offer, where rotations A and B have a K_{sat} of the same magnitude and rotations C and D have a much lower value. At Röbäcksdalen, rotation B has the highest K_{sat} while rotation A has an unexpectedly low value.

BULK DENSITY IN THE TOPSOIL

Bulk densities at the different sites and crop rotations are shown in Table 8.7. At Offer and Ås, bulk density was lower in rotation A than in the other treatments. At Röbäcksdalen, treatments A and B have a lower bulk density than treatments C and D. This confirms our visual judgment in the field, where rotations A and B were easily separated from rotations C and D by their appearance. Although this difference is clear only at Röbäcksdalen, the tendency at the other sites is the same. Soils in plots from rotations C and D also seemed to have a greater tendency to form a crust. This was especially evident at Offer, where the soil type is susceptible to this kind of phenomenon. This can also be seen in the aggregate analyses (see below).

TABLE 8.8
Fractions of Different Aggregate Sizes (%) After Dry Sieving of the Topsoil at Offer and Röbäcksdalen; Numbers Followed by the Same Letter Are Not Significantly Different ($p = 05$; Fisher's LSD Test)

| | Aggregate size distribution (%) | | | | | |
| | Offer | | | Röbäcksdalen | | |
Treatment	>5 mm	2 to 5 mm	<2 mm	>5mm	2 to 5 mm	<2 mm
A	24 a	19 a	57 a	37 a	23 a	40 a
B	37 b	15 a	48 b	45 ab	19 ab	36 ab
C	34 b	19 a	47 b	52 bc	19 ab	29 bc
D	33 b	17 a	50 b	59 c	18 b	23 c

TABLE 8.9
Number of Roots at Different Depths in Different Treatments (Röb = Röbäcksdalen)

	Rotation											
	A			B			C			D		
Depth	Site											
(cm)	Offer	Ås	Röb	Offer	Ås	Röb	Offer	Ås	Röb	Offer	Ås	Röb
10	11	16	22	13	10	14	19	8	14	14	14	9
20	17	11	19	17	6	13	9	7	10		6	2
30	4	13	5	7	8	4	4			5		
40	4	8	6	7	2		2			6		
50	1	7	2	7			1			4		
60	1									4		
70	1									1		

AGGREGATE SIZE DISTRIBUTION

Results from the dry sieving are presented in Table 8.8. At both sites the amount of small aggregates is significantly higher in rotation A. At Röbäcksdalen there is a trend towards bigger aggregates from rotation A to rotation D, while at Offer rotations B to D seem to have the same aggregate size distribution. There are only small differences in the amount of aggregates in the fraction 2 — 5 mm between rotations. This is in accordance with findings of Broersma et al. (1997).

ROOT DEPTH

In Table 8.9 the number of roots at different depths is listed. Our method can only give a superficial idea of the situation. The pits that were used for the observations were also used to investigate the soil profile and to take samples from each 10-cm level down to 100 cm.

Roots at Ås appear to be much less affected by the treatment than those at the other sites. The soil at Ås, being poorly graded, has a much more favorable structure than the highly graded soils at the other sites, explaining this result. At the two other sites there is a trend, however, where root depths are greatest in rotation A and smallest in rotation D.

SOIL ORGANIC CARBON

Results of the analysis of SOC are shown in Figures 8.2 through 8.4. At all sites differences between SOC in the different rotations have increased, with the highest content in rotation A and the lowest

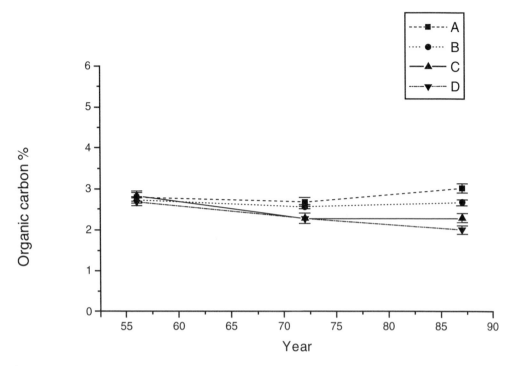

FIGURE 8.2 Soil organic carbon at Offer sampled in 1956, 1972, and 1987; means of 6 samples for each rotation and year; error bars indicate ± standard error.

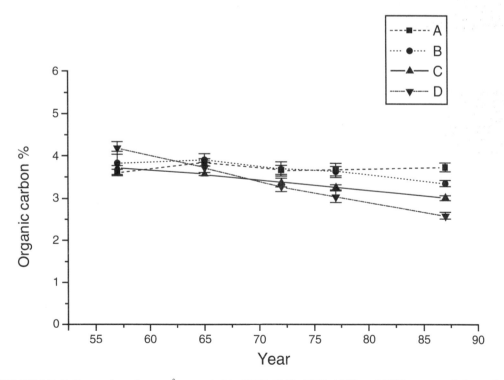

FIGURE 8.3 Soil organic carbon at Ås sampled in 1957, 1965, 1972, 1977, and 1987; means of 6 samples for each rotation and year; error bars indicate ± standard error.

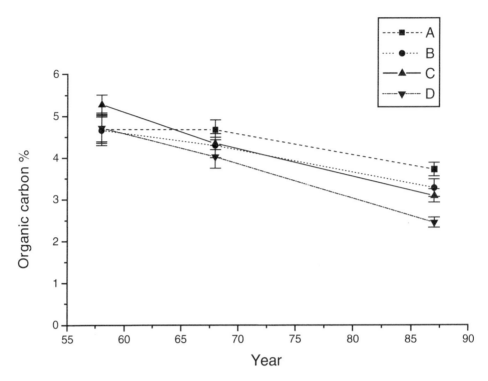

FIGURE 8.4 Soil organic carbon at Röbäcksdalen sampled in 1958, 1968, and 1987; means of 6 samples for each rotation and year; error bars indicate ± standard error.

in rotation D. At Röbäcksdalen there has been a decrease in all rotations, while at Ås SOC has decreased in all rotations except in A, and at Offer it has decreased in rotation C and D, but remained unchanged in rotation B, and has increased somewhat in rotation A.

Calculations of the total amount of SOC in the soil are reported in Table 8.10. As can be seen, the difference in SOC corresponds to as much as 690 kg ha^{-1} year^{-1}. The value for rotation C at Röbäcksdalen is unexpected; likely resulting from the considerable unevenness at this site.

C/N Ratio

In Tables 8.11 to 8.13 the C/N ratios for the different sites are shown. There are no differences between treatments at the different sites. Only at Röbäcksdalen is there a change over time in the C/N ratio, which seems plausible as there is a decrease in organic carbon in all treatments. This indicates that humus in this soil was young at the start of the experiment. Before the experiment here was started, the site was a poorly drained grassland.

Trends in Barley Yield

The calculated slopes for barley yield over the rotations are listed in Table 8.14. At Röbäcksdalen there were no statistically significant interactions between rotation and rotation period, indicating that there was no significant trend in yield at this site. Therefore, no slopes are calculated for this site. There is a positive trend in barley yield for rotation A at Ås and for rotation A and B at Offer. Rotation B at Ås has a small negative trend, while rotations C and D both have an unchanged level at Offer but a negative trend at Ås.

There seems to be a relatively strong correlation between crop rotation, decrease in SOC, changes in physical properties of the soil, and barley yields.

TABLE 8.10

The Amount of Carbon, Bulk Density, Difference to Treatment A, and the Change in SOC Per Year in Different Treatments After 30 Years of Cropping

Treatment	C%, 1986	Bulk density Mg m^{-3}	Mg C ha^{-1}	A-X Mg C ha^{-1}	A-X kg C ha^{-1} yr^{-1}
		Offer site			
A	3.06	1.05	64.26	—	—
B	2.71	1.16	62.87	1.39	46
C	2.32	1.23	57.07	7.19	240
D	2.06	1.25	51.50	12.76	425
		Ås site			
A	3.77	1.33	100.28	—	130
B	3.37	1.43	98.38	3.90	343
C	3.02	1.49	90.00	10.28	691
D	2.60	1.53	79.56	20.72	
		Röbäcksdalen			
A	3.87	1.18	91.30	—	—
B	3.64	1.19	86.60	4.70	157
C	3.27	1.39	90.90	0.40	13
D	2.64	1.50	79.20	11.70	390

TABLE 8.11

The C/N Ratio of the Soil Organic Matter in Different Treatments at the Ås Site; Figures Followed by the Same Letter Are Not Significantly Different (p = 05, Fisher's LSD Test)

Treatment	1957	1965	1972	1977	1987
A	11.4a	14.1a	11.3a	11.6a	12.6a
B	11.2a	14.6a	12.2a	12.1a	12.2a
C	11.6a	13.6a	11.3a	11.5a	12.5a
D	12.2a	14.0a	11.3a	11.6a	12.9a

TABLE 8.12

The C/N Ratio of the Soil Organic Matter in Different Treatments at the Offer Site; Figures Followed by the Same Letter Are Not Significantly Different (p = 05, Fisher's LSD Test)

Treatment	1956	1972	1987
A	11.1a	10.9a	12.3a
B	11.0a	10.9a	11.5b
C	11.0a	10.3a	11.5b
D	10.2a	10.8a	11.1b

TABLE 8.13
The C/N Ratio of the Soil Organic Matter in Different Treatments at the Röbäcksdalen Site; Figures Followed by the Same Letter Are Not Significantly Different ($p = 05$, Fisher's LSD Test)

Treatment	1958	1968	1987
A	19.1a	19.3a	15.4a
B	19.5a	19.7a	17.7a
C	19.5a	18.9a	16.0a
D	18.8a	18.5a	15.7a

TABLE 8.14
Slopes for Barley Yield at Offer and Ås (kg/Rotation Period); Treatments Followed by the Same Letter Are Not Significantly Different ($p = 05$; Fisher's LSD Test)

Treatment	Experiment Site	
	Offer	Ås
A	200.7 a	110.8 a
B	147.8 a	−2.9 b
C	−16.7 b	−92.5 b
D	7.8 b	−93.5 b

GENERAL DISCUSSION

The studies reported in this chapter show the great impact that cropping can have on SOC and also on the physical properties of the soil. It is interesting to note the differences in response in SOC that is found at the different sites in Study 3. To be able to evaluate the impact of a cropping system on SOC at a certain site, one has to know the former cropping history at the site. Crop rotation A has led to a raise in SOC at Offer, kept SOC unchanged at Ås, but led to a reduction in SOC at Röbäcksdalen. This reflects very clearly the effects of former cropping. The field at Röbäcksdalen had for a long time before the experiment started been poorly drained grassland. This means that drainage and cropping per se will reduce SOC, even in the rotation that consists of five years of ley and only one year of barley.

The effect of crop rotation on soil physical properties is clear in Study 3. However, a raise in SOC will lead to a higher mineralization of nutrients from the soil, which will affect yield. Although fertilizer levels in the cited experiments have been adjusted to meet these differences, one can not be sure that it is sufficient to compensate. Calculations on nutrient balances in the different rotations, not reported here, indicate that phosphorus and potassium have not been limiting factors. The effect of nitrogen on yield mineralization from SOC is much more difficult to estimate. The differences in trends observed in barley yield can not be distinctly linked to differences in the physical properties of the soil. However, they will give an idea of the sustainability of the different systems in the long run. One would expect a positive trend for yields in all systems as new, better varieties are introduced and fertilizer levels are raised according to common practice during the experimental period. The poor, and in some cases negative, trend in barley yield in rotations C and D is therefore a strong evidence that the sustainability of those systems are not good. It is also clear that soil physical properties resulting from these systems play an important role for yield, although it is not possible to separate that from other factors.

Results from Study 2 give a brief idea of the effect of manure on SOC, which is not possible to estimate in Study 3. Unfortunately we have little data on SOC from the experiments in Study 2. We have just started a project were we will use stored samples from these experiments to estimate SOC levels in the different treatments over the years. We are also planning to study soil mineralization.

CONCLUSIONS

1. Crop rotation has a profound effect on SOC and also on the physical properties of a soil.
2. Whether a certain crop rotation will lead to an increase or a decrease in SOC depends on the former crop rotation at the site.
3. A decrease in SOC, which has led to less favorable physical properties in the soil, also affects trends in crop yields negatively.
4. In a monoculture with cereals (barley/oats) approximately 17% of the carbon from incorporated straw is found after 16 years of cropping.
5. The difference in SOC after 30 years of cropping can be as much as 20 Mg ha^{-1} between different crop rotations.

REFERENCES

Andersson, A., Andersson, R., and Eriksson, J., Tillståndet i svensk åkermark (Current status of Swedish arable soils, in Swedish). Swedish Environmental Protection Board, report no 4778. Stockholm, 1997.

Andersson, S., Markfysikaliska undersökningar i odlad jord. VIII. En experimentell metod. Grundförbätttring nr 8. Specialnummer 2, 98 pp. (Soil physical studies in cultivated soils. VIII. An experimental method. In Swedish), 1995.

Angers, D.A. and Girroux, M., Recently deposited organic matter in water-stable aggregates, *Soil Sci. Soc. Am. J.,* 60, 1547, 1996.

Broersma, K., Robertson, J.A., and Chanasyk, D.S., The effects of diverse cropping systems on aggregation of a Luvisolic soil in the Peace River region, *Can. J. Soil. Sci.,* 77(2), 323-329, 1997.

Ericson, L., Soil physical properties, organic carbon and trends in barley yield in four different crop rotations. In: C.A.S. Smith (ed.), Proceedings of the 1st Circumpolar Agricultural Conference. Whitehorse, YT, Canada. Agricultural Canada, Research Branch, Centre for Land and Biological Resources Research, Ottawa, 1994, p. 189.

Fisher, R.H. and Yates, Y., *Statistical Tables for Biological, Agricultural and Medical Research,* Oliver and Boyd, Edinburgh, 963.

Håkansson, I., A method for characterising the state of compactness of the plough layer, *Soil Tillage Res.,* 16, 105, 1990.

Johansson, W., Kolbindning och kolflöden vid odling. Sammanfattning av en analys rörande inverkan av växtföljd/odlingssystem och av restprodukttillförsel till marken. (Carbon flows in agriculture. Summary of an analysis about the effect of crop rotation/growing system and addition of organic residues to the soil. In Swedish). Swedish University of Agricultural Sciences. Department of Soil Sciences. Uppsala, 1994.

Kritz, G., Såbäddar för vårstråsäd – en stickprovsundersökning. (Physical conditions in cereal seedbeds – a sampling investigation in Swedish spring-sown fields. In Swedish). Swedish Univ. of Agric. Sciences. Dept of Soil Sciences. Reports from the Division of Soil Management, no 65. Uppsala, 1983.

Oades, J.M., Soil organic matter and structural stability; mechanisms and implications for management, *Plant Soil* 76, 319, 1984.

SCB, *Yearbook of Agricultural Statistics,* SCB, Statistics Sweden, Örebro, 1997.

Schnitzer, M., Bedeutung der organischen Bodensubstanz für die Bodenbildung, Transport-prozesse in Böden und die Bodenstruktur. (Significance of soil organic matter in soil formation, transport processes in soils and in the formation of soil structure. In German), *Ber. Landwirtsch. Sonderh.,* 206, 63, 1992.

Tisdall, J.M. and Oades, J.M., The effect of crop rotation on aggregation in a red-brown earth. *Austr. J. Soil Res.,* 18, 423, 1980.

J.M. Tisdall and Oades, J.M., Organic matter and water stable aggregates in soils, *J. Soil Sci.,* 33, 141, 1982.

9 Impact of Potential Global Warming on Erosion and Water Quality of Some Cryosols

R.M. Bajaracharya, R. Lal, and J.M. Kimble

CONTENTS

INTRODUCTION

The bulk of the terrestrial carbon pool occurs in world soils, of which an estimated 25 to 33% is contained in soils of the northern ecosystems, i.e., tundra and boreal forest regions (Ping et al., 1998; Tarnocai and Ballard, 1994; Post et al., 1982). Global climate change, therefore, poses major consequences for water quality, environmental, and health-related issues due to the potential changes in soil C dynamics and erosion, particularly in organic C-rich northern (high latitude) ecosystems (Lal, 1995; Malcolm et al., 1995; Ping et al., 1995). Most climate change models show the largest temperature changes at the high latitudes. A warming trend in the global climate is expected to influence high-latitude soils the most due to their high stored C status and predominantly frozen soil conditions. Increased temperatures would favor increased SOC decomposition rates due to an increase in the biologically active zone of these soils (Ping et al., 1998; Ping et al., 1995). This in turn could set up a positive feedback system where increased CO_2 emission of northern ecosystems further increase radiatively active atmospheric trace gases, hence further accelerating the global warming process (Bouwman, 1990; Kirschbaum, 1995; Lal et al., 1995; Ping et al., 1995). The objective of this study was, therefore, to evaluate the effect of simulated rainfall under increased temperature conditions on soil erosion and nitrate and DOC concentrations in runoff and leachate water from some Alaska soils.

MATERIALS AND METHODS

Pedons 1 and 2 are located on the North Slope of Alaska's Prudhoe Bay in the tundra in the area of continuous permafrost. Pedons 1 and 2 were formed in loess with an aquic moisture regime. Pedon number 3 was sampled in an area cleared of forest in the discontinuous permafrost zone near Delta Junction south of Fairbanks. Pedons 4, 5, and 6 were sampled near Smith Lake next to the campus of the University of Alaska, Fairbanks. This area is in the zone of discontinuous

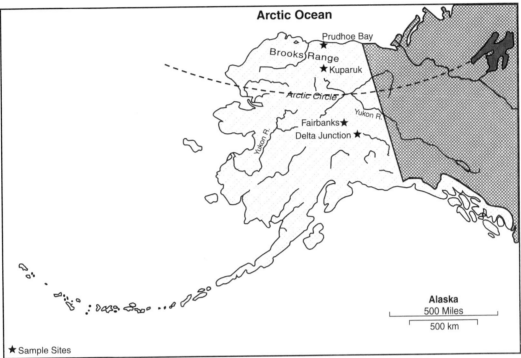

FIGURE 9.1 Location of the pedons.

TABLE 9.1
Pedon Classification and Horizon Description for Six Soils from Alaska

Soil number	Horizon	Depth (cm)	Texture	Pedon classification
1	O_e	13–23	Muck	Loamy, mixed Typic
1	Bg	23–37	Silty clay loam	Histoturbel add for location 69°24'16.8" N latitude,
1	2O/Ab	37–50	Silt loam/muck	148°16'6.4" W longitude
2	Oe	0–10	Peat	Coarse-loamy, mixed Typic Histoturbel add for location
2	Oa	10–22	Peat/muck	70°16"02.8" N latitude, 148°53'02" W longitude
2	Oaf	30–50	Muck	
3	Ap	0–10	Loamy fine sand	Coarse-loamy over sandy skeletal, mixed Aquic Haplorthel
3	Bw1	10–20	Loamy fine sand	
4	Oa	15–27	Peat	Coarse-loamy, mixed Typic Histoturbel add for location
4	Bg1	27–42	Silt loam	4°52'10.5" N latitude, 147°51'27.4" W longitude
5	Ap1	0–12	Silt loam	Coarse-loamy, mixed Typic Histoturbel add for location
5	Ap2	12–29	Silt loam	64°51'56.3" N latitude, 147°51'12.9" W longitude
5	A1b	29–46	Silt loam	
6	Bg	16–23	Silt loam	Coarse-loamy, mixed Typic Histoturbel add for location
6	Oa	23–37	Muck	64°51'56.3" N latitude, 147°51'12.9" W longitude

permafrost. Perdons 4 and 6 had their native vegetation and pedon 6 was cleared forest. Figure 9.1 shows the locations of the pedons.

Samples from the top two or three horizons of six Alaskan Cryosols were transported to Ohio State University for laboratory rain simulation. Pedon classification, horizon, depth, and texture for each soil are described in Table 9.1. Soil textures ranged from very high organic (30-40%) peat and muck to loamy fine sand (LFS).

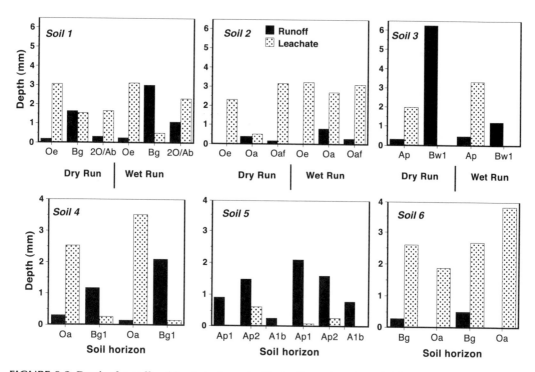

FIGURE 9.2 Depth of runoff and leachate from six Alaska Cryosols under rainfall simulation.

The soil samples were air-dried and clods broken up with a mallet and sieved through an 8-mm mesh. The sieved soil was then packed carefully into micro soil trays of $150 \times 200 \times 30$ mm. Two simulated rain events of about 50 mm h^{-1} intensity were applied 1 h apart. The trays were inclined at a 10% slope to allow surface runoff generation. The first rain simulation event was over air-dried soil, while the second represented wet soil at near field capacity moisture content. Each soil sample was run in duplicate.

Runoff, leachate, and soil splash were collected and total water volumes as well as sediment concentration was measured by oven-drying and weighing wet samples. The splashed soil and sediment were analyzed for soil organic carbon (SOC) using the dry combustion technique with a muffle furnace. Runoff and leachate water samples were analyzed for nitrate, DOC, and pH. The data means were compared and correlation analyses performed for relevant parameters.

RESULTS AND DISCUSSION

Peat or muck horizons containing very high SOC had little or no surface runoff, but had high amounts of leachate water, ranging from about 23 to 38 mm h^{-1} (Figure 9.2). This could be attributed to the low bulk density of the organic-rich material and high porosity. The opposite, however, was generally true for silty clay loam (SiCL), loam of fine sand (LFS), and silt loam (SiL) horizons, which had considerably less SOC and higher bulk densities than the organic horizons and were highly erodible. Soil splash was the predominant mode of erosion (up to 650 g m^{-2}) on these microplots, although silty horizons also had considerable soil loss (20 to 228 g m^{-2}) in runoff water (Figure 9.3). Wet rain simulation runs expectedly had more runoff or leachate once the surface became smooth, due to raindrop impact, and/or the profile became saturated (especially in the case of organic horizons).

Nitrate showed a trend of relatively higher concentration in runoff for peat/muck horizons than for silty or loamy soil, while the opposite was generally true for leachate water (Figure 9.4). Leachate tended to have higher nitrate in A, Bg, and Bw than for organic horizons. The concentrations of DOC in runoff water (0 to 35.7 ppm) were similar or slightly higher that nitrate (0.7 to 27.9 ppm),

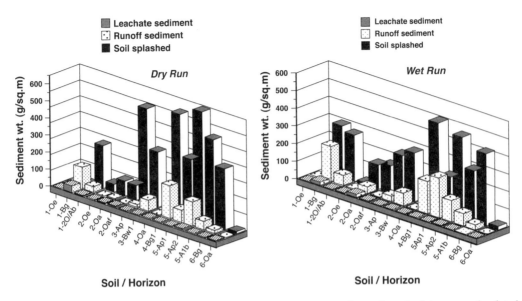

FIGURE 9.3 Sediment in runoff, leachate, and splash from six Alaska Cryosols under laboratory simulated rainfall.

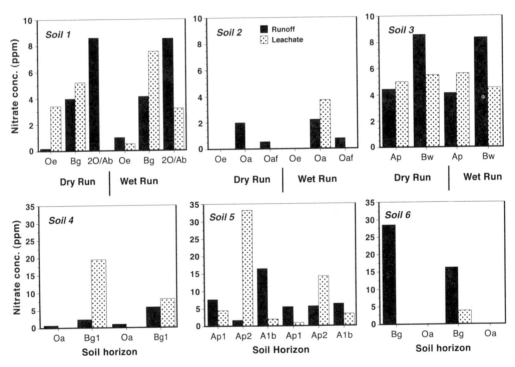

FIGURE 9.4 Nitrate concentration in runoff and leachate during rainfall simulation runs for six dry and wet samples of Alaska Cryosols.

but DOC in leachate water (10 to 186 ppm) was substantially higher than nitrate (0 to 32.5 ppm) (Figure 9.5). The high DOC contents in the leachate were apparently due to high SOC and partially decomposed organic matter in these soils. The pH of runoff and leachate waters ranged from about 5.9 to 7.6, with organic horizons having a typically somewhat lower pH in leachate (Figure 9.6) compared to the runoff.

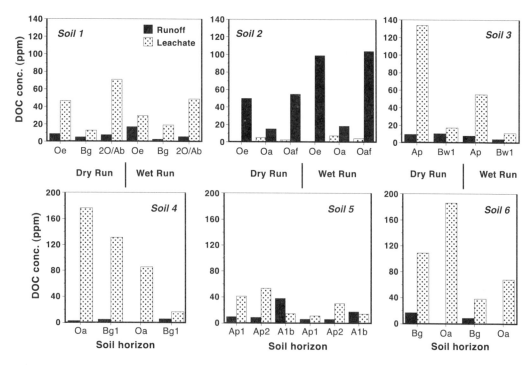

FIGURE 9.5 Dissolved organic carbon (DOC) in runoff and leachate during rainfall simulation runs for six dry and wet samples of Alaska Cryosols.

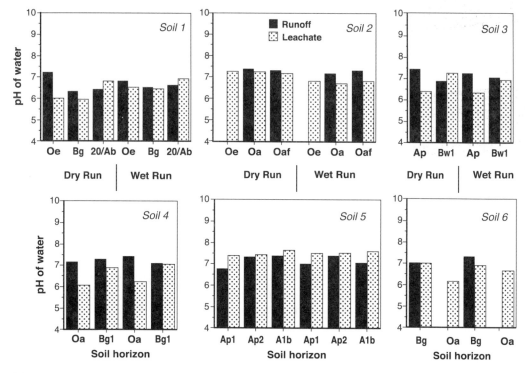

FIGURE 9.6 pH of runoff and leachate water during rainfall simulation runs for six dry samples of Alaska Cryosols.

TABLE 9.2
Soil Organic Carbon (SOC) Content of Sediment in Runoff,
Leachate, and Splash from Six Alaska Cryosols

Soil/horizon	Runoff Dry run	Runoff Wet run	Leachate Dry run	Leachate Wet run	Splash Dry run	Splash Wet run
			SOC (% by weight)			
1 – O$_e$	—	—	—	—	28.0	31.2
1 – Bg	3.7	3.7	8.4	—	2.4	2.1
1 – 2O/Ab	16.7	19.5	14.0	—	18.6	18.1
2 – Oe	—	—	—	—	5.5	14.2
2 – Oa	10.4	14.7	12.3	15.4	8.4	11.7
2 – Oaf	26.1	—	—	27.4	23.9	23.2
3 – Ap	6.2	5.7	—	—	5.6	5.0
3 – Bw1	3.0	3.4	—	—	0.9	1.0
4 – Oa	—	—	—	—	30.4	36.6
4 – Bg1	2.8	2.5	—	—	2.0	1.7
5 – Ap1	1.5	2.1	—	—	2.1	2.2
5 – Ap2	1.9	1.5	10.2	—	1.1	1.0
5 – A1b	6.8	3.4			2.7	2.3
6 – Bg	4.3	4.2	—	—	2.5	2.4
6 – Oa	—	—	—	—	33.3	38.2

The SOC contents of sediment reflected the SOC status of the soil horizon. Organic horizons had the highest SOC contents in sediment, ranging from 5.5 to 38.2% (Table 9.2). Nitrate and DOC were significantly positively correlated for runoff but not for leachate. Both nitrate and DOC in runoff were also positively correlated with each other and between dry and wet runs (Table 9.3). On the other hand, nitrate and DOC in leachate water were not well correlated or were negatively correlated (wet run). Runoff and leachate were negatively correlated, r = –0.803 with each other (Table 9.4). Also, all parameters were generally positively correlated between dry and wet runs, i.e., high amounts of leachate, sediment, or splash in the dry run meant a likewise result in the wet run and vice versa (Table 9.4).

CONCLUSIONS

A number of conclusions can be drawn from this simulated rainfall study. These include the following:

- Soil from organic horizons had little or no runoff, but high amounts of leachate (23 to 38 mm h^{-1}).
- Silt loam and loamy fine sand horizons typically had high amounts of runoff (3 to 30 mm h^{-1}) and little or no leachate, as well as, highest amounts of soil splash (250 to 650 g m^{-2}).
- Organic soil horizons yielded considerably higher amounts of DOC that other horizons.
- Runoff water contained low amounts of DOC (0 to 35.7 ppm), but higher amounts of nitrate (0.7 to 27.9 ppm) compared to leachate water.
- Runoff volume was significantly negatively correlated (r = -0.803) with leachate amount.
- Nitrate and DOC were positively correlated for runoff, and between dry vs. wet runs, but this did not hold true for leachate water.

In general, the results of this study suggest that warming of and increased precipitation on SOC-rich, high-latitude Cryosols could result in DOC and nitrate pollution of groundwater. This

TABLE 9.3
Correlation of Relevant Chemical Properties of Runoff and Leachate Water

Parameter	R-NO$_3$-D	R-pH-D	R-DOC-D	L-NO$_3$-D	L-pH-D	L-DOC-D	R-NO$_3$-W	R-pH-W
R-DOC-W	0.775**							
R-DOC-D	−0.153ns							
L-NO$_3$-D		0.257ns						
L-pH-D								
L-DOC-D				−0.335ns	0.233ns			
R-NO$_3$-W	0.685**							
R-pH-W		0.745**						
R-DOC-W			0.720**				0.022ns	
L-NO$_3$-W					0.372ns		0.460ns	
L-pH-W						0.820**		0.065ns
L-DOC-W							0.455ns	

Note: R = runoff; L = leachate; D = dry run; W = wet run; NO$_3$ = nitrate; DOC = dissolved organic carbon; ** indicates significance at the 1% probability levels, and ns indicates nonsignificance.

TABLE 9.4
Correlation of Relevant Runoff, Leachate, and Sediment Parameters for Six Alaska Cryosols

Parameter	RO-D	L-D	R-SED-D	L-SED-D	SPL-D	RO-W
L-D	−0.493ns					
R-SED-D	0.369ns					
L-SED-D		0.057ns	0.352ns			
SPL-D		−0.561*	0.582*			
RO-W	0.376ns					
L-W		0.818*				−0.803**
R-SED-W			0.836**			0.952**
L-SED-W				0.761**		
SPL-W					0.864**	0.544*

Note: R, RO = runoff; L = leachate; D = dry run; W = wet run; SED = sediment; SPL = soil splashed; *, ** indicates significance at the 5% and 1% probability levels, respectively, and ns indicates nonsignificance.

would result from high concentrations of DOC in water percolating through the soil profile, as well as high nitrate concentrations in runoff water. Exposure of silty and loamy subsurface soil layers could lead to high rates of erosion.

REFERENCES

Bouwman. A.F., Exchange of greenhouse gases between terrestrial ecosystems and the atmosphere, in *Soil and the Greenhouse Effect*, Bouwman, A.F., Ed., John Wiley & Sons, Chichester, U.K., 1990.
Kirschbaum, M.U.F., The temperature dependence of soil organic matter decomposition, and the effect of global warming on soil organic C storage, *Soil Biol. Biochem.*, 27, 753, 1995.

Lal, R., Kimble, J., and Stewart, B.A., World soils as a source or sink for radiatively-active gases, in *Soil Management and the Greenhouse Effect,* Lal, R., Kimble, J., Levine, E., and Stewart, B.A., Eds., Lewis Publishers, Boca Raton, FL, 1995, p. 1.

Lal, R., Global soil erosion by water and carbon dynamics, in *Soil Management and the Greenhouse Effect,* Lal, R., Kimble, J., Levine, E., and Stewart, B.A., Eds., Lewis Publishers, Boca Raton, FL, 1995, p. 131.

Malcolm, R.L., Kennedy, K., Ping, C.L., and Michaelson, G.J., Fractionation, characterization, and comparison of bulk soil organic substances and water-soluble soil interstitial organic constituents in selected Cryosols of Alaska, in *Soils and Global Change,* Lal, R., Kimble, J., Levine, E., and Stewart, B.A., Eds., Lewis Publishers, Boca Raton, FL, 1995, p. 315.

Ping, C.L., Michaelson, G.J., and Malcolm, R.L., Fractionation and carbon balance of soil organic matter in selected Cryic soils in Alaska, in *Soils and Global Change,* Lal, R., Kimble, J., Levine, E., and Stewart, B.A., Eds., Lewis Publishers, Boca Raton, FL, 1995, p. 307.

Ping, C.L., Michaelson, G.J., Loya, W.M., Chandler, R.J., and Malcolm, R.L., Characteristics of soil organic matter in Arctic ecosystems of Alaska, in *Soil Processes and the Carbon Cycle,* Lal, R., Kimble, J., Follett, R.F., and Stewart, B.A., Eds., Lewis Publishers, Boca Raton, FL, 1998, p. 157.

Post, W.M., Emanuel, W.R., Zinke, P.J., and Stangenberger, G., Soil carbon pools and world life zones, *Nature,* 298, 156, 1982.

Tarnocai, C. and Ballard, M., Organic carbon in Canadian soils, in *Soil Processes and the Greenhouse Effect,* Lal, R., Kimble, J., and Levine, E., Eds., Soil Conservation Service, National Soil Survey Center, U.S. Department of Agriculture, Lincoln, NE, 1994, p. 31.

10 The Soil Organic Carbon Dynamics in High Latitudes of Eurasia Using ^{14}C Data and the Impact of Potential Climate Change

S.V. Goryachkin, A.E. Cherkinsky, and O.A. Chichagova

CONTENTS

INTRODUCTION

Soil organic carbon (SOC) is one of the main pools of terrestrial carbon (Post et al., 1982; Kobak, 1988; Schlesinger, 1991; Eswaran et al., 1995). Its total store is evaluated about 1500 Pg ($1Pg = 10^{15}g$). It actively exchanges with the atmosphere by gases (CO_2, CH_4) which can control the greenhouse effect (Schlesinger, 1991; Houghton et al., 1996; Trumbore et al., 1996). That is why it is of primary importance to evaluate SOC dynamics and characterize the flows between the pedosphere and atmosphere.

One of the main components of global soil carbon pool (23-30%) is SOC of northern ecosystems (arctic, boreal forests, and northern peatlands) which contain an estimated 350 to 455 Pg of C (Billings, 1987; Post et al., 1982; Gorham, 1991; Oechel and Billings, 1992). Arctic tundra contains about 13% of the global soil pool or 192 Pg (Billings, 1987). Northern ecosystems, besides Amazon rain forest, are one of the most important sources of oxygen on the Earth — "lungs of the planet"

(Zavarzin, 1994). These ecosystems include high latitudes where the most pronounced warming in the twenty-first century is expected (Houghton et al., 1996). For these reasons many specialists are attracted to the sparsely populated regions of the Arctic to study carbon dynamics by different methods.

There are many methods scientists can use to investigate the store and dynamics of SOC. The method based on chronosequences (Schlesinger, 1990) supposed that SOC accumulation is a permanent (or very long-term) process and it does not reach steady state; this method strongly underestimates the role of soil in carbon turnover. For the long-term change the paleoecological method is effective (Billings, 1987; Kremenetski et al., 1997; Eisner, 1998) but it is not good enough for the more exact estimations needed to model current flows of carbon. Calculations based on rates of SOC accumulation and plant residue inputs are rather precise but they can be done only in the case of long-term experiments on the stations (Jenkinson and Raynor, 1977; Andren and Kätterer, 1997) — it is not possible, at present, in the Arctic. Direct measurements of greenhouse gas emanation (mainly CO_2 and CH_4) from the soils, including those from the arctic regions (Oechel and Vourlitis, 1995; Christensen et al., 1995) are popular now. These methods are very effective and precise. But their application poses many problems related to interpretation of the data. Gas emanation is very dynamic and variable in spatial characteristics. It is the result of dynamics not only of SOC but also respiration of plants, invertebrates, and microorganisms. The soil in these measurements is often taken into account as a "black box". So, the application using only this method cannot characterize the mechanisms and parameters of SOC dynamics.

This report considers one of the best methods to study SOC dynamics, including sparsely populated arctic regions, to be radiocarbon analysis. Since the early years of its application in soil science the specialists proposed to use it for the investigations of the dynamics of soil humus (Paul et al., 1964) giving birth to the concept of "mean residence time" of soil humus. For some time specialists in radiocarbon analysis of soil were more focused on the problems of soil age (Scharpenseel, 1971; Gerasimov and Chichagova, 1971; Gerasimov, 1974), but meanwhile the investigation of different aspects of carbon dynamics also took place (Martel and Paul, 1974; Zavel'skiy, 1975). On the basis of this background the up-to-date radiocarbon studies of soils are mainly focused on the following aspects of investigations of SOC dynamics: (1) general aspects of radiocarbon dating, turnover rates, databases organization (Hsieh, 1993; Sharpenseel and Becker-Heidmann, 1994; Paul et al., 1995; 1997; Becker-Heidman et al., 1996); (2) using "bomb" ^{14}C for estimating SOC turnover (Harkness et al., 1986; 1991; Scharpenseel et al., 1989); (3) estimation of slow- and fast-cycling SOC pools (Harrison et al., 1993; Leavitt et al., 1996; Harrison, 1998); (4) dynamics of SOC in different fractions of soil humus (Trumbore et al., 1990; Chichagova and Cherkinsky, 1993; Ping et al., 1998); and (5) paleoecological aspects of SOC dynamics (Chichagova, 1985, 1996). There is some experience on comparison of different methods of SOC dynamics study (Kobak, 1988; Demkina and Zolotareva, 1997), which found good correspondence of the radiocarbon method with other procedures.

Unfortunately, the main volume of radiocarbon data does not include much information on the soils of high latitudes. There are only few papers containing these data (Cherkinsky and Goryachkin, 1993; Cherkinsky, 1996; Ping et al., 1998). One of the causes for this may be that many surface soil horizons of tundra have more than 100% of modern carbon (pmC) because of "bomb effect" and it is difficult to interpret. We suppose that radiocarbon analysis of soils having such a wide spectrum of investigations on SOC dynamics can be successfully applied in high latitudes as an integrated parameter. It can be harmoniously combined with the direct measurement of soil respiration and gas emanations.

The purpose of this chapter is to propose the model of interpretation of radiocarbon data (including those with high pmC) and on the basis of it evaluate renovation rates of SOC (except the fast-cycling one) in soils of high latitudes of Eurasia, and their possible change due to global warming.

THEORY

On the basis of the previous works (Cherkinsky, 1996; Cherkinsky and Brovkin, 1991, 1993; Cherkinsky and Paul, 1995) this chapter suggests a summarized and revised model of carbon dynamics in a soil being developed under stable conditions of pedogenesis in a steady state (input of carbon is equal to its loss due to mineralization and leaching out of soil). In these conditions humus and ^{14}C accumulation can be represented by simple differential Equations 10.1 and 10.2:

$$d^{12}C_s/dt = K_h{}^{12}C_{pd} - K_r{}^{12}C_s \tag{10.1}$$

$$d^{14}C_s/dt = K_h{}^{14}C_{pd} - K_r{}^{14}C_s - \lambda^{14}C_s \tag{10.2}$$

where $^{12}C_{pd}$ and $^{14}C_{pd}$ are amounts of plant detritus input of ^{12}C and ^{14}C, respectively; $^{12}C_s$, and $^{14}C_s$, are the content of carbon in the soil, respectively; and K_h, K_r and λ are coefficients of humification, renovation (including mineralization and leaching from a soil horizon), and radioactive ^{14}C decay constant, respectively. Given the specific activity of humus, these equations for dynamic equilibrium, where input = ^{14}C decay constant, which allows us to calculate the coefficients of humus renovation. These equations define the correlation of humus and ^{14}C accumulation with time; using the coefficients of correlation, it is possible to establish the duration of the process of humus accumulation.

Using Equations 10.1 and 10.2, and assuming the first approximation that input and output (mineralization + leaching) of organic material are constant values, we arrive at the expressions below that show correlation of humus and ^{14}C accumulation with time (Equations 10.3 and 10.4):

$$^{12}C_s = (K_h/K_r)^{12}C_{pd}(1 - expK_rt) \tag{10.3}$$

$$^{14}C_s = [K_h/(K_r + 1)]^{14}C_{pd}[1 - exp(Kr + \lambda)t] \tag{10.4}$$

The proposed model works sufficiently well if it assumes constant concentration of ^{14}C in the atmosphere and, correspondingly, in the biota that are the primary material for depositing soil organic matter. However, as many researchers have shown, atmospheric ^{14}C concentrations are not constant, and are affected by explosions of supernovae, variations in solar activity, oscillations of the geomagnetic field, industrial activity, and atmospheric nuclear testing. Nuclear power plants also contribute to increased ^{14}C concentration in humus; they generate a significantly higher specific radioactivity for biota than the specific activity of the NIST (National Institute of Standards) oxalic acid standard. According to our data for European Russia, this surplus depends little on the type of vegetation and ranges from 130 to 135% of modern carbon (pmC). This was most pronounced in the humus of soils with short regeneration periods (several decades).

On the basis of the proposed mathematical model of change of specific humus activity, I(t), it is possible to correlate the mean rate of its renovation with the I_o index that is stable for I(t). This correlation can be expressed as

$$I(t)/I_o = K_r/(K_r + \lambda) \tag{10.5}$$

Atmospheric ^{14}C fluctuations make it impossible to calculate, with Equation 10.5, the rates of humus renovation and ^{14}C dates of soils that undergo rapid rejuvenation of carbon. The I_o index in this equation ceased to be constant in the time interval, 1956–1982. Instead, two stationary levels of specific activity are found: $I_o{}'$ — "pre-nuclear" activity (100 pmC), and $I_o{}''$ — atmospheric activity after 1982 (ca. 135 pmC). If the measured level of activity in a soil is, for example, 110 pmC, how

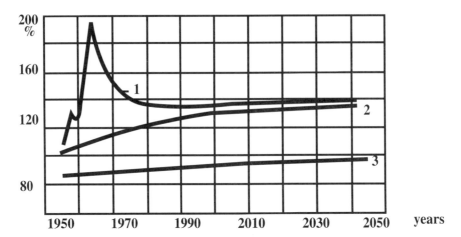

FIGURE 10.1 (1) Calculated change of specific carbon activity (% from NBS) in the atmosphere; (2) soil humus with $K_r = 22$ g kg⁻¹ C yr⁻¹; and (3) $K_r = 1$ g kg⁻¹ C yr⁻¹.

can this result be interpreted? Calculation of ¹⁴C age with Equation 10.5, if I_o equals I_o', it gives a negative age. Measurements on the European part of Russia after 1982 to 1990 were about 135 pmC. Perhaps this is a result of local emissions. If $I_o = I_o''$, the measurement provides a date of about 1500 years, which disagrees with other ¹⁴C data for the soil. The specific activity of the soil is not constant, but gradually increases with time, which disagrees with the assumptions of Equation 10.5. These difficulties explain why soils with specific activity above 100 pmC are described by specific activity of a sample in percent of the modern standard of prenuclear time, or by the designation "modern" without a date.

This chapter proposes a technique for calculating the ¹⁴C age of soils and the rates of humus renovation that takes into account the whole curve of change of specific activity of atmospheric carbon from 1956. It is assumed that the humus content in a soil is stable, i.e., meeting the condition of equilibrium between humification and mineralization+leaching. Then, the dynamics of specific ¹⁴C activity in soil humus, I(t), is calculated as

$$I(t) = I(t - 1) - (K_r + \lambda)I(t - 1) + K_r I_o(t) \tag{10.6}$$

where $I(t - 1)$ is the specific activity of humus in the preceding year, and $I_o(t)$ is specific activity of atmosphere in the year, t. It is assumed that by 1955 the amount of ¹⁴C in soil stabilized; then, according to Equation 10.6, $I(1955) = I_o [K_r/(K_r + \lambda)]$. Thus, given the dynamics of $I_o(t)$ and the original value of $I(1955)$, Equation 10.6 can be used to calculate the pattern of change of I(t). If the specific activity was measured in a year, t_o, and equals $I(t_o)$, it is possible by changing K_r to select the rate of humus renovation that agrees with $I(t_o) = \hat{I}(t_o)$ (if $\hat{I}(t_o) < I_o''$). It can be shown that the problem has a single solution: each $\hat{I}(t_o)$ corresponds to a single value of the rate of renovation. Thus calculated, K_r in Equation 10.5 can be used to measure the specific activity of a soil up to 1955. Thus, the suggested technique allows one to calculate the rate of humus renovation with the specific activity of humus in the year t, using equation

$$K_r = \lambda \, [I(t)/(I_o(t) - I(t))exp\lambda(t - 1950)]1000 \text{ g kg⁻¹ C yr⁻¹} \tag{10.7}$$

For example, the dynamics of the specific humus activity, I(t), in a soil with rapid renovation of organic material, 22 g kg⁻¹ Cyr⁻¹, is shown in Figure 10.1, which also demonstrates the curve

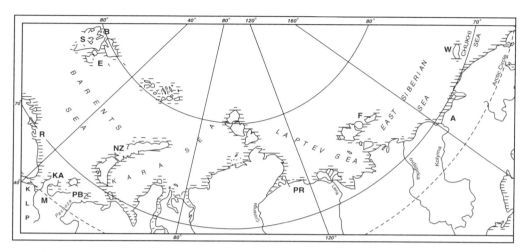

FIGURE 10.2 Location of study sites in Eurasian Arctic and Subarctic; B - Bockfjord, Spitzbergen; S - West Spitsbergen; E - Edge; NZ - Novaya Zemlya (Pankova Land); PR - Pronchishev Ridge; F - Faddeevsky Island; A - Anui Ridge; W - Wrangel Island; R - Rybachiy Peninsula; KA - Kanin Peninsula; PB - Pechora Bay Coast; K - Koida; L - Laka; M - Mezen; P - Pinega.

of change of the specific activity of atmospheric ^{14}C after Bolin (1986). The original specific activity of soil is 99.8%, and rapidly increases because of the high activity of the newly formed humus. In three years, I(t) becomes more than 100%, prohibiting calculations with Equation 10.5. The new stationary value of the specific activity of soil is 133.8%. In a soil with a low rate of humus renovation ($K_r = 1$ g kg^{-1} C yr^{-1}), the specific activity changes much more slowly (Figure 10.1). Thus, from 1950 to 1985, I(t) changed from 89.0 to 90.2%, and by 2000, it will be up to 90.7%. The stationary value, $I_o(\infty)$, is 119.7%. Because of the slow change of I(t), Equation 10.5 can be used for calculations, although it will produce an error towards the increased rate of renovation.

STUDY SITES AND METHODS

LOCATIONS OF STUDY SITES, ENVIRONMENTAL CHARACTERISTICS, AND SOILS

The locations of study sites are shown in the Figure 10.2. Three study sites from the northern part of boreal forests zone were used for comparison to the data obtained from high latitudes. The brief characteristics of the study sites' environment and soils are in the Table 10.1.

Greater details were partly published earlier (Goryachkin et al., 1994; Cherkinsky, 1996; Kremenetski et al., 1997). It should be added that despite different annual precipitation all the sites could be characterized as having a humid climate. The study sites are representative for many types of Eurasian arctic landscapes with varying climate, relief, substrates, and vegetation.

All the soil profiles had the morphological features of mature soils of the studied regions except the Gelic Regosol of site 9 on the solifluction slope, where the parallel sampling of immature and stable soils took place. The soils of study sites were classified in accordance with Soil Map of the World revised legend (FAO, 1990).

The weak correlation between the latitude and mean annual temperature is the common phenomenon for Eurasia (Table 10.1). It is because of warm Atlantic influence in the West and very cold Siberian winters in the East (ultra-continental climate). In this condition the more informative climatic parameter for our purposes is the mean temperatures of the warmest month (July). That is why we add these data in Table 10.1.

TABLE 10.1
Location, Environment, and Soils of Study Sites

Site no	Area	Lat. (N) Long. (E)	Mean annual and July t (°C)	Landform/vegetation	Soil horizons and classification (FAO, 1990)
1	Bockfjord, Spitzbergen	79° 32' 13° 21'	−5 4	Marine terrace/saxifraga, mosses	O-AO-C-Ci Haplic Arenosol
2	West Spitzbergen	78° 42' 16° 30'	−4 4	Marine terrace, calcareous/ saxifraga, mosses	AO-Ck-Cki Gelic Leptosol
3	Edge, Spitzbergen	77° 35' 20° 56' 77° 27' 21° 01' 77° 27' 21° 01'	−4 5	Marine terrace/polar willow, bog mosses	H-Cg-Cgi Gelic Gleysol
				Marine terrace, calcareous/ sparse lichens, grass	AO-Ck-Cki Calcaric Arenosol
				Bog in depression/cotton grass, mosses	H-Hi Gelic Histosol
4	Faddevsky isl. (E. Sib)	75° 30' 143° 15'	−15 3	Patterned surface/sedges, forbs	A-Bg-Cgi Gelic Gleysol
5	Pronchishev Ridge, NW Yakutia	73° 18' 116° 56'	−14 4	Flat patterned surface, crack/grass, mosses Foothill/sedge meadow	O-AO-A-Ci Gelic Regosol H-Bg-Hb-Cgi Gelic Gleysol
6	Pankova Land, Novaya Zemlya	73° 03' 53° 10'	−7 6	Marine terrace, calcareous/ grass meadow Patterned surface/polar willow, mosses Peat mound/bog mosses, sedges, polar willow	Ak-ACk-Ck-R Rendzic Leptosol O-Bg-Cgi Gelic Gleysol O-H-Hi Gelic Histosol
7	Wrangel Island	71° 02' 179° 33'	−11 4	Ridge slope, calcareous/grass, willow	Ak-Ack-Ck-R Rendzic Leptosol
8	Rybachiy Peninsula	69° 38' 32° 22'	1 10	Bog in depression/bog mosses, sedges	H-Cr Dystric Histosol
9	Anui, W Chukotka	69° 21' 163° 35'	−13 7	Solifluction slope/grass, mosses, polar willows	AO-C-Ci Gelic Regosol
10	Pechora Bay	68° 31' 52° 48'	−6 9	Sandy hill/lichens, dwarf shrubs, grass	AO-E-Bhs-C-Ci Gelic Podzol
11	Kanin Peninsula	68° 27' 45° 25'	−2 7	Hummocky surface/mosses, dwarf shrubs	O-OH–Bg-Cgi Gelic Gleysol
12	Koida, European North of Russia (ENR)	66° 25' 42° 34'	−1 12	Hummocky depression/dwarf birch, mosses Sandy moraine hill/ lichens, mosses Foothill/ dwarf birch, mosses	O-H-Bg-Cgi Gelic Gleysol AO-E-Bhs-C Carbic Podzol O-E-Bh-Bhs-C Carbic Podzol
13	Laka (ENR)	65° 17' 43° 11'	−0.4 15	Calcareous residual hill/ spruce-larch forest, grass Slope of residual hill/ spruce-larch forest, grass	O-A-AB-Ck Rendiz Leptosol A-AB-B-BCk-Ck Eutric Cambisol
14	Mezen (ENR)	65° 03' 45° 34'	−0.6 15	Hill, red loam/larch forest, shrubs, mosses	O-AO-AB-B-Ck Eutric Cambisol
15	Pinega (ENR)	64° 36' 42° 55'	−0.2 16	Slope of moraine hill/spruce forest, mosses	OH-E-Bt-Btg-Cg Gleyic Podzoluvisol

Sampling, Methods of Chemical Pretreatment, and Radiocarbon Dating

Soil samples were taken from the genetic horizons and air dried. All obvious fragments of roots and other unhumified organic material were discarded by hand-picking from peat, litter, and mineral soil samples.

The peat and litters were then digested in 2.0 M HCl (at 96°C for 2 h), 0.5 M NaOH (at 96°C for 0.5 h) and 2.0 M HCl (at 96°C for 0.5 h). Samples were washed free of reagent with distilled water after each treatment and dried.

Small roots and plant residues were removed by flotation and then the soil sample was washed calcium ion free in 0.1 M HCl. After this, humus acids were separated by 0.1 M NaOH, repeatedly. Then humic acid (HA) was precipitated from the separated solution by HCl or H_2SO_4 in pH 1 to 2. The sediment of HA was washed acid free with distilled water and dried. All reactions were at the room temperature.

Pretreatment of fossil bones consists in acid-alkali extraction of the collagen and purification from humus contamination. The insoluble residue from the bone apatite dissolution procedure was filtered and washed. It was then boiled in slightly acid distilled water to solubilize any collagen present. The broth was filtered and the filtrate was evaporated to dryness to recover collagen as bone gelatin. Rootlets, humic acids, and other contaminants would have been removed by the filter and discarded. The recovered bone gelatin was combusted and the carbon dioxide was recovered and used for the analysis.

After treatment, the received samples were converted to benzene using the standard technique of Gupta and Polach (1985). Then ^{14}C-dating was performed by liquid scintillation counting in spectrometers ("IGAN" and "Mark-11-Nuclear Chicago" in [14]C- laboratory in Institute of Geography, Moscow). Bone gelatin from the reindeer horn (site 5, profile of Gelic Regosol) was prepared and dated in Geochron Laboratory, Cambridge, MA.

Total organic carbon contents (C_{tot}) were determined by wet oxidation with potassium dichromate and concentrated sulfur acid in humus soil mineral horizons. In the organic horizons (peat, litter) organic carbon contents were determined by dry combustion. Values were expressed as weight percent of the dry sample.

Data Analysis

To define the renovation rate of organic carbon we proposed to use the coefficient of renovation K_r, which is the integral figure of organic carbon renovation which results both in the biochemical reactions of mineralization and of its leaching from the soil profile. This coefficient of renovation was calculated according to Equation 10.7. Percent of modern carbon (pmC) was calculated according to the definition of Stuiver and Polach (1977).

All ^{14}C data are expressed at the 2-sigma interval for overall analytical confidence. Conventional ^{14}C ages were reported in years B.P.

The coefficient of SOC renovation was then correlated to the depth of soil horizons (boreal soils were excluded), to latitude, to mean annual, and to mean July temperature of the study sites. The main trends with the highest mean square deviation (R^2) and coefficients of correlation (R) were calculated with the help of MS Excel software. Those soils and soil horizons that did not reach steady state (immature ones) were excluded from this analysis.

To assess the possible climate-induced change of SOC renovation rate we use the proposed algorithm of geographic analog approach (Goryachkin and Targulian, 1990, 1993). It is based on the assumption that spatial climate-induced soil change is approximately similar to the possible temporal transformation of soils in changed climate, and on the concept of different characteristic response times of various soil characteristics.

TABLE 10.2

Data on Soils of High Arctic of Eurasia — to 70° N Latitude

Site	Latitude	July t (°C)	Soil (FAO)	Depth (cm)	OM type	pmC, %	Age, yr B.P.	Kr, g kg⁻¹ C y⁻¹	C_{tot}, %	Sampling
Bockfjord,	79.5°	4	Haplic	0–5	HA	99.5 ± 1.2	Modern	2.82	20.4	1989
Spitzbergen			Arenosol	5–15	HA	83.4 ± 0.6	1500 ± 65	0.59	10.0	
				18–28	HA	71.4 ± 0.6	2780 ± 80	0.30	7.5	
				28–32	HA	66.3 ± 0.7	3400 ± 120	0.24	4.5	
West Spitzbergen	78.7°	4	Gelic	0–3	HA	105.3 ± 0.9	Modern	6.04	4.8	1988
			Leptosol	3–8	HA	95.7 ± 1.2	360 ± 70	1.77	2.6	
				8–16	HA	87.2 ± 0.7	1130 ± 80	0.77	1.4	
Edge, Spitzbergen	77.6°	5	Gelic	0–4	HA	98.9 ± 1.5	Modern	2.65	19.3	1988
			Histosol	4–11	HA	92.6 ± 1.3	640 ± 90	1.24	14.3	
				11–21	HA	89.0 ± 1.0	960 ± 80	0.90	11.7	
				21–31	HA	70.9 ± 0.6	2840 ± 60	0.30	11.9	
	77.4°	5	Calcaric	0–2	Bone	54.6 ± 0.9	5000 ± 250	n.d.	n.d.	1988
			Arenosol	0–2	HA	95.9 ± 1.0	350 ± 70	1.80	2.3	
				2–11	HA	84.7 ± 0.9	1370 ± 80	0.65	0.9	
	77.4°	5	Gelic	0–6	Peat	87.7 ± 1.4	1090 ± 100	0.80	43.8	1988
			Gleysol	6–15	Peat	75.4 ± 0.8	2330 ± 70	0.37	42.1	
				15–26	Peat	65.1 ± 0.7	3550 ± 60	0.23	44.2	
				26–30	Peat	64.0 ± 0.7	3690 ± 60	0.22	43.5	
Faddevsky Island	75.5°	3	Gelic Gleysol	0–2	HA	97.1 ± 2.0	240 ± 120	2.00	5.0	1994
Pronchishev Ridge,	73.3°	4	Gelic	0–4	Litter	116.5 ± 0.6	Modern	14.94	60.7	1994
Northwest Yakutia			Regosol	4–7	HA	95.6 ± 0.5	360 ± 50	1.62	6.9	
				26–29	HA	n.d.	1720 ± 60	n.d.	46.8	
	73.3°	4	Gelic	0–4	Litter	116.9 ± 3.3	Modern	15.50	60.0	1994
			Regosol	4–8	HA	99.5 ± 1.0	Modern	2.69	8.33	
				8–22	HA	91.2 ± 1.0	690 ± 80	1.44	6.18	
				30–32	Bone	n.d.	790 ± 80	n.d.	n.d.	
Pankova Land,	73°	6	Rendzic	0–10	HA	92.3 ± 0.4	Modern	1.18	10.1	1995
Novaya Zemlya			Leptosol							
	73°	6	Gelic	5–10	HA	95.6 ± 0.3	350 ± 30	1.66	3.5	1995
			Gleysol							
	73°	6	Gelic	0–10	Peat	97.1 ± 0.3	230 ± 20	1.97	52.4	1995
			Histosol	10–20	Peat	91.9 ± 0.4	680 ± 30	1.14	59.6	
Wrangel Island	71°	4	Rendzic	0–20	HA	96.7 ± 0.9	260 ± 90	1.91	16.2	1994
			Leptosol							

Note: OM — organic matter; HA — humic acids; n.d. — not determined.

RESULTS AND DISCUSSION

The obtained and calculated data are shown in Tables 10.2 to 10.4.

APPLICABILITY OF THE PROPOSED MODEL TO THE SOC RENOVATION ASSESSMENT

One of the key questions of our investigation is whether it is possible to apply the proposed model to the studied soil profiles. As it is widely known, arctic soils are characterized by periodic cryoturbations and substrate renovation (Tedrow, 1977). That is why it is disputable if processes of SOC accumulation in tundra soils reach steady state (basic condition for applicability of the model) or not.

TABLE 10.3

Data on Soils of the Southern Part of the Tundra Zone of Eurasia — South to 70° N Latitude.

Site	Latitude	July t (°C)	Soil (FAO)	Depth (cm)	OM type	pmC, %	Age, yr B.P.	Kr, g kg⁻¹ C y⁻¹	C_{tot}, %	Sampling
Rybachiy	69.6°	10	Dystric	0–10	Peat	92.8 ± 0.5	630 ± 50	1.26	n.d.	1990
Peninsula,			Histosol	10–20	Peat	81.6 ± 0.4	1630 ± 40	0.53	n.d.	
North of Kola				20–30	Peat	64.7 ± 0.3	3510 ± 40	0.22	n.d.	
Peninsula				30–40	Peat	64.3 ± 0.3	3530 ± 40	0.22	n.d.	
				40–50	Peat	57.7 ± 0.3	4430 ± 40	0.17	n.d.	
Anui, West	69.4°	7	Gelic	0–5	HA	120.5 ± 1.2	Modern	19.9	n.d.	1994
Chukotka			Gleysol[a]							
	69.4°	7	Regosol	0–9	HA	101.1 ± 0.9	Modern	3.28	17.8	1994
				12–20	HA		140 ± 40	2.31	12.3	
Pechora Bay	68.5°	9	Gelic	0–2	HA	99.4 ± 0.7	Modern	2.65	6.8	1994
			Podzol							
Kanin Peninsula	68.4°	7	Gelic	0–6	Litter	106.4 ± 0.5	Modern	6.16	90.4	1994
			Gleysol	6–13	HA	99.3 ± 0.8	Modern	2.61	87.7	
Koida, Russian	66.4°	12	Gelic	0–9	Peat	109.0 ± 0.5	Modern	9.41	40.7	1986
European North			Gleysol	33–39	Peat	53.4 ± 0.5	5190 ± 60	0.14	38.6	
				39–42	HA	56.2 ± 0.5	4770 ± 50	0.16	3.8	
				39–42	FA	54.8 ± 0.5	4980 ± 60	0.15	3.8	
				42–47	HA	56.3 ± 0.5	4750 ± 60	0.16	2.6	
				42–47	FA	71.1 ± 1.0	2820 ± 90	0.30	2.6	
				47–55	HA	61.9 ± 0.8	3960 ± 100	0.20	2.3	
				47–55	FA	63.8 ± 1.0	3710 ± 170	0.22	2.3	
Koida, Russian	66.4°	12	Haplic	0–3	HA	99.6 ± 1.6	Modern	2.99	28.3	1986
European North			Podzol	3–6	HA	93.7 ± 0.4	540 ± 30	1.42	6.5	
				15–22	HA	72.5 ± 0.4	2660 ± 30	0.32	1.8	
				15–22	FA	80.8 ± 1.1	1760 ± 130	0.50	1.8	
				22–27	HA	71.3 ± 0.6	2800 ± 50	0.30	1.6	
				22–27	FA	78.8 ± 0.7	1970 ± 70	0.45	1.6	
	66.4°	12	Carbic	0–4	HA	115.7 ± 0.7	Modern	16.65	9.4	1986
			Podzol	4–19	HA	78.3 ± 0.5	2020 ± 40	0.43	11.1	
				19–38	HA	72.9 ± 0.4	2610 ± 40	0.33	3.4	
				38–72	HA	69.6 ± 0.5	3000 ± 70	0.28	1.2	

Note: OM — organic matter; HA — humic acids; FA — fulvic acid; n.d. — not determined.

[a] Immature soil profile (not used in following calculations).

Morphologically mature soils were used in this study. To assess whether the organic profiles of these soils approach the steady state is possible by comparing the ^{14}C age of soil organic matter belonging to open system and the ^{14}C age of some organic material (charcoal, wood, bone) in a soil horizon not exchanging with atmosphere carbon (Zavel'skiy, 1975; Scharpenseel and Becker-Heidman, 1994). The problem is to find material of such kind. However, a whale bone in the surface horizon of Calcaric Arenosol in Edge, Spitzbergen and remnants of reindeer horn at the depth of 30–32 cm in Gelic Regosol of Pronchishev ridge were found. In both cases the ^{14}C ages of soil organic matter were less than that of organic relics (Table 10.2) — evidence for steady state of the soil. This gives the possibility to evaluate the time that organic profiles of arctic soils need to approach steady state. Organic horizons of well-drained soils need n∗100–1000 years to reach equilibrium with the environment.

TABLE 10.4
Data on Soils of Boreal Forest of the European North of Russia

Site	Latitude	July t (°C)	Soil (FAO)	Depth (cm)	OM type	pmC, %	Age, yr B.P	Kr, g kg⁻¹ C y⁻¹	Ctot, %	Sampling
Laka	65.3°	15	Rendzic Leptosol	0–1	Litter	117.2 ± 0.8	Modern	18.96	28.9	1985
				1–5	HA	107.0 ± 0.4	Modern	7.79	7.5	
				5–10	HA	99.1 ± 0.4	Modern	2.83	2.4	
				10–17	HA	91.9 ± 0.4	650 ± 30	1.18	0.9	
	65.3°	15	Eutric Cambisol	1–8	HA	106.8 ± 0.7	Modern	7.62	2.7	1985
				8–16	HA	93.2 ± 0.5	580 ± 50	1.35	1.4	
				16–31	HA	87.0 ± 0.7	1150 ± 100	0.77	1.8	
Mezen	65°	15	Eutric Cambisol	—	Grass	126.1 ± 0.5	—	—	—	1986
				0–2	HA	115.6 ± 1.6	Modern	16.45	17.1	
				2–8	HA	112.4 ± 0.8	Modern	12.74	3.0	
Penega	64.6°	16	Gleyic Podzoluvisol	0–6	Litter	126.9 ± 0.6	Modern	37.16	41.4	1984
				6–12	Litter	98.3 ± 0.4	140 ± 30	2.56	34.7	
				12–25	HA	79.7 ± 0.4	1880 ± 50	0.47	4.4	
				33–42	HA	57.6 ± 0.7	4760 ± 110	0.17	3.2	

Note: OM — organic matter; HA — humic acids; n.d. — not determined.

These facts showed that majority of investigated soils are mature not only morphologically but also in concern to organic carbon dynamics. That is why SOC dynamics of these mature soils can be studied with the help of the proposed model.

Since recognizing the immature character of the soil horizon in the field (for example, 0-5 cm horizon of Gelic Regosol on the solifluction slope of Anui ridge) it fully confirmed by radiocarbon analysis the highest pmC values — 120.5% (Table 10.3). Such soils should not be included for SOC dynamics analysis on the basis of ^{14}C data.

TREND OF ^{14}C AGE AND K_R CHANGE WITH DEPTH IN SOIL PROFILES

As can be seen from the Tables 10.2 and 10.3, the ^{14}C age in Eurasian arctic soils has a pronounced trend to increase with depth. The only exception is Gelic Gleysol of Koida, where the alluvial horizons are younger than overlaid peat. Such inversion of ^{14}C dates is an often occurring phenomena in many soils with alluvial processes (Chichagova, 1985).

In respect to the general trend of getting older (or more accurately — increasing mean residence time) with depth, the organic matter of Eurasian arctic soils has a typical behavior for almost all soils of the world (Chichagova, 1985; Becker-Heidman et al., 1996; Scharpenseel and Pfeifer, 1998). And these data closely correlate to the trends shown for North America arctic soils (Tedrow, 1977; Ping et al., 1998).

Being reciprocal of the ^{14}C age the coefficient of renovation of SOC has inverse distribution in soil profiles. And Figure 10.3 shows that according to radiocarbon analysis the essential input in the exchange of carbon with atmosphere contributes only to the upper layer of soils with a thickness less than 10 cm (two profiles with high K_r in deeper layers are the soils of solifluction slope and foothill, and accretion of fresh material could cause such magnitudes of coefficient of renovation). It is also relevant to take into consideration that because of periodic turbation a large amount of organic carbon is buried in tundra soils (Tedrow, 1977; Glazovskaya, 1997). The admixture of this old buried organic matter (often morphologically invisible) can essentially decrease the mean value of radiocarbon activity in soil horizons (Scharpenseel and Becker-Heidmann, 1994) and consequently minimize the K_r magnitudes. There is a need for the parallel control of SOC behavior in

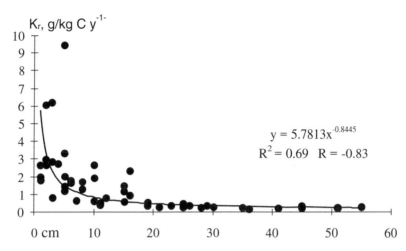

FIGURE 10.3 Relationship between the coefficient of renovation of SOC (K_r) and sampling depth in soils of the Eurasian Arctic.

arctic soil beneath 10 cm by other methods to confirm or to reject this interpretation of ^{14}C data, but in this chapter the priority for SOC dynamics is study of the upper 10 cm.

It is also noted that in spite of the large amplitude of different factors (climate, drainage, substrate) the radiocarbon data for all mature soils of Eurasian Arctic show the obvious order in distribution with the depth. It indicates the general unity of arctic soils in respect to carbon dynamics.

SPATIAL DISTRIBUTION OF SOC RENOVATION RATES IN NORTHERN EURASIA

As is known, the main parameters of soil organic turnover in mature soils depends on climatic, drainage, substrate, and biotic factors (Kobak, 1988; Cherkinsky and Goryachkin, 1993; Schimel et al., 1994). On the continental scale the first-order factor is climate. Arctic climate especially relevant to the SOC cycle is that of the summer time, related to the latitude.

As is seen from Figure 10.4 on the basis of the data of the surface horizons and the layers at 10 cm depth it can be concluded that the climate is the most important factor for SOC dynamics in Eurasian Arctic. Despite all the differences of other factors (drainage, substrate, and local microclimate) we see the regular distribution of the data related to the latitude.

In the layer, at the depth of 10 cm the latitude difference of 15° (>1600 km) and principal change of the life zone (arctic tundra - boreal forests) does not play the key role in carbon cycle rate. Only slight and gradual enhancement of SOC renovation can be observed while moving from north to south. It should be related to the commonly low soil temperatures at this depth in the Arctic and Subarctic and resulting weak biological activity (Goryachkin et al., 1994).

SOC in surface horizons of Eurasian arctic soils behave in different ways. The data in Figure 10.4 show small oscillations of coefficient of carbon renovation in high latitudes up to the arctic circle, related to change of local factors. But not far to the north from the tree line there is a pronounced change of the carbon cycling rate. This is because of the principal change of the vegetation type from tundra to forest-tundra and then to boreal forest.

The limited data collected on the litter radiocarbon dating show that in high latitudes the renovation rate of this component of soil organic profile can be greater than in southern tundra. It may take place because of prevalence of well-decomposed remnants of gramen, sedges, and other herbs in the High Arctic vs. a high portion of hard dwarf shrub fragments in the litter of the southern part of the tundra zone. But the tree leaves and needles in the litter of the forest regions together with enhancement of the thermic conditions make the K_r so high (19.0 and 37.16 g kg^{-1} C per year) that it is out of the scale of Figure 10.4.

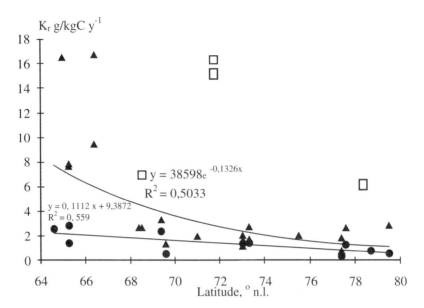

FIGURE 10.4 Spatial distribution of the SOC renovation rate in upper soil horizons: relationship between K_r and latitude; triangles are data on the surface horizon and the upper line is the consequent trend, circles are K_r values in the layer at the depth of 10 cm and the lower line is the trend; empty boxes are out of trend analysis data on renovation rates in litters and on the very thin upper horizon of highly calcareous soil of Spitzbergen.

On five study sites the spatial distribution of SOC renovation rates can be characterized by radiocarbon data on the local scale (Tables 10.2 and 10.3). The sequence is usual for the soil landscapes of humid territories — the lowest rates are in relief depressions in the Histosols with impeded drainage, and the highest ones are on the top of a hill in perfectly drained soils, especially in those with the influence of calcareous rocks. In respect to SOC renovation rates the principal difference is between Histosols and other soils — even Podzols, Regosols, and Gleysols can have the similar magnitudes of K_r in the upper horizons. But while a bog system develops in permafrost conditions there can be the pronounced change of the carbon turnover — on Novaya Zemlya the highest rate (1.97 g kg^{-1} C y^{-1}) is on the top of a peat mound (result of the former frost upheaving). It is because the upper peat horizons after upheaving occurred in well-drained conditions and consequent biomass growth and intensification of mineralization it results in higher K_r values.

POSSIBLE CLIMATE-INDUCED CHANGE OF SOIL CARBON CYCLES IN NORTHERN EURASIA

In order to use the geographical analog approach (Goryachkin and Targulian, 1990, 1993) to evaluate climate-induced soil changes, it is necessary to find the linkage between climatic and relevant soil parameters. In this study these are temperature (precipitation is not the factor of limitation) and coefficient of SOC renovation.

There exists low correlation between mean annual temperature and latitude in northern Eurasia. The correlation between mean annual temperature and K_r of SOC in surface horizons and at the depth of 10 cm was only $R^2 = 0.28$ for the surface horizon and $R^2 = 0.02$ for the 10-cm depth. It means that there is almost no relationship between SOC renovation rates and mean annual temperatures.

The correlation between a more informative climatic parameter — mean July temperature and K_r, is shown in Figure 10.5. To avoid drainage- and substrate-induced "noise", overmoistened Histosols and soils with calcareous upper horizons were excluded from the analysis. This is an exponential curve depicting the trend of the SOC renovation rate with increase in July temperatures for the surface soil horizons and the linear correlation between two parameters for the layer at the depth of 10 cm.

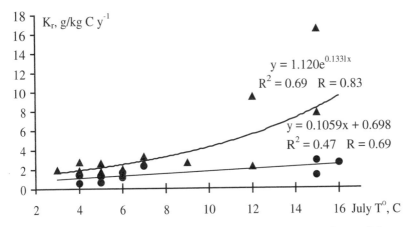

FIGURE 10.5 The relationship between coefficient of SOC renovation (K_r) and mean July temperature for well-drained noncalcareous soils of the Eurasian Arctic and Subarctic; triangles are data on the surface horizon and circles are K_r values in the layer at the depth of 10 cm.

The exponential character of the upper curve is related to the pronounced increase of SOC renovation rate in the territories with a mean July temperature of 12°C and more. This is in the zone of transition between tundra and forest life zones, and the K_r threshold takes place because of the change of carbon turnover character (tundra cycle for forest cycle).

On the basis of the correlation it was possible to evaluate the climate-induced change of SOC dynamics because of the climate warming. As suggested by Houghton et al. (1996) there will be mainly a later fall and winter warming in the high latitudes with very slight increase (especially in the scenario with aerosol content increase) of the summer temperatures. In this case the enhancement of carbon dynamics in the tundra zone may not be observed although there will be some change in the vicinity of the tree line. But the geographical analogue approach offers to consider the characteristic response time and possible feedbacks of the changing processes. The forest plays a great role in sustaining the environment of the ecosystem (for example, it levels the sun radiation input, gas content, and the microclimate). This system has to be formed over a rather long time, as not only low temperature but winds, high air humidity, imperfect drainage, and sometimes shallow permafrost are the causes of absence of trees in tundra zone of Eurasia (Puzachenko, 1988).

So, the essential temperature-induced enhancement of humus mineralization rate in high latitudes of Eurasia is possible only in case of a noticeable northward shift of the tree line, as has taken place in the Middle Holocene (Kremenetski et al., 1997) because of more pronounced warming than that expected by experts during the 21st century.

The comparison of ^{14}C data on litter (Tables 10.2 and 10.3, Figure 10.4) shows that in case of stable position of the tree line the climate warming would even slow down the renovation of litter if the proportion of dwarf shrubs and grasses increases in vegetation cover.

But there could be some change of carbon turnover in tundra zone that is not directly coupled with climatic warming (Gorham, 1991; Eisner, 1998). It relates to transformation of northern peatlands. As the data show there is pronounced intensification of SOC turnover in the case of palsa and peat mound formation (compare Histosols of Novaya Zemlya and those of Edge and more southern Rybachiy, Tables 10.2 and 10.3). It is known that palsa and pingo are mainly features of the arctic sectors with relatively mild falls and winters where the gradual freezing of overmoistened peat takes place (Tedrow, 1977; Goryachkin et al., 1994). In case of expected warming, especially during fall and winter, palsa growth in Siberian Arctic can result in intensification of SOC mineralization in this region.

This manuscript considers only the relative renovation of SOC (g kg^{-1} y^{-1}) without reckoning of the soil-atmosphere flows (g m^{-2} y^{-1}) and taking into account all the problems connected with determination of SOC store in highly cryoturbated soils (Kimble et al., 1993). There are also complicated problems of extrapolation of point data on the vast space of arctic regions (Arnold, 1995). Thus special studies are needed for northern Eurasia to make relatively reliable calculations. The evaluation of carbon sequestration caused by biomass growth was also not used in this study because evidence of it is very poor (Sweda, 1996) and its prediction is recognized as complicated and contradictionary (Oechel et al., 1993; Oechel and Billings, 1992; Oechel and Vourlitis, 1995). These data show low enhancement of SOC renovation in case of warming — the biomass growth would result in the well-drained soils of arctic territories to be the sink of atmospheric carbon. The role of peatlands are more complicated — palsa formation would enlarge CO_2 emanation because of the peat mineralization but at the same time the productivity of the vegetation of peat mounds would also increase. These oppositely directed carbon flows should be thoroughly calculated.

One more indirect consequence of global warming in tundra zone is the enhancement of the fossilizing function (Glazovskaya, 1997) of permafrost-affected soils in carbon cycle. It is known that cryoturbations bury a large amount of carbon from litter, peat, and humus horizons which occur in deep and even permafrost layers. Radiocarbon datings of deep horizons confirmed it for North America (Ping et al., 1998) and also for Eurasia (by our data). Every tundra specialist knows that cryoturbations are more intensive in milder sectors of Arctic (Tedrow, 1977). That means that due to expected climate milding the role of arctic soils as a fossilizing sink of carbon would increase.

CONCLUSIONS

The following conclusions are based on the results of the study:

1. The model for calculating the rates of humus renovation of soils in a steady state is proposed. It takes into account the whole curve of change of specific activity of atmospheric carbon since 1956, including pmC magnitudes >100%.
2. Organic horizons of well-drained soils of Eurasian Arctic need n*100–1000 years to reach equilibrium with the environment. The SOC dynamics of these mature soils can be studied with the help of the proposed model.
3. According to radiocarbon analysis the essential input in exchange of carbon with atmosphere contributes only to the upper layer of arctic soils with a thickness less than 10 cm.
4. In respect to the spatial distribution there are only small oscillations of the coefficient of SOC renovation in high latitudes up to the Arctic Circle, related to change of local factors. But near the tree line there is pronounced change of the carbon cycling rate because of the principal change of the vegetation type from tundra to forest-tundra, and then to boreal forest.
5. The spatial sequence of renovation rates in Eurasian Arctic on the local scale is as follows: the lowest rates are in relief depressions in the Histosols with impeded drainage, and the highest ones are on the top of a hill in perfectly drained soils, especially those with the influence of calcareous rocks.
6. There is almost no relationship between SOC renovation rates and mean annual temperatures for high latitudes. But it is exponential for the relationship between renovation in the surface soil horizons and mean July temperatures and the linear correlation for that of the layer at the depth of 10 cm.
7. An essential temperature-induced enhancement of humus mineralization rate in high latitudes of Eurasia is possible only in case of a noticeable northward shift of the tree line. In case of expected climate warming, we can suppose the appearance of palsa in Siberian Arctic which can result in intensification of SOC mineralization in this region.
8. Due to expected climate milding the role of arctic soils as a fossilizing sink of carbon would increase because of the intensification of cryoturbations.

ACKNOWLEDGMENTS

Project funding was provided by Russian Foundation for Basic Research, and Russian State Scientific and Technical Programs "Global Change…" and "World Ocean." Special thanks to Swedish Polar Research Secretariat and P.V. Boyarski (leader, Integrated Marine Arctic Expedition) for arrangement of field investigations. We are grateful to R.I. Zlotin, V.A. Odintsov, A.E. Roslyakov, and T.A. Vostokova for assistance in sampling and to E.N. Subbotina, G.E. Kadantseva (chemical pretreatment), K.E. Pustovoitov, L.D. Sulerzhitsky, and V.M. Alifanov (radiocarbon dating) for the help in the analytical part of the study. We also appreciate very much the support of J. Kimble, R. Lal, and L. Everett in preparation of the chapter.

REFERENCES

Andren, O. and Kätterer, T., ICBM: the introductory carbon balance model for exploration of soil carbon balances, *Ecol. Appl.,* 7, 1226, 1997.

Arnold, R.W., Role of soil survey in obtaining a global carbon budget, *Soils and Global Change,* Lal, R., Kimble, J.M., Levine, E., and Stewart, B.A., Eds., Lewis Publishers, Boca Raton, FL, 1995.

Becker-Heidman, P., Sharpenseel, H.W., and Wiechmann, H., Hamburg radiocarbon thin layers database, *Radiocarbon,* 38, 295,1996.

Billings, W.D., Carbon balance of Alaskan tundra and taiga ecosystems: past, present and future, *Q. Sci. Rev.,* 6, 165, 1987.

Bolin, B., How much CO_2 will remain in the atmosphere? The carbon cycle and projections for the future, *The Greenhouse Effect, Climatic Change and Ecosystems,* SCOPE Rep. 29, Bolin, B., Doos, B.R., Jager, J., and Warrick, R.A., Eds., John Wiley & Sons, New York, 1986, p. 157.

Cherkinsky, A.E., ^{14}C dating and soil organic matter dynamics in arctic and subarctic ecosystems, *Radiocarbon,* 38, 241, 1996.

Cherkinsky, A.E. and Brovkin, V.A., A model of humus formation in soils based on radiocarbon data of natural ecosystems, *Radiocarbon,* 33, 186, 1991.

Cherkinsky, A.E. and Brovkin, V.A., Dynamics of radiocarbons in soils, *Radiocarbon,* 35, 363, 1993.

Cherkinsky, A.E. and Goryachkin, S.V., Distribution and renovation time of soil carbon in boreal and subarctic ecosystems of European Russia, in *Carbon Cycling in Boreal Forest and Subarctic Ecosystems,* Kolchugina, T. and Vinson, T., Eds., Oregon State University Press, Corvallis, OR, 1993, p. 65.

Cherkinsky, A.E. and Paul, Y.M., Application of radiocarbon method for evaluating the mineralization rate and humus-acids renewal time in Krasnozems (Ustox Suborder), *Eurasian Soil Sci.,* 27, 81, 1995.

Chichagova, O.A., *Radiocarbon Dating of Soil Humus,* Nauka, Moscow, 1985 (in Russian).

Chichagova, O.A., Modern trends in radiocarbon studies of soil organic matter, *Eurasian Soil Sci.,* 29, 89, 1996.

Chichagova, O.A. and Cherkinsky, A.E., Problems in radiocarbon dating of soils, *Radiocarbon,* 35, 351, 1993.

Christensen, T.R., Jonasson, S., Callaghan, T.V., and Havströem, M., Spatial variation in high-latitude methane flux along a transect across Siberian and European tundra environments, *J. Geophys. Res.,* 100, 21,035-21,045, 1995.

Demkina, T.S. and Zolotareva, B.N., Determination of the rate on mineralization of humic substances in soils, *Eurasian Soil Sci.,* 30, 1085, 1997.

Eisner, W.R., Arctic paleoecology and soil processes: developing new perspectives for understanding global change, in *Soil Processes and the Carbon Cycle,* Lal, R., Kimble, J.M., Follett, R.F., and Stewart, B.A., Eds., Lewis Publishers, Boca Raton, FL, 1998, p. 127.

Eswaran, H., Van den Berg, E., Reich, P., and Kimble, J., Global soil carbon resources, in *Soils and Global Change,* Lal, R., Kimble, J.M., Levine, E., and Stewart, B.A., Eds., Lewis Publishers, Boca Raton, FL, 1995, p. 27.

FAO, Soil Map of the World, revised legend, Food and Agricultural Organization of the United Nations, Rome, 1990.

Gerasimov, I.P., The age of recent soils, *Geoderma,* 12, 17, 1974.

Gerasimov, I.P. and Chichagova, O.A., Some problems of radiocarbon dating of soil humus, *Pochvovedenie,* N10, 3, 1971 (in Russian).

Glazovskaya, M.A., Fossilizing functions of the pedosphere in continental cycles of organic carbon, *Eurasian Soil Sci.,* 30, 240, 1997.

Gorham, E., Northern peatlands: role in the carbon cycle and probable response to climate warming, *Ecol. Appl.,* 2, 182, 1991.

Goryachkin, S.V. and Targulian, V.O., Climate-induced changes in the boreal and subpolar soils, *Proc. Int. Workshop Effects of Expected Climate Change on Soil Processes in the Tropics and Subtropics,* Scharpenseel, H.W., Schomaker, M., and Ayoub, A., Eds., Feb. 12-14, Nairobi, 1990.

Goryachkin, S.V. and Targulian, V.O., Soil cover evolution as a result of global climate change: approaches to modelling, Soil Cover Structure, *Proc. Int. Soil Sci. Soc. Symp.,* Moscow, Sept. 6-11, 1993.

Goryachkin, S.V., Zlotin, R.I., and Tertitsky, G.M., *Diversity of Natural Ecosystems in the Russian Arctic,* Lund University, Sweden, 1994.

Gupta, S.K. and Polach, H.A., Handbook of Radiocarbon Dating Practices at ANU, Australian National University, Canberra, 1985.

Harrison, K.G., Using bulk soil radiocarbon measurements to estimate soil organic matter turnover times, in *Soil Processes and the Carbon Cycle,* Lal, R., Kimble, J.M., Follett, R.F., and Stewart, B.A., Eds., Lewis Publishers, Boca Raton, FL, 1998, p. 549.

Harrison, K.G., Broecker, W.S., and Bonani, G., A strategy for estimating the impact of CO_2 fertilization on soil carbon storage, *Global Biochem. Cycles,* 7, 69, 1993.

Harkness, D.D., Harrison, A.F., and Bacon, P.J., The temporal distribution of "bomb" [14]C in a forest soil, *Radiocarbon,* 28, 328, 1986.

Harkness, D.D., Harrison, A.F., and Bacon, P.J., The potential of bomb ^{14}C measurements for estimating soil organic matter turnover, in *Advances in Soil Organic Matter Research,* Wilson, W.S., Ed., Royal Chemical Society, Cambridge, U.K., 1991, p. 239.

Houghton, J.T., Meira Filho, L.G., Callander, B.A., Harris, N., Kattenberg, A., and Maskell, K., Eds., Climate change 1995, in *The Science of Climate Change,* IPPC, Cambridge University Press, Cambridge, U.K., 1996.

Hsieh, Y.P., Radiocarbon signatures of turnover rates in active soil organic carbon pools, *Soil Sci. Soc. Am. J.,* 57,1020, 1993.

Jenkinson, D.S. and Raynor, J.H., The turnover of organic matter in some of the Rothamsted classical experiments, *Soil Sci.,* 123, 298, 1977.

Kimble, J.M., Tarnocai, C., Ping C.L., Ahrens, R., Smith, C.A.S., Moore, J., and Lynn, W., Determination of the amount of carbon in highly cryoturbated soils, Joint Russian-American Seminar on Cryopedology and Global Change, Post-seminar Proceedings, Gilichinsky, D.A., Ed., Puschino, Russia, 1993, p. 277.

Kobak, K.I., *Biotic Components of the Carbon Cycle,* Hydrometeoizdat, Leningrad, 1988 (in Russian).

Kremenetski, C., Vaschalova, T., Goriachkin, S., Cherkinsky, A., and Sulerzhitsky, L., Holocene pollen stratigraphy and bog development in the western part of the Kola Peninsula, Russia, *Boreas,* 26, 91, 1997.

Leavitt, S.W., Follett, R.F., and Paul, E.A., Estimation of slow- and fast-cycling soil organic carbon pools from 6*N* HCl hydrolysis, *Radiocarbon,* 38, 231, 1996.

Martel, Y.A. and Paul, E.A., Use of radiocarbon dating of organic matter in the study of soil genesis, *Soil Sci. Soc. Am. Proc.,* 38, 501, 1974.

Oechel, W.C. and Billings, W.D., Anticipated effects of global change on carbon balance of arctic plants and ecosystems, in *Arctic Physiological Processes in a Changing Climate,* Chapin, F.S., III, Jefferies, R.L., Shaver, G.R., Reynolds, G.F., and Svobada, J., Eds., Academic Press, San Diego, 1992, p. 139.

Oechel, W.C. and Vourlitis, G.L., Effects of global change on carbon storage in cold soils, in *Soils and Global Change,* Lal, R., Kimble, J.M., Levine, E., and Stewart, B.A., Eds., Lewis Publishers, Boca Raton, FL, 1995, p. 117.

Oechel, W.C., Hastings, S.J., Vourlitis, G.I., Jenkins, M.A., Riechers, G., and Grulke, N., Recent change of arctic tundra ecosystems from a carbon sink to a source, *Nature,* 361, 520, 1993.

Paul, E.A., Campbell, C.A., Rennie, D.A., and McCallum, K.J., Investigations of the dynamics of soil humus utilizing carbon dating techniques, Vol. 3, in Trans. 8th Int. Congr. Soil Sci., Bucharest, Aug. 31 to Sept. 9, 1964, Rompresfilatelia, Bucharest, 1964, p. 201.

Paul, E.A., Horwath, W.R., Harris, D., Follett, R., Leavitt, S., Kimball, B.A., and Pregitzer, K., Establishing pool sizes and fluxes of CO_2 emissions from soil organic matter turnover, in *Soils and Global Change,* Lal, R., Kimble, J.M., Levine, E., and Stewart, B.A., Eds., Lewis Publishers, Boca Raton, FL, 1995, p. 297.

Paul, E.A., Follett, R.F., Leavitt, S.W., Halvorson, A., Peterson, G.A., and Lyon, D.J., Radiocarbon dating for determination of soil organic matter pool sizes and dynamics, *Soil Sci. Soc. Am. J.,* 61, 1058, 1997.

Ping, C.L., Michaelson, G.J., Cherkinsky, A.E., and Malcolm, R.L., Characterization of soil organic matter by stable isotopes and radiocarbon ages of selected soils in Arctic Alaska, Proc. 8th Int. Humic Substances Conf., Sept. 9-14, Wroclaw, Poland, 1996.

Post, W.M., Emanuel, W.R., Zinke, P.J., and Stangenberger, A.G., Soil carbon pools and world life zones, *Nature,* 298, 156, 1982.

Puzachenko, Y.G., Climate-induced southern boundary of tundra, in *Communities of the Far North and a Man,* Chernov, Y.I., Ed., Nauka, Moscow, 1988 (in Russian), p. 22.

Scharpenseel, H.W., Radiocarbon dating of soils — problems, troubles, hopes, in *Paleopedology,* Yaalon, D.H., Ed., University of Jerusalem Press, 1971, p. 77.

Scharpenseel, H.W. and Becker-Heidman, P., ^{14}C dates and ^{13}C measures of different soil species, in *Soil Processes and the Greenhouse Effect,* Lal, R., Kimble, J.M., and Levine, E., Eds., SCS, NSSC, U.S. Department of Agriculture, Lincoln, NE, 1994, p. 72.

Scharpenseel, H.W. and Pfeiffer, E.M., Carbon turnover in different climates and environments, in *Soil Processes and the Carbon Cycle,* Lal, R., Kimble, J.M., Follett, R.F., and Stewart, B.A., Eds., Lewis Publishers, Boca Raton, FL, 1998, p. 577.

Scharpenseel, H.W. Becker-Heidman, P., Neue, H.U., and Tsutsuki, K., Bomb carbon ^{14}C-dating and Δ^{13}C measurements as tracers of organic matter dynamics as well as of morphogenetic and turbation processes, *Sci. Total Environ.,* 81/82, 99, 1989.

Schimel, D.S., Braswell, B.H., Holland, E.A., McKeown, R., Ojima, D.S., Painter, T.H., Parton, W.J., and Townsend, A.R., Climatic, edaphic, and biotic controls over storage and turnover of carbon in soils, *Global Biogeochem. Cycles,* 8, 273, 1994.

Schlesinger, W.H., Evidence from chronosequence studies for a low carbon storage potential of soils, *Nature,* 348, 232, 1990.

Schlesinger, W.H., *Biogeochemistry, An Analysis of Global Change,* Academic Press, New York, 1991.

Stuiver, M. and Polach, H.A., Discussion: reporting of ^{14}C-data, *Radiocarbon,* 19, 355, 1977.

Sweda, T., Some vegetation indications of climate warming as detected on the forest-tundra border in the continental Canadian Arctic, Proc. Int. Symp. Environ. Res. in the Arctic, July 19-21, Watanabe, O., Ed., National Institute for Polar Research, Tokyo, 1996.

Tedrow, J.C.F., *Soils of the Polar Landscapes,* Rutgers University Press, New Brunswick, NJ, 1977.

Trumbore, S.E., Bonani, G., and Wölfli, W., The rates of carbon cycling in several soils from AMS C-14 measurements of fractionated soil organic matter, in *Soils and the Greenhouse Effect,* Bouwman, A.F., Ed., John Wiley & Sons, Chichester, U.K., 1990, p. 407.

Trumbore, S.E., Chadwik, O.A., and Amudson, R., Rapid exchange between soil carbon and atmospheric carbon dioxide driven by temperature change, *Science,* 272, 393, 1996.

Zavarzin, G.A., Carbon cycle in natural ecosystems of Russia, *Priroda,* N7, 15, 1994 (in Russian).

Zavel'skiy, F.S., Radiocarbon dating and theoretical models of carbon turnover in soils, *Bull. Acad. Sci. USSR, Geogr. Ser.,* N1, 27, 1975 (in Russian).

11 Influence of Overgrazing by Reindeer on Soil Organic Matter and Soil Microclimate of Well-Drained Soils in the Finnish Subarctic

G. Broll

CONTENTS

INTRODUCTION

Large areas of northern Scandinavia are damaged by reindeer overgrazing (Helle and Aspi, 1983). Comparing different countries in Northern Scandinavia, the differences in the extent of ecosystem damage become apparent. The worst situation can be observed in Norway (Oksanen et al., 1995). Within the Subarctic, characterized by dwarf shrubs, lichens, and mosses, semi-domestic reindeer can cause serious soil erosion (Evans, 1995; Kashulina et al., 1997). Lichens are the main winter forage for reindeer. Therefore, lichen cover on windblown heaths with a thin snow cover is especially influenced by grazing (Wielgolaski, 1997). This effect of overgrazing can also be seen in the boreal forest of Scandinavia (Väre et al., 1995). In addition, reindeer can destroy vegetation cover and soil by trampling (Oksanen and Virtanen, 1995).

    The objective of this study is to show the effects of reindeer grazing on soil organic matter and soil microclimate at a representative well-drained tundra site in northern Scandinavia. A fenced site without grazing is compared with an unfenced site with grazing.

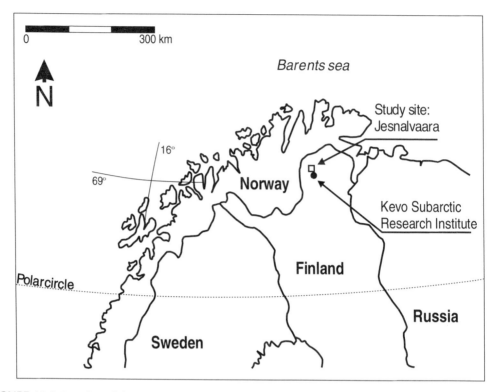

FIGURE 11.1 Location of the study site "Jesnalvaara" in northernmost Finland.

MATERIALS AND METHODS

SITE DESCRIPTION

The study area is situated in northernmost Finland at 69°46' N and 26°57' E close to the Baltic Sea and the Norwegian border (Figure 11.1). The study sites are located on the "Jesnalvaara" mountain at 327 m a.s.l. The climate is slightly continental with a mean annual air temperature of −3.5°C and an anuual precipitation of 420 mm (Seppälä, 1976). The study sites, located on the top of the mountain, are exposed to strong winds. Therefore, winter snow cover is thin. The bedrock consists of granite gneiss belonging to the Precambrian Baltic Shield. On the "Jesnalvaara" only small remnants of till are left, most of the till on the top of the mountains in this area was eroded after deglaciation (Hirvas, 1991). The soils are seasonally frozen, showing a cryic temperature regime (Soil Survey Staff, 1998). Permafrost does not occur at this study site. The shallow soils with a high percentage of skeletal material are podzolized (Typic Haplocryod) (Hinneri et al., 1975), which is typical for well-drained sites in Finnish Lapland (Räisänen, 1994). At the study site podzolization is enhanced by the sandy texture (loamy sand), the low pH value of about 4.0, and the wide C/N ratio of the litter. The oligotrophic low alpine heath just above the timberline is characterized by a mosaic of dwarf shrubs and lichens. Beneath the dwarf shrubs, soil organic matter accumulates, and a thick O horizon develops. On the other hand, only small amounts of organic carbon occur under the lichen cover (Broll, 1994).

The study site is divided into a grazed plot and an ungrazed plot. The ungrazed plot is a former study site of the International Biological Programme (IBP). The site was fenced in 1968 (Kallio, 1975). Thus, natural succession continued for over 30 years. The size of the plot is about 4000 m². The area outside the fence is grazed by reindeer. The grazed area which was sampled has the same size as the ungrazed plot. Both plots are characterized by patchy vegetation of lichens and dwarf shrubs (lichen microsites and dwarf shrub microsites). Without grazing, the boundaries between the vegetation patches increasingly become diffuse.

FIELD MEASUREMENTS, SOIL SAMPLING, AND SOIL ANALYSIS

The vegetation was mapped according to Braun-Blanquet (1964) with special regard to the height of the vegetation cover. Soil sampling (mixed samples) for analysis of organic carbon and total nitrogen was carried out only at the lichen microsites of the grazed plot and the ungrazed plot. Ten mixed samples were taken at each microsite. Sampling depth was 0–2 cm of the mineral soil. Organic carbon and total nitrogen were determined with an elemental analyser (Carlo Erba NA 1500). At the lichen microsites as well as at the dwarf shrub microsites bulk density samples were taken with a metal cylinder in a depth of 0-4 cm. Soil water content was measured with TDR (TRIME FM-2, Eijkelkamp) in the A horizon of the lichen microsites and in the O horizon of the dwarf shrub microsites. Soil temperature was recorded at a depth of 2.5 cm from August 1996 to August 1997 (logger — DT 3, Elpro-Fuchs).

RESULTS AND DISCUSSION

VEGETATION

The vegetation of the oligotrophic low alpine heath on the top of the "Jesnalvaara" mountain can be classified as Arctic-Empetrum-Cetraria nivalis type (Haapasaari, 1988) or as a Empetrum-Cetraria nivalis type according to Oksanen and Virtanen (1995). This type is characteristic for wind-exposed sites where snow cover is thin or even missing in winter. The plants of this type are adapted to extreme microclimatic conditions such as very low winter temperatures and dry conditions. At some parts of the study site another type of vegetation occurs: the Hemiarctic Empetrum Lichens type (Haapasaari, 1988) or the Betula nana-Cladina-type (Oksanen and Virtanen, 1995), respectively.

Plant cover of the two investigated plots is different. On the grazed plot the plant cover is discontinuous with bare soil and spots with fragments of lichens. Disturbances of the lichen cover by reindeer enhance the occurrence of *Juncus trifidus* as was also found on other sites by Helle and Aspi (1983). Grasses are able to generate faster than lichens, and sometimes grazing can even stimulate the growth of grasses, as in the case of *Festuca ovina*. Also, a slight increase of mosses was found on the grazed plot. At the grazed plot the coverage of *Polytrichum piliferum* and *Polytrichum juniperinum*, for example, were higher than on the ungrazed plot. Reindeer avoid consuming *Polytrichum piliferum* (Oksanen and Virtanen, 1995), and *Polytrichum juniperinum* prefers sites which are disturbed like grazed sites (Helle and Aspi, 1983).

On the ungrazed plot the plant cover is continuous and very dense. Lichens and dwarf shrubs are growing side by side. *Cetraria nivalis* is dominating within the lichens (50–75% coverage). On the grazed plot, the coverage of *C. nivalis* is extremely reduced to 1–5% coverage. *C. nivalis* does not belong to the favorite diet for reindeer, but it is consumed in case the food supply is limited. Also, reindeer destroy the lichen cover by trampling (Haapasaari, 1988). Besides *C. nivalis, C. islandica* and *Cladonia stellaris* grow better on the study site when reindeer are excluded. *C. stellaris* and *C. islandica* are favorite food for reindeer (Helle and Aspi, 1983; Haapasaari, 1988; Oksanen et al., 1995). Decreasing lichen cover on grazed sites compared to ungrazed sites was also recognized by Väre et al. (1996) and by Manseau et al. (1996) in Canada due to grazing by caribou. In general, only when the food supply is very bad do reindeer consume dwarf shrubs (Danell et al., 1994). Comparing the height of the plant cover inside and outside the enclosure on the "Jesnalvaara" (Figure 11.2) even the dwarf shrubs, in this case mainly *Empetrum hermaphroditum*, are very likely to be influenced by grazing. The same result was found by Väre et al. (1996) in Finland and by Crete and Doucet (1998) and Manseau et al. (1996) in Canada. The decrease of the height of the lichen cover from about 4 cm without grazing is conspicuous when compared to

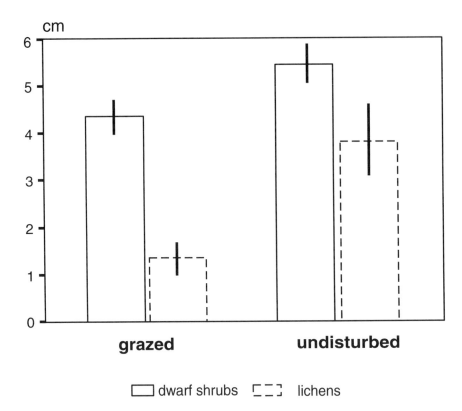

FIGURE 11.2 Height of the plant cover at the grazed plot and the ungrazed plot (n = 20; I = SD; lichens ***p <001, dwarf shrubs ***p <001).

the grazed plot where the height of the lichens is about 1.5 cm. The reduced lichen biomass on the grazed plot influences the soil organic matter and the microclimatic conditions.

SOIL

Soil Organic Matter

Grazing not only affects the vegetation but also the soil organic matter. There are distinct differences in the carbon content between the lichen microsites at the grazed and the ungrazed plot (Figure 11.3). On the grazed plot about 1.8 kg m² organic carbon was found. On the ungrazed plot in addition to this 1.1 kg m² organic carbon accumulated in 0–2 cm depth as a result of the plant succession. For example, Morris and Jensen (1998) found an increase of 0.55 kg m² organic carbon in a depth of 0–10 cm after 25 years without grazing. At the study site in Finnish Lapland an O horizon has developed, where more than 20% Corg accumulated (Figure 11.3). This gives evidence for a changing of the humus composition. Lower soil organic matter content in the case of grazing is reducing field capacity and enhancing the lack of water supply for the plants at this well-drained and wind-exposed heath site.

Parallel to the increase of carbon after exclusion of grazing, the nitrogen content also increased (Figure 11.4). However, on the "Jesnalvaara" study site the extent of accumulation of nitrogen is not high enough to maintain the C/N ratio at the same level. At the grazed plot the C/N ratio is about 29, whereas at the ungrazed plot it is about 34. An influence of reindeer grazing on nitrogen mineralization can be expected (Stark, 1999). The potential mineralization of organic nitrogen is already low at this dry alpine heath because of the high soil acidity and the wide C/N ratio of the litter (Broll, 1994). Under field conditions the nitrogen mineralization would be reduced further

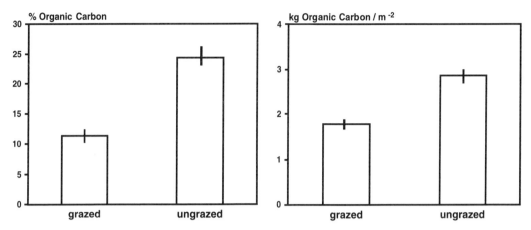

FIGURE 11.3 Organic carbon content of the lichen sites at the grazed plot and the ungrazed plot (n = 10; I = SD; ***p <001).

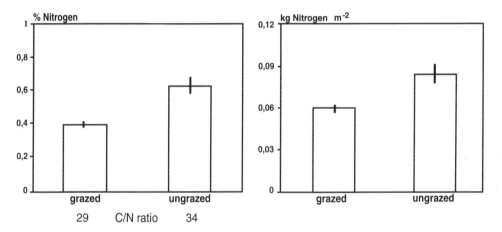

FIGURE 11.4 Total nitrogen content of the lichen sites at the grazed plot and the ungrazed plot (n = 10; I = SD; ***p <001).

due to the low soil moisture during summer, as was evidenced for comparable dry heath sites in Swedish Lapland (Weih, 1998). If the C/N ratio is widened during succession the nitrogen mineralization will be hampered even more. On the other hand, increasing soil moisture during plant succession would enhance nitrogen mineralization on the ungrazed plot.

Soil Microclimate

On the lichen microsites, reindeer grazing heavily influences the soil microclimate. Besides the indirect effect through the changes of the soil organic matter content, grazing opens the vegetation cover which protects the soil from extreme heating and cooling of the surface. In case of heavy grazing the soil temperature in the 2.5 cm depth is higher during summer and lower during winter compared to the undisturbed plot (Figures 11.5 and 11.6). On the lichen microsites the greatest difference in summer is about 6°C and about 4°C in winter. Also, the daily amplitude between the minimum and the maximum soil temperature is higher at the grazed plot. The temperature drops below 0°C earlier on the grazed plot. Väre et al. (1996) also measured lower soil temperatures in the 5 cm depth, comparing grazed and ungrazed sites in an oligotrophic pine heath in northeastern Fennoscandia. Compared to forest sites with a thick snow cover in the Finnish Subarctic the soil

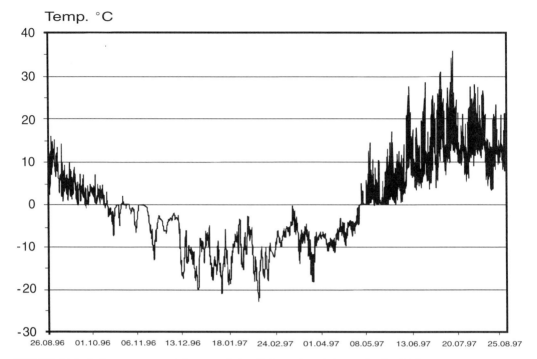

FIGURE 11.5 Soil temperature of the grazed plot (lichen microsite) at a depth of 2.5 cm.

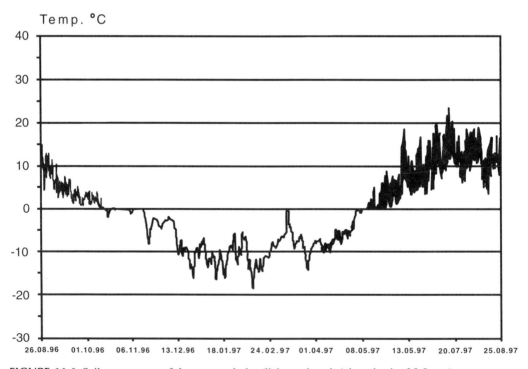

FIGURE 11.6 Soil temperature of the ungrazed plot (lichen microsite) in a depth of 2.5 cm.)

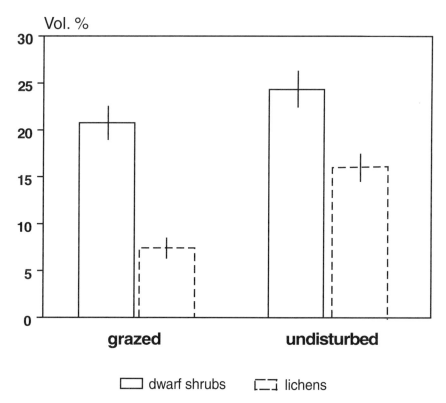

FIGURE 11.7 Soil water content (vol. %) recorded August 24, 1996 between 2 and 4 p.m. of the dwarf shrub microsites and the lichen microsites at the grazed plot and the ungrazed plot (dwarf shrubs: organic layer; lichens: 0–4 cm depth) (n = 20, I = SD; lichens **p <01, dwarf shrubs NS).

temperatures on the study site "Jesnalvaara" are very low. Measured in eastern Finnish Lapland in a Scots pine (*Pinus sylvestris*) forest with a snow cover of about 1 m during winter, soil temperatures not lower than about 0°C were found in a depth of 5 cm.

During summer the high soil temperature may increase evaporation, followed by exhaustion of soil moisture. As an example of lower water content at the grazed plot, Figure 11.7 shows the results of soil moisture measurement during one summer day of 1996. The water content in the O layer of the dwarf shrub microsites did not show any significant difference between the plots, but both plots were higher compared to the lichen microsites.

Bulk Density

In the event of intense reindeer grazing and trampling one would expect changes of the bulk density of the soil (Greenwood et al. 1998). On the one hand, the compaction due to trampling could increase bulk density. Morris and Jensen (1998) measured five times greater bulk density at a grazed site compared to an ungrazed site. On the other hand, grazing could lower bulk density by removal of the belowground biomass together with the aboveground biomass. However, it depends on soil texture, in particular in which way bulk density is affected. On soils with loamy soil texture grazing would lead to compaction, while on soils with sandy soil texture the grazing would enhance a loosening of the soil surface. At the heath site "Jesnalvaara" with a loamy sand, the bulk density on the lichen microsites was slightly higher on the grazed plot compared to the ungrazed plot (Figure 11.8). Compared to the organic layer beneath the dwarf shrubs the bulk density of the lichen microsites is higher in any case, of course.

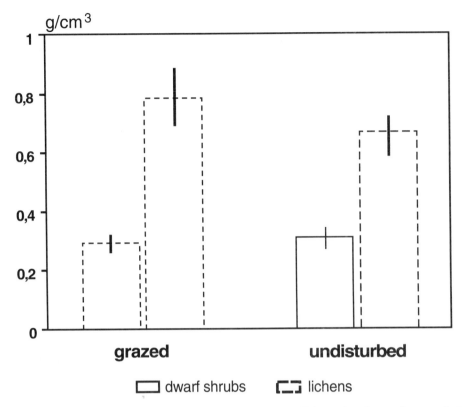

FIGURE 11.8 Bulk density of the dwarf shrub microsites and the lichen microsites at the grazed plot and the ungrazed plot (dwarf shrubs: organic layer; lichens: 0–4 cm depth) (n = 20, I = SD; lichens *p <05, dwarf shrubs NS).

CONCLUSIONS

Reindeer grazing at well-drained alpine heath sites in northernmost Finland influences the vegetation cover and both the physical and chemical properties of the soil. The height and the coverage of lichens are reduced by grazing. Organic carbon content and nitrogen content of the soil decrease on grazed sites. Due to lack of soil organic matter, field capacity and soil moisture decline. The grazing further effects the microclimate by increasing the soil temperature during summer and decreases the soil temperature during winter. An increasing daily amplitude of soil temperature was also observed. During summer high soil temperatures may increase evaporation, which also would cause a decline of soil moisture.

In view of global change, the question is whether the rise of soil temperature caused by grazing exceeds the increase of soil temperature attributed to predicted global change (Heal et al., 1998). Moreover, it would be of interest whether the impact of grazing on decomposition of soil organic matter at well-drained soils in the Subarctic is possibly higher than the predicted impact due to global warming (Press et al., 1998). There is no question that regional studies on changing land use, in this case grazing intensity, should be more thoroughly considered in discussions on global change. At the long-term perspective the slow regeneration of lichen cover after heavy reindeer grazing is of paramount importance for the ecological carrying capacity of tundra ecosystems (Wielgolaski, 1997; Leader-Williams et al., 1987).

ACKNOWLEDGMENTS

I would like to thank the staff of the Kevo Subarctic Research Institute for support and the opportunity to work on the study site on "Jesnalvaara" mountain. Many thanks also go to Roland Ossenbrink who did the vegetation mapping and to Gerald Mueller who assisted in the field.

REFERENCES

Braun-Blanquet, J., *Pflanzensoziologie,* Springer, Berlin, 1964.

Broll, G., Influence of the soil mosaic on biodiversity at heath sites in the European Subarctic, in Trans. 15th World Congr. Soil Sci., Vol 4a, Mexico City, 1994, p. 220.

Crête, M. and Doucet, G.J., Persistent suppression in dwarf birch after release from heavy summer browsing by caribou, *Arctic Alpine Res.,* 30, 126, 1998.

Danell, K., Utsi, P.M., Palo, T., and Eriksson, O., Food plant selection by reindeer during winter in relation to plant quality, *Ecography,* 17, 153, 1994.

Evans, R., Impacts of reindeer grazing on soils and vegetation in Finnmark. A report for NORUT Information Technology, Anglia Polytechnic University, Cambridge, England, 1995.

Greenwood, K.L., MacLeod, D.A., Scott, J.M., and Hutchinson, K.J., Change of soil physical properties after grazing exclusion, *Soil Use Manage.,* 14, 19, 1998.

Haapasaari, M., The oligotrophic heath vegetation of northern Fennoscandia and its zonation, *Acta Bot. Fenn.,* 135, 1988.

Heal, O.W., Broll, G., Hooper, D.U., McConnell, J., Webb, N.R., and Wookey, P.A., Impacts of global change on tundra soil biology, in *Global Change in Europe's Cold Regions,* Heal, O.W., Callaghan, T.V., Cornelissen, J.H.C., Körner, C., and Lee, S.E., Eds., Rep. of the Arteri Workshops at the Danish Polar Centre, Copenhagen, Nov. 4-10, 1996; Ecosystems Res. Rep. 27, European Communities, Luxembourg, 1996, p. 65.

Helle, T. and Aspi, J., Effects of winter grazing by reindeer on vegetation, *Oikos,* 40, 337. 1983.

Hinneri, S., Sonesson, M., and Veum, A.K., Soils of the Fennoscandian IBP tundra ecosystems, in *Fennoscandian Tundra Ecosystems, Part 1: Plant and Microorganisms,* Springer, New York, 1975, p. 31.

Hirvas, H., Pleistocene stratigraphy on Finnish Lapland, *Geogr. Surv. Finl. Bull.,* 354, 1, 1991.

Kallio, P., Kevo, Finland, in *Structure and Function of Tundra Ecosystems, Ecological Bulletin 20,* Rosswall, T. and Heal, O.W., Eds., Kevo, Finland, 1975, p. 193.

Kashulina, G., Reimann, C., Finne, T.E., Halleraker, J.H., Äyräs, M., and Chekushin, V.A., The state of the ecosystems in the central Barents Region, scale, factors, and mechanism of disturbance, *Sci. Total Environ.,* 206, 203, 1997.

Leader-Williams, N., Smith, R.I.L., and Rothery, P., Influence of introduced reindeer on the vegetation of south Georgia, results from a long-term exclusion experiment, *J. Appl. Ecol.,* 24, 801, 1987.

Manseau, M., Huot, J., and Crête, M., Effects of summer grazing by caribou on composition and productivity of vegetation: community and landscape level, *J. Ecol.,* 84, 503, 1996.

Morris, J.T. and Jensen, A., The carbon balance of grazed and non-grazed *Spartina anglica* salt marshes at Skallingen, Denmark, *J. Ecol.,* 86, 229, 1998.

Oksanen, L. and Virtanen, R., Topographic, altitudinal, and regional patterns in continental and suboceanic heath vegetation of northern Fennoscandia, *Acta Bot. Fenn.,* 153, 1, 1995.

Oksanen, L., Moen, J., and Helle, T., Timberline patterns in northernmost Fennoscandia — relative importance of climate and grazing, *Acta Bot. Fenn.,* 153, 93, 1995.

Press, M.C., Potter, J.A., Burke, M.J.W., Callaghan, T.V., and Lee, J.A., Responses of a subarctic dwarf shrub heath community to simulated environmental change, *J. Ecol.,* 86, 315, 1998.

Räisänen, M.L., Reflection of bedrock on the soil geochemistry and weathering in the tundra region of northernmost Finland, in Proc. Meet. Classification, Correlation Manage. of Permafrost-Affected Soils, Kimble, J.M. and Ahrens, R.J., Eds., Soil Conservation Service, National Soil Survey Center, U.S. Department of Agriculture, Lincoln, NE, 1994, p. 99.

Seppälä, M., Periglacial character of the climate of the Kevo region (Finnish Lapland) on the basis of meteorological observations 1962-1971, *Rep. Kevo Subarctic Res. Stn.,* 13, 1, 1976.

Soil Survey Staff, Keys to Soil Taxonomy, U.S. Department of Agriculture, Washington, D.C., 1998.

Stark, S., The effect of reindeer on soil nitrogen and carbon dynamics, Sustainable Development in Northern Timberline Forests, Proc. Timberline Workshop, Whitehorse, Canada, May 10-11, 1998, Kankaanpää, S., Tasanen, T., and Sutinen, M.-L., Eds., Res. Pap. 734, Kolari Res. Stn., Finnish Forest Research Institute, Helsinki, 1999.

Sutinen, M.-L., Ritari, A., Holappa, T., and Kujala, K., Seasonal changes in soil temperature and in the frost hardiness of Scots Pine roots under subarctic conditions, Int. Symp., Phys. Chem. Ecol. of Seasonally Frozen Soils, Faitbanks, AK, June 10-12, Iskandar, I.K., Wright, E.A., Radke, J.K., Sharratt, B.S., Groenevelt, P.H., and Hinzman, L.D., Eds., Spec. Rep. 97-10, Cold Regions Research and Engineering Laboratory, Hanover, NH, 1997, p. 513.

Väre, H., Ohtonen, R., and Mikkola, K., The effect and extent of heavy grazing by reindeer in oligotrophic pine heaths in northeastern Fennoscandia, *Ecography,* 19, 245, 1996.

Väre, H., Ohtonene, R., and Oksanen, J., Effects of reindeer grazing on understorey vegetation in dry *Pinus sylvestris* forests, *J. Vegetation Sci.,* 6, 523, 1995.

Weih, M., Seasonality of nutrient availability in soils of subarctic mountain birch woodlands, Swedish Lapland, *Arctic Alpine Res.,* 30, 19, 1998.

Wielgolaski, F.E., Fennoscandian tundra, in *Polar and Alpine Tundra,* Wielgolaski, F.E., Ed., Ecosystems of the World, 3, Amsterdam, 1997, p. 27.

Section III

Methods

12 An Approximation of the Organic Carbon Content of the Soils of Alaska's Arctic Tundra

L.R. Everett and K.R. Everett*

CONTENTS

INTRODUCTION

During the last several decades a number of estimates have been developed in an attempt to quantify the amount of carbon storage in various components of the global terrestrial environment (Schlesinger, 1977, 1980; Woodwell et al., 1978; Ajtay et al., 1979; Post et al, 1982; Eswaran et al., 1995, among others). Such estimates are critical in defining the potential sources and sinks for carbon in the global carbon cycle as it is today or as it may be altered by global change. As with all such estimates, there is a significant range depending on how the areas are delineated: extent of individual biomass; standing crop (volume of carbon per unit area) and how it is partitioned between biomass; above- and/or belowground and dead organic matter which may include standing dead and litter or only organic and mineral soil horizons. Substantial sources of error lie in the lack of detailed ground truth because of the requirement of extrapolation of data over extended areas. The lack of ground truthing is probably of greatest concern in the estimation of soil organic carbon (SOC) because of variation in soil C content and bulk density.

The following summary focuses on the quantity of SOC, estimates of which, on a global scale, range between 1515 and 1552 Pg (Whittaker and Likens 1973, 1975; Woodwell et al., 1978; Schlesinger, 1980; and cited in Miller, 1981). Current global change scenarios (Mitchell et al, 1990; Post et al, 1982; Billings, 1989; Oechel et al., 1993) indicate the polar regions, especially the north, will see the largest increase in mean annual temperatures. The past several decades have already seen an increase of approximately 1°C per decade (Oechel et al., 1993) not only in much of Alaska, but likewise in northwestern Canada and Siberia.

It is within the northern, permafrost-dominated ecosystems (Tundra and Taiga) that significant volumes of SOC are sequestered in cold, commonly saturated, anaerobic soils or soils which are

---

* Deceased.

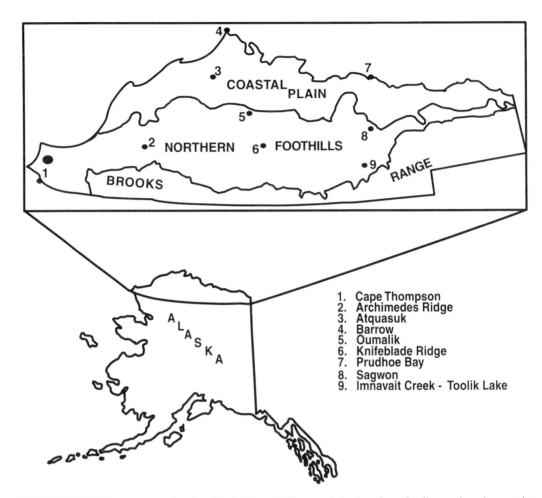

FIGURE 12.1 Reference map showing Alaska's North Slope and the location of soil mapping sites on the Coastal Plain and Foothills region used for carbon determination.

perennially frozen below the shallow (0.2–1.5 m), seasonally thawed, soil active layer. In 1981, Miller indicated this amount was approximately 10% of the total globally sequestered SOC. While more recent work by Tarnocai and Smith (1992) and Whalen et al. (1996) suggest the total globally sequestered amount in these northern ecosystems is closer to 27%. It is for this reason that the Tundra and Taiga ecosystems have been regarded traditionally as carbon sinks, and under the various scenarios of global warming may well be converted to carbon sources (CO_2 and CH_4) contributing to the greenhouse effect (Oechel, 1993).

Kane et al. (1991) studied the thermal characteristics of soils at Imnavait Creek, a Foothills site near the Brooks Range (Figure 12.1). This study indicated that an increase in mean annual air temperature of 2–4°C would result, over a 50-year period, in the summer active layer deepening by ~40 cm, thus exposing significant volumes of now perennially frozen organic carbon to anaerobic and/or aerobic decomposition. Chapin and Shaver (personal communication, 1994) have noted recent changes in vegetation composition within the Foothills area and attributed these changes to increased summer temperature. Consequently, it is important and necessary to determine: (a) if the Foothills tundra is a net source of greenhouse gases, (b) the potential carbon sequestered in the active layer and at least the upper 30-40 cm of the perennially frozen soil, and (c) the carbon of the northern Foothills and Coastal Plain in northern Alaska as one unit.

Estimates of SOC for extensive, often remote, areas such as northern Alaska or the circumarctic area in general are often based on widely separated point measurements. These measurements are then extrapolated regionally to similar vegetation and/or land unit types (Moraes and Khalil, 1992; Whiting et al., 1987). It is only in agricultural regions that sufficient closely spaced analyses can be related to detailed soils maps, e.g., federal and state soil surveys in the contiguous states. This chapter refines the estimates of SOC for northern Alaska using a series of detailed small-scale soils (geo-botanical) studies (e.g., Everett 1981; Everett and Parkinson, 1977; Parkinson, 1978; Walker et al. 1980; Walker and Walker, 1996; Figure 12.1), as well as various unpublished contract reports (Walker et al., 1987), analytical databases (Everett et al., 1981; J. Brown, 1965; Parkinson, 1978), and maps that are then related to a more generalized soil map covering Northern Alaska (USDA, 1979).

SAMPLING AND ANALYSES

Individual soil profiles for each soil map unit recognized within a particular unit were described and characterized. Numerous profiles were described and sampled over a period of years with some sites, such as those with polygons, having several cores taken from them. Typically those sites that were located on polygon landforms had three cores taken: one from the rim, one from the center, and one from the trough. A site description and complete pedon description were done for each profile. Most sampling was done using a 3" coring device to a depth of 1 m. Landforms and vegetation were also identified. Sampling was done in late July to early August to coincide with the time when the active layer depth was at its greatest.

Organic carbon content was determined for diagnostic horizons, i.e., A, O, B, C (including frozen materials below the active layer). This was done by dry combustion and corrected for carbonates (Total Organic Carbon – 0.12 x Soil Inorganic Carbon = SOC). Soil classification follows *Soil Taxonomy* (1975) and horizon nomenclature follows Attachment 1 of the revised *Soil Survey Manual* 430-v-SSM (May, 1981). The values are reported as percent organic carbon. Horizons were combined as A and/or O, B, and C, frozen (Bf, Cf, Of). All calculations include bulk density and that value depends upon the decomposition state of the organic material. If no value is reported a median value is assumed using published relationships between decomposition state and bulk density (Boeltier, 1969; Everett, 1983).

Bulk density measurements were taken from each horizon within the individual pedons. Cores of a known volume were taken and each core was then placed into a tared moisture (tin) can. A numbering system was used on the tins to identify the pedon as well as the horizon within the pedon. Each tin was sealed, kept cool, and shipped back to Ohio State University for immediate processing. All samples were then oven-dried at 105°C, reweighed, and the bulk density calculated according to standard USDA procedure (Soil Survey Staff, 1996) and reported on an oven-dry basis.

Calculation took the following form:

> Soil type (e.g., Cryaquept); mean horizon thickness (A + O) in meters $\times$ % organic carbon content = meters of SOC $\times$ combined area of the soil type = M^3 SOC $\times$ 1 $\times$ 10^6 = cm^3 SOC $\times$ bulk density (g cm^3) = g $\div$ 1000 = kg OC $\div$ combined area in meters = kg M^{-2}. The same calculation is repeated for B + C, horizons and frozen layers. Calculations are extended to a regional scale using the exploratory soil map (1979). Mean organic and horizon thickness values were used for specific taxonomic units.

RESULTS AND DISCUSSION

ACTIVE LAYER CARBON CONTENT

The mean value for stored carbon in the Foothills Province of Northern Alaska is 19.7 kg C* m^{-2}. This compares favorably with the estimate for world tundra of 21.60 kg C* m^{-2} (Schlesinger, 1977),

TABLE 12.1
Soil Organic Matter Content

	Active layer A/O	B/C	% of area mapped	Upper 30 cm	% of area mapped
	kg C m^{-2}			kg C m^{-2}	
Foothills					
Cape Thompson 111 km^2	11.9	8.1	92	38.7 (3.5)[c]	9
Sagwon 5.3 km^2	12.5	7.5	80	18.0 (0.54)	3
				76.1 (3.80)	5
Imnavait Creek 2.1 km^2	17.0	6.8	77	95.0 est. (13.3)	14
Atqasuk 32.8 km^2	14.5	4.5	89	13.1 (2.4)	18
Archimedes Ridge 38.4 km^2	7.8	6.9	87	1.98 (2.2)	55
Mean	13.7	6.1	85	40.5 (6.9)	17
Toolik[a] 1.1 km^2	25.4	2.9	—	—	—
Coastal Plain					
Barrow	33.5	9.2	90 est.	41.0	80 est.
Prudhoe Bay	13.6	[b]	72	[d]	69
Mean	24.5	—	81	29.6	74

[a] Values for Toolik Lake are based on transects only and over-represent the amount of soil carbon in the geographically broader landscape.
[b] Data analysis are incomplete.
[c] Weighted by % of map area represented.
[d] At selected sites data from cores extending well below 30 cm (33) and Everett data (field notes) indicates carbon storage values similar to or somewhat higher than Barrow (60.12 kg C* m^{-2}).

and 21.80 kg* m^{-2} for the Tundra Life Zone described by Billings (1989). The values in Table 12.1 are probably significantly lower than the 29.0 kg C* m^{-2} given in the report on circumarctic carbon storage (Miller, 1981). In the Foothills landscapes the quantity of carbon in the active layer is related closely to the surface geometry and hence to vegetation and soil moisture conditions. This is shown in Figure 12.2 and Table 12.2 that represents a 2.1 km^2 surface in the Imnavait Creek region and is representative of much of the Foothills Province. The dry and probably much of the poorly drained area may well be the source of carbon to the atmosphere. Some researchers (Barnett and Schell, 1991; Schell and Barnett, 1990; Oechel and Billings, 1992) argue that at least the Foothills region of the Alaskan tundra is a source of carbon now after being a sink for perhaps the last 2000 years. These opinions are based upon ^{14}C sequences (Oswood et al., 1996) and studies of net primary productivity (Murray et al., 1989).

Carbon storage in the active layer of the Coastal Plain sites (mean 24.5 kg C* m^{-2}, Table 12.1) is substantially greater than the 13.4 kg C* m^{-2} given in the wet sedge tundra (Miller, 1981) report. The value for Barrow, however, is close to 33.8 kg C* m^{-2} (Gersper et al., 1980).

There is probably little argument that the Coastal Plain region is a net sink for carbon, but supporting data are few (Schell and Ziemann, 1983). Chapin et al. (1980) calculated a detailed carbon capture/loss budget for the wet tundra at Barrow and found a small positive balance of approximately 0.11 kg m^{-2} yr^{-1}. Most of the carbon in the soil active layer A/O horizons in both Coastal Plain and Foothills Provinces has accumulated in the last 4000 to 5000 years (Everett, 1979,1981; Walker et al, 1980). However, carbon moved into the B/C horizons as a result of cryogenic or pedogenic processes may be older.

Just as in the Foothills, active layer carbon content is closely related to surface geometry that reflects the thaw lake cycle (Britton, 1967; Walker et al, 1980), ice wedge formation and evolution (Everett, 1979), and wind transport of sand and loess.

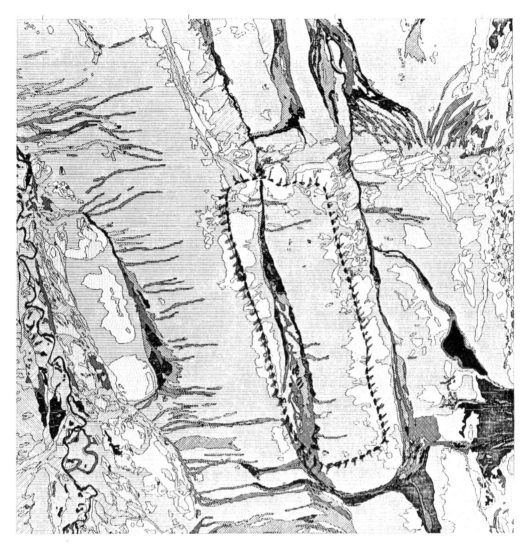

FIGURE 12.2 Excerpt from a vegetation map of the headwater basin of Imnavait Creek, defined by dashed line (Site 9 in Figure 12.1) and adjacent areas of the upper Kuparuk River Basin. (From Walker, D.A., Webber, P.J., Lederer, N.D., and Walker, M.D., U.S. Army Cold Regions Research and Engineering Laboratory, Hanover, NH, 1987.)

TABLE 12.2
Soil Organic Carbon (SOC) Content for Imnavait Creek, Alaska

Drainage	A/O	B/C	Total
	kg C m^{-2}		
Poor	9.6[a]	4.7	14.3
Wet	n.a.	n.a.	30.4
Dry	10.5	8.8	19.3
			Mean = 21.3

[a] Values developed from data in Walker et al., 1987.

TABLE 12.3
Total Soil Organic Carbon in the Active Layer and
Upper Permafrost Zone in Northern Alaska

	Area (10⁶ ha)	SOC pools	
		Active layer[a] (Pg)	Permafrost (Pg)
Foothills	11.3	2.24	0.78
Coastal Plain	4.3	1.06	2.40
Circum Arctic[b]	3180	55.13 (Total SOC pool)	

[a] Includes A/O and B/C horizons (Table 12.1).
[b] Wet sedge, tussock, and low shrub tundra (modified from Miller, 1981.)

PERENNIALLY FROZEN CARBON

Because soils studied in permafrost regions were largely confined to the active (seasonally thawed) layer, relatively little information is available on the SOC in the permafrost. A 30 cm thickness was chosen for the permafrost carbon determination because most of the existing data are in this depth range and it is the range most likely to be influenced by climatic change (warming). The amount and distribution of these reserves depends, as does that in the active layer, on the geometry and geomorphic history of the landscapes.

It is clear from (Table 12.1) that in the Foothills large volumes of presently frozen carbon are found in a relatively small percent of the landscape: the wetter areas or Histosols (see also Figure 12.2, Table 12.2). Whereas on the Coastal Plain considerably larger volumes of frozen carbon occupy a greater area of the landscape, as a greater percentage of this physiographic province is occupied by Histosols.

In order to extrapolate the pedon data to larger landscape units The Exploratory Soil Survey of Alaska (1979) was used for the Foothills and Coastal Plain physiographic provinces (Wahrhaftig, 1965) together with enhanced satellite imagery in specific areas. The values given in Table 12.3 were derived from area summaries of soils in those regions known to have significant organic carbon contents (the percent organic carbon and bulk densities used in the site surveys [Table 12.1] were applied in the calculation of total carbon).

Although the Coastal Plain accounts for only 28% of the Alaskan Arctic slope it contains approximately the same volume of SOC as the Foothills. However, in the Foothills 76% of the SOC is in the active layer while in the Coastal Plain only 34% is in the active layer, with the rest found in the permafrost layer. In aggregate the two physiographic regions account for approximately 12.0% of the carbon sequestered in the circumarctic tundra.

CONCLUSIONS

Significant volumes of organic carbon are sequestered in soils of the Tundra and Tiaga ecosystems. It has been estimated that approximately 27% of the total global soil carbon is in the perennially frozen soils of the circumarctic Tundra (Whalen et al., 1996; Post et al., 1982). This zone has been traditionally regarded as a sink for carbon in the global cycle (Billings, 1987; Oberbauer, 1996) as plant organic remains are slowly incorporated into the cold, commonly anoxic soils and/or are perennially frozen. More recent work by Oechel et al. (1993) indicates the area has definitely changed from a sink to a source. Global climate models of the last decade predict that the temperature increase resulting from climatic warming will be largest in the arctic regions of the northern hemisphere. This creates the potential for a progressively deeper seasonal thaw allowing stored soil organic carbon (SOC) to be more susceptible to release into the atmosphere as CO_2 or CH_4, increasing the greenhouse gases thus amplifying the greenhouse effect. Accurate assessment of this potential problem depends upon the knowledge of both the volume of SOC which is sequestered in the northern ecosystems and its current dynamics. These questions can be answered in part by careful analysis of the soils and their distribution on northern landforms. Such analysis indicates the permafrost (perennially frozen) landscape of northern Alaska contains approximately 4.5 Pg or 8% of the circumarctic Tundra's SOC and that its distribution is closely tied to surface geometry.

The SOC sequestered in the active layer and upper 30-40 cm of the underlying perennially frozen soil represents 12.0% of the total SOC of the circumarctic tundra. It is this resource that will be vulnerable to decomposition and conversion to CO_2 and CH_4 as a result of increased active layer thickness and changes in moisture balance due to projected global climate change over the next 50+ years.

It should be noted that in addition to the 1979 USDA report (Soil Survey Staff, 1979) a series of GIS maps exist for the North Slope region of Alaska (North Slope Borough) that can be used for additional refinements in estimates of soil carbon.

ACKNOWLEDGMENTS

This chapter represents the final work done by Dr. K.R. Everett prior to his death in 1994. Dr. Everett had done research in the Arctic for nearly 30 years with the majority of his work involving numerous aspects of research on the permafrost-rich soils of northern Alaska and Russia. Had this paper been published several years ago, as he originally intended, it would have been one of the first to recognize that more detailed studies were needed to derive a more accurate assessment of the amount of carbon that exists within the Northern Alaska ecosystems. With all of the global change scenarios that were developing he felt that this assessment was necessary to provide a more realistic idea of the potential impact of carbon (C) release into the atmosphere and whether or not the Arctic was changing from a net sink to a net source. It has been with the assistance and encouragement of several of Dr. Everett's colleagues that I have attempted to fill in the missing gaps and present this paper for publication.

Numerous agencies provided support for studies that have contributed data between 1979 and 1992 on which this work is based. They include the U.S. Army, U.S. Army Research Office, Research Triangle Park, N.C.; U.S. Army Cold Regions Research and Engineering Laboratory, Hanover, N.H.; National Science Foundation, Division of Polar Programs and the U.S. Department of Energy, R4D Program. This chapter is Byrd Polar Research Center Contribution Number 1159.

REFERENCES

Ajtay, G.L., Ketner, P., and Durigneaud, P., Terrestrial primary production and phytomass, in *The Global Carbon Cycle*, Bolin, B., Kenye, S., and Ketner, P., Eds., SCOPE B, John Wiley & Sons, New York, 1979, p. 129.

Barnett, B. and Schell, D.M., Peat carbon accumulation rates in Arctic Alaska, abstr, AAAS Conf., Fairbanks, Alaska, 1991.

Billings, W.D., Carbon balance of Alaska tundra and Taiga ecosystems: past, present and future, *Q. Sci. Rev.*, 6, 165, 1989.

Boelter, B.D., Physical properties of peats as related to the degree of decomposition, *Soil Sci. Soc. Am. Proc.*, 33, 606, 1969.

Britton, M.E., Vegetation of the arctic tundra, in *Arctic Biology*, 2nd ed., Hansen, H.P., Ed., Oregon State University Press, Corvallis, OR, 1967.

Brown, J., Radiocarbon dating, Barrow, Alaska, *Arctic*, 18, 36, 1965.

Chapin, F.S., III and Shaver, G.R., Changes in soil properties and vegetation following disturbance of Alaskan arctic tundra, *J. Appl. Ecol.*, 18, 605, 1981.

Chapin, F.S., Miller, P.C., Billings, W.D., Coyne, P.I. et al., *Arctic Ecosystem: the Coastal Tundra at Barrow, Alaska,* Brown, J., Miller, P.C., Tieszen, L.L., and Bunnell, T.L., Eds., Dowden, Huchinson & Ross, Inc., Stroudsburg, PA, 1980, pp. 458-482.

Eswaran, H., Van den Berg, E., Reich, P., and Kimble, J., Global soil carbon resources, in *Soils and Global Change, Advances in Soil Science,* Lal, R., Kimble, J., Levine, E., and Stewart, B., Eds., CRC Press, Boca Raton, FL, 1995, p. 27.

Everett, K.R., Histosols, in *Pedogenesis and Soil Taxonomy II. The Soil Orders,* Wilding, L.P., Smeck, N.E., and Hall, G.F., Eds., Elsevier, New York, 1983, chap. 1, p. 1.

Everett, K.R., Soil-landscape relations at selected sites along environmental gradients in northern Alaska, Rep. to the U.S. Department of the Army, Research Office, Triangle Park, NC, 1983.

Everett, K.R., Evolution of the soil landscape in the sand region of the Arctic Coastal Plain as exemplified at Atkasook, Alaska, *Arctic,* 32, 207, 1979.

Everett, K.R. and Parkinson, R.J., Soil and landscape associations, Prudhoe Bay area, Alaska, *Arctic Alpine Res.,* 9, 1, 1977.

Gersper, P.L., Alexander, V., Barkley, S.A., Barsdate, R.J., and Flint P.S., The soils and their nutrients, in *An Arctic Ecosystem, The Coastal Tundra at Barrow, Alaska,* Brownin, J., Miller, P.C., Tiezen, L.L., and Bunnell, F.L., Eds., Dowden, Hutchinson & Ross, Stroudsburg, PA, 1980, chap. 7, p. 219.

Kane, D.L., Hinzman. L.D., and Zarling, J.P., Thermal response of the active layer to climatic warming in a permafrost environment, *Cold Reg. Sci. Technol.,* 19, 111, 1991.

Miller, P.C., Ed., Carbon Balance in Northern Ecosystems and the Potential Effect of Carbon Dioxide-Induced Climatic Change, Carbon Dioxide Effects Research and Assessment Program Report 015, U.S. Department of Energy, Washington, D.C., 1981.

Mitchell, J.F.B., Manake, S., Meleshenko, V., and Tolioka, T., Equilibrium climate change and its implications for the future, in *Climate Change, IPCC Scientific Assessment,* Cambridge University Press, New York, 1990, p. 135.

Moraes, F. and Khalil, M.A.K., The effect of permafrost on atmospheric concentrations of CH_4, CO_2, H_2O, and CO, Abstr. AGU Spring Meet., Boston, 822A(562), 1992.

Murray, K.J., Tenhunen, J.D., and Kummerow, J., Limitations on Sphagnum growth and net primary production in the Foothills of the Philip Smith Mountains, Alaska, *Oecologia,* 80, 256, 1989.

Oberbaurer, S.F., Cheng, W., Gillespie, C.T., Ostendorf, B., Sala, A., Gebauer, R., Virginia, R.A., and Tenhunen, J.D., Landscape patterns of carbon dioxide exchange in Tundra ecosystems, in *Landscape Function and Disturbance in Arctic Tundra, Ecological Studies,* Reynolds, J.F. and Tenhunen. J.D., Eds., Springer-Verlag, Berlin, 1996, p. 223.

Oechel, W.C. and Billings, W.D., Effects of global change on the carbon balance of Arctic plants and ecosystems, in *Arctic Ecosystems in a Changing Climate: An Ecophysiological Perspective,* Chapin, F.S., III et al., Eds., Academic Press, New York, 1992, p. 139.

Oechel, W.C., Hastings, S.J., Vourlitis, G., Jenkins, M., Reichers, G., and Grulke, N., Recent change of Arctic tundra ecosystems from a net carbon dioxide sink to a source, *Nature,* 361, 520, 1993.

Oswood, M.W., Irons, J.G., III, and Schell, D.M., Dynamics of dissolved and particulate carbon in an arctic stream, in *Landscape Function and Disturbance in Arctic Tundra, Ecological Studies,* Reynolds, J.F. and Tenhunen, J.D., Eds., Springer-Verlag, Berlin, 1996, p. 276.

Parkinson, R.J., Genesis and classification of Arctic Coastal Plain soils, Prudhoe Bay, Alaska, Institute of Polar Studies, Rep. No. 68, Ohio State University, Columbus, OH, 1978.

Post, W.M., Emanuel, W.R., Zinke, P.J., and Stangenberger, J., Soil carbon pools and world life zones, *Nature,* 298, 156, 1982.

Schell, D.M. and Barnett, B., Peat accumulation rates in Arctic Alaska: responding to recent climate change?, Abst. Int. Conf. Role of Polar Regions in Global Change, June 11-15, University of Alaska, Fairbanks, 1990, p. 117.

Schell, D.M. and Ziemann, P.J., Accumulation of peat carbon in the Alaskan Arctic Coastal Plain and its role in biological productivity, *Permafrost,* 4th Int. Conf. Proc., Fairbanks, Alaska, National Academy of Sciences Press, Washington, D.C., 1983, p. 1105.

Schlesinger, W., Carbon balance in terrestrial detritus, *Annu. Rev. Ecol. Syst.,* 8, 51, 1977.

Schlesinger, W., The world carbon pool in soil organic matter: a source of atmospheric CO_2?, *The Role of Terrestrial Vegetation in the Global Carbon Cycle: Methods for Appraising Changes,* Woodwell, G.M., Ed., John Wiley & Sons, New York, 1980.

Soil Survey Staff, Soil Survey Laboratory Methods Manual, Version 3.0, NRCS, Lincoln, NE, 1996.

Soil Survey Staff, Revised Soil Survey Manual, 430-V.SSM, Soil Conservation Service, U.S. Department of Agriculture, Washington, D.C., 1981.

Soil Survey Staff, Exploratory Soil Survey of Alaska, Soil Conservation Service, U.S. Department of Agriculture, Washington, D.C., 1979.

Soil Survey Staff, Soil Taxonomy, a Basic System of Soil Classification for Making and Interpreting Soil Surveys, Handbook No. 436, Soil Conservation Service, U.S. Department of Agriculture, Washington, D.C., 1975.

Tarnocai, C.S. and Smith, C.A.S., The formation and properties of soils in the permafrost regions of Canada, in First Int. Conf. Cryopedology Workshop Proc., Puschino, Russia, 1992, p. 21.

Wahrhaftig, C., Physiographic Divisions of Alaska, Prof. Pap. 482, U.S. Geological Survey, Washington, D.C., 1965.

Walker, D.A. and Walker, M.D., Terrain and vegetation of the Imnavait Creek, in *Landscape Function and Disturbance in Arctic Tundra: Ecological Studies 120,* Reynolds, J.F. and Tenhunen, J.D., Eds., Springer-Verlag, Berlin, 1996, p. 73.

Walker, D.A., Everett, K.R., Webber, P.J., and Brown, J., Geobotanical Atlas of the Prudhoe Bay Region, Alaska. Rep. 80-14, U.S. Army Cold Regions Research and Engineering Laboratory, Lincoln, NE, 1980.

Walker, D.A., Webber, P.J., Lederer, N.D., and Walker, M.D., (Unpublished) Terrain and vegetation of the Department of Energy R4D Research Site, Imnavait Creek, Alaska. Classification and mapping, draft manuscript, 1987.

Whalen, S.C., Reeburgh, W.S., and Reimers, C.E., Control of tundra methane emission by microbial oxidation, in *Landscape Function and Disturbance in Arctic Tundra, Ecological Study,* Reynolds, J.F. and Tenhunen, J.D., Eds., Springer-Verlag, Berlin, 1996, p. 257.

Whiting, G.J., Crill, P.M., Cheautou, J.P., Bartlett, K.B., and Bartlett, D.B., The effect of light, dark, and CO_2 on short-term measurements of methane flux, *EOS Trans.,* 68, 1225, 1987.

Whittaker, R.H. and Likens, G.E., Carbon in the biota, Brookhaven Symp., Biology 24, 1973, p. 281.

Whittaker, R.H. and Likens, G.E., The biosphere and man, in *Primary Production of the Biosphere*, Lieth, H. and Whittaker, R.H., Eds., Springer-Verlag, New York, 1975.

Woodwell, G.M., Whittaker, R.H., Reiners, W.A., and Likens, G.E., The abiota and the world carbon budget, *Science,* 199, 141, 1978.

13 Carbon Storage and Accumulation Rates in Alpine Soils: Evidence from Holocene Chronosequences

J.G. Bockheim, P.W. Birkeland, and W.L. Bland

CONTENTS

INTRODUCTION

Alpine ecosystems (i.e., those ecosystems above altitudinal treeline) comprise 2.9×10^{12} m² or about 2.2% of the biosphere. Compared with other soils, very little is known about carbon storage in alpine soils. Because of low decomposition rates and an allocation of biomass to the root systems of alpine plants, alpine soils could constitute a substantial reservoir of organic C. Concern over the possible effects of global warming has sparked interest in C dynamics of arctic and alpine ecosystems. There is considerable uncertainty as to whether these ecosystems will act as a CO_2 sink, thereby ameliorating the problem, or as a CO_2 source, thereby exacerbating the problem (Billings, 1987; Oechel et al., 1993).

Soil chronosequence studies have the potential to determine whether or not there have been changes in C accumulation rates during the Holocene and late Wisconsin Period (Schlesinger, 1990; Harden et al., 1992). Soil chronosequences have been examined in numerous alpine areas, including the Rocky Mountains (Mahaney, 1974, 1978; Birkeland et al., 1987) and Sierra Nevada (Shroba and Birkeland, 1983) of the western U.S.A., the Southern Alps of New Zealand (Birkeland, 1984; Rodbell, 1990), Mt. Everest in Nepal (Birkeland et al., 1989), and Mt. Elbrus in Russia (Gennadiyev and Sokolova, 1977; Gennadiyev, 1979).

1-56670-459-6/00/$0.00+$.50
© 2000 by CRC Press LLC

In this chapter, C pools in alpine soils are summarized and compared with data from other life zones, C accumulation rates for different periods within the Holocene are estimated, and the issue as to whether or not alpine soils will act as a source or a sink for atmospheric CO_2 during global warming is addressed.

METHODS AND MATERIALS

STUDY AREAS

The soil chronosequences chosen for inclusion in this study are all of the post-incisive type (Vreeken, 1975) and are summarized in Table 13.1. Most of the chronosequences are in areas strictly defined as alpine (Zwinger and Willard, 1972); however, the chronosequence on Mt. Cook, New Zealand is in the alpine-subalpine ecotone. The soils at all of the sites except Mt. Cook are derived from till or rock-glacier deposits composed of granites and gneisses; most of the soils have a thin (3–29 cm) loess mantle.

The soils constitute four general age groups (Birkeland et al., 1989): late Holocene (up to several centuries old), late-to-middle Holocene (1000 to 2000 years old), middle Holocene (about 3500 to 5500 years), and early Holocene to late Pleistocene (about 8000 to 10,000 years old). These ages are based partially on radiocarbon dates but mostly from relative-age data such as lichenometry, dendrochronology, rock weathering, and soil properties (Birkeland et al., 1989).

SOIL SAMPLING AND LABORATORY ANALYSIS

All of the soils were described using methods of the Soil Survey Staff (1993). Samples were collected from each horizon and passed through a 2-mm screen. The skeletal fraction was either weighed, or the volume of the horizon accounted for by coarse fragments was estimated. Organic C was determined using the Walkley-Black chromic acid digestion procedure (Nelson and Sommers, 1982). The work to this point was carried out by the authors cited in Table 13.1.

Profile thicknesses ranged from 18 cm on the youngest surface to 100 cm on the oldest surface. Because bulk density was not reported in any of the studies, bulk density was estimated from percent organic C using Equation 13.1 of Alexander (1989).

Profile C density (kg m^{-2}) was determined by taking the product of horizon thickness, percent organic C, and estimated bulk density; these values were summed for the entire profile. Corrections were made for coarse fragments, which commonly accounted for about 50 to 75% of the profile volume or mass in the non-loessal materials.

The long-term rate of accumulation of profile C was predicted using a reservoir model (Jenny et al., 1949; Harden et al., 1992) that incorporates a zero-order input of C and a first-order rate of loss:

$$dC/dt = I - kC \tag{13.1}$$

where C is the mass of organic C per unit area of land surface in the fine-earth fraction (i.e., profile C density), I is the rate of organic C input, and k is the first-order rate constant for organic matter oxidation. The time integral of this equation is

$$C = I/k \times (1 - e^{-kt}) = C_e(1 - e^{-kt}) \tag{13.2}$$

where C_e is the long-term equilibrium C soil inventory. The model assumes a constant rate of C input.

Interval-specific accumulation rates were calculated from the difference in profile C density between two soil members of the same chronosequence, divided by the difference in their times of exposure (Schlesinger, 1990).

TABLE 13.1
Soil Forming Factors for the Areas from Which Chronosequences Were Selected

Location	Latitude	Longitude	Altitude range (m)	Mean annual precipitation (mm)	Mean annual temperature (°C)	Vegetation	Parent materials	Approx. age span (yr)	Dating method[a]	Ref.
Wind River Mountains, Wyoming	43°45'N	110°15'W	3400–3500	1000	–3.5	Alpine tundra, lichens	Granitic, gneiss; Till, loess	~100–10,000	L, W, S	Mahaney, 1978; Birkeland et al., 1989
Front Range, Colorado	40°03'N	105°36'W	3400–3800	1020	–3.8	Alpine tundra	Granitic, gneiss; Till, rock–glacier deposits	~100–10,000	L, W, S, C	Mahaney, 1974; Birkeland et al., 1987
Mt. Cook, Southern Alps, New Zealand	43°40'S	170°05'E	600–900	4000	7 to 8	Subalpine forest, low scrub	Graywacke, low-grade metamorphics; Till, loess	~100–9,000	W, C, S	Birkeland, 1984; Birkeland et al., 1989
Ben Ohau Range, Southern Alps, New Zealand	43°55'S	170°05'E	1200–1900	1000	4	Tussock grasses	Graywacke, low-grade metamorphics; Till, loess	~100–9,000	W, C, S	Birkeland, 1984; Birkeland et al., 1989

[a] Dating method: L = lichenometry; W = rock weathering ratios, rinds, pits; C = radiocarbon; S = soil properties; D = dendrochronology.

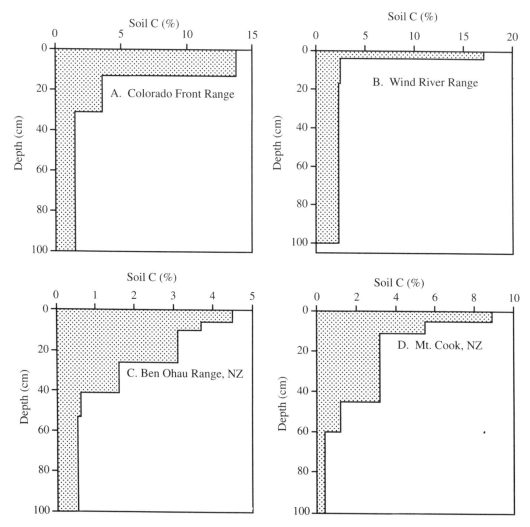

FIGURE 13.1 Depth distribution of organic C in early Holocene alpine soils representative of (A) the Colorado Front Range (Birkeland et al., 1987), (B) the Wind River Range (Birkeland et al., 1989), (C) the Ben Ohau Range, New Zealand (Birkeland, 1984), and (D) Mt. Cook, New Zealand (Birkeland, 1984).

RESULTS AND DISCUSSION

ORGANIC C DISTRIBUTION IN ALPINE SOILS

In mature (3000 to 10,000-year-old) alpine soils, organic C generally is concentrated in the surface 40 cm and decreases sharply with depth (Figure 13.1). This depth corresponds with the rooting depth of alpine plants which contribute organic matter by turnover of fine roots (Bliss, 1956; Webber, 1974). Although there is considerable variation in the data, the amount of organic C in the surface mineral horizon increases with time of exposure for a given chronosequence. As an example, the relationship between organic C in the surface horizon and time of exposure for alpine soils in the Colorado Front Range is illustrated in Figure 13.2 using Equation 13.2.

PROFILE C DENSITY FOR ALPINE SOILS

Profile C densities range from <1 kg m^{-2} for late Holocene soils to 12 ± 1.4 kg m^{-2} for early Holocene to late Pleistocene soils (Table 13.2). Although time is an important factor controlling

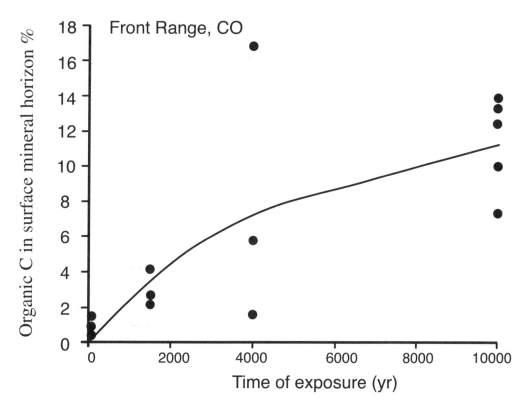

FIGURE 13.2 Relation between organic C concentration of the surface mineral horizon and time of exposure for alpine soils in the Colorado Front Range (calculated from data by Birkeland et al., 1987 and Mahaney, 1974, and Equation 13.2).

the distribution of C in alpine soils, microclimate is also important. For example, Burns (1980) illustrated the importance of topographic position on both snow distribution and soil formation in the alpine areas of the Colorado Front Range. He divided the alpine into seven topographic-climatic units, including extremely windblown, windblown, minimal snow cover, early melting snowbank, late melting snowbank, perennial snowbank, and wet meadow.

Profile C density was calculated for 19 soils representing 6 of the topographic units from data provided by Burns (1980). The soils occur on surfaces of unknown age but, based on profile C density, are likely of middle to early Holocene age. There is a threefold greater amount of profile C in wet meadow soils in topographic depressions than in dry soils on windblown ridges (Figure 13.3). These trends likely are related to differences in available moisture and its effect on productivity of alpine plant communities and on decomposition rates (Webber, 1974; Burns, 1980).

The relation between profile C density and time of exposure was tested using the reservoir model. The equilibrium C content of the soils can be ranked: Mt. Cook (17.0 kg m^{-2}) > Colorado Front Range (15.0) > Ben Ohau Range (12.6) > Wind River Mountains (5.3 kg m^{-2}) (Table 13.3; Figure 13.4). These rankings may be related to climatic differences among the sites. Of the sites included in the study, Mt. Cook has the greatest current mean annual precipitation and temperature (Table 13.1). However, the remaining C_e values were not as obviously related to precipitation or temperature.

RATES OF PROFILE C ACCUMULATION IN ALPINE SOILS

The best-fit annual rate of organic C input (I) in the soils was calculated using the reservoir model and ranged from 1.8 g m^{-2} yr^{-1} in the Wind River Range to 10.0 g m^{-2} yr^{-1} at Mt. Cook (Table 13.3).

TABLE 13.2
Soil Carbon Density and Accumulation Rates for Alpine Chronosequences

Profile No.	Approximate time of exposure (yr)	Soil C density (kg m^{-2})
Wind River Mountains, Wyoming (Mahaney, 1978)		
TB15	100	0.31
TB5	1,500	4.1
TB6	1,500	3.2
TB14	10,000	3.7
Wind River Mountains, Wyoming (Birkeland et al., 1989)		
S48	100	0.54
S19	1,500	1.2
S11	4,000	3.6
S5a	10,000	7.2
Front Range, Colorado (Birkeland et al., 1987)		
CO2	100	0.10
CO1	1,500	1.1
CO3	4,000	5.8
CO4	10,000	8.2
CO5	10,000	9.7
CO6	10,000	12.6
CO7	10,000	11.5
CO8	10,000	9.2
Front Range, Colorado (Mahaney, 1974)		
A1	100	0.24
H3	100	0.86
A11	1,500	1.3
H1	1,500	1.9
T2	4,000	5.0
H5	5,000	11.2
T3	5,000	4.5
Mt. Cook, New Zealand (Birkeland, 1984)		
C13	100	0.90
C14	500	2.9
C8	1,200	9.5
C19	5,500	16.5
C1	8,000	19.6
C20	8,000	21.8
C18	9,000	13.0
Ben Ohau Range, New Zealand (Birkeland, 1984)		
B11	100	0.67
B9	3,000	2.1
B19	4,000	12.6
B10	4,000	12.9
B27	4,000	13.3
B4	9,000	8.8
B5	9,000	13.9

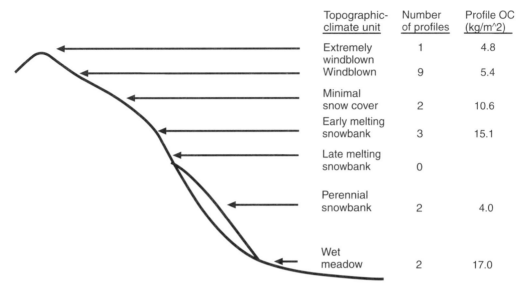

Topographic-climate unit	Number of profiles	Profile OC (kg/m^2)
Extremely windblown	1	4.8
Windblown	9	5.4
Minimal snow cover	2	10.6
Early melting snowbank	3	15.1
Late melting snowbank	0	
Perennial snowbank	2	4.0
Wet meadow	2	17.0

FIGURE 13.3 Relation between profile carbon content in alpine soils and topographic position in the Colorado Front Range (calculated from data by Burns, 1980).

TABLE 13.3
Best-Fit Values of Rate of Carbon Input (I) and Inverse of the Rate Constant (k) for Soil C Accumulation Modeled as dC/dt = 1 – kC; the Estimated Equilibrium C (C_e) Is Found as I/k

Chronosequence	I (g m^{-2} yr^{-1})	1/k (yr)	C_e (kg m^{-2})
Wind River Mountains, Wyoming (Mahaney, 1978; Birkeland et al., 1989)	2.4	2,200	5.3
Front Range, Colorado (Mahaney, 1974; Birkeland et al., 1987)	1.8	8,570	15.0
Mt. Cook, New Zealand (Birkeland, 1984)	10.0	1,670	17.0
Ben Ohau Range, New Zealand (Birkeland, 1984)	5.2	2,140	12.6

Because the soils likely did not evolve under steady state conditions of C input and organic matter decay, interval-specific accumulation rates in profile C density were determined. The greatest interval-specific C accumulation rates generally were during the late-to-middle and middle Holocene (ca. 1500–4000 years B.P.) and the lowest rates commonly occurred during the early Holocene to late Pleistocene (ca. 8000 to 10,000 years B.P.) (Figure 13.5). The accumulation rate from the reservoir model is also shown in Figure 13.5. The accumulation rate is initially the C input rate, and then declines as oxidation of the C pool increases (Equation 13.1). This model assumes constant rates of C input and of decay (k), which were not the case.

Based on Holocene paleoenvironmental reconstruction from geological, botanical, and historical data, together with critical radiocarbon dates (Elias, 1985), the early Holocene to late Pleistocene in the central Rocky Mountains featured a cool-moist climate. The period between 4500 and 9000 years B.P. was the climatic optimum, followed by climatic deterioration from 4500 to 3000 years B.P. Climate warming occurred during the period from 3000 to 1800 years B.P., followed by cooling from 1800 years B.P. to the present. Therefore, data from the four sites included in this study suggest that the rate of accumulation of organic C in alpine soils is greatest during the warmer mid-to-late and middle Holocene and lowest in the cooler and moister early Holocene to late Pleistocene. These findings are consistent with those of Marion and Oechel (1993) who examined Holocene carbon balance in arctic tundra of northern Alaska.

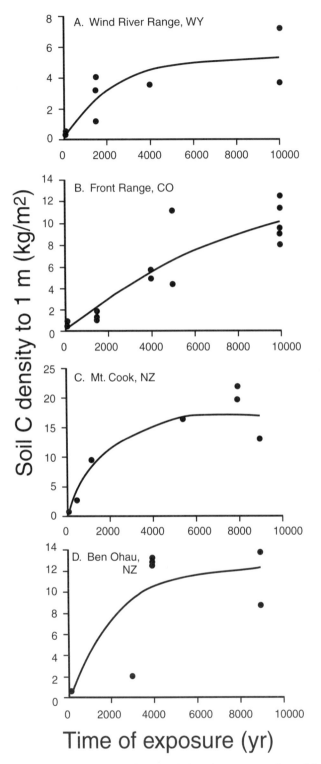

FIGURE 13.4 Long-term rate of accumulation of profile C, based on a reservoir model (Equation 13.2) and data from (A) the Wind River Range (Birkeland et al., 1989; Mahaney, 1974); (B) the Colorado Front Range (data from Birkeland et al., 1987; Mahaney, 1978); (C) the Ben Ohau Range, New Zealand (data from Birkeland, 1984); and (D) Mt. Cook, New Zealand (data from Birkeland, 1984).

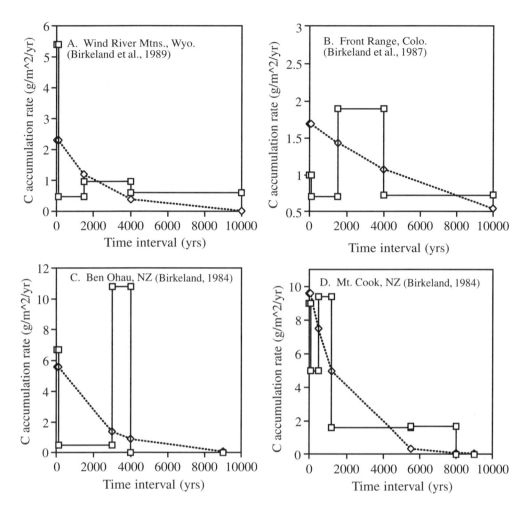

FIGURE 13.5 Interval-specific rates of profile C accumulation in alpine soils of (A) the Wind River Range, Wyoming (data from Birkeland et al., 1989); (B) the Colorado Front Range (data from Birkeland et al., 1987); (C) the Ben Ohau Range, New Zealand; and (D) Mt. Cook, New Zealand (data from Birkeland, 1984). The long-term C accumulation from Equation 13.1 is shown with dashed lines.

COMPARISON WITH OTHER STUDIES

Although the C concentrations of alpine soils often are greater than other mineral soils at an equivalent depth, organic C pools in middle Holocene to late Pleistocene alpine soils are similar to those of other mineral soils (Table 13.4). The reason for this similarity is that alpine soils contain an abundance of coarse fragments (ca. 50–75%) that "dilutes" the soil.

The long-term rate of organic carbon accumulation was calculated by dividing the profile C density (kg C m^{-2}) by the accumulation interval (yr) for soils between the ages of 3,000 and 10,000 (Schlesinger, 1990). Based on this approach, the long-term rate of C accumulation in the alpine soils included in this study ranges between 0.7 and 2.5 g C m^{-2} yr^{-1} (Table 13.4). This is comparable to the 2.4 g ± 0.70 C m^{-2} yr^{-1} value reported for 16 middle Holocene to late Wisconsin (3,000 to 10,000 years old) soils from a variety of ecosystems (Schlesinger, 1990).

Assuming that the alpine soils considered in this study are representative of those worldwide, alpine soils contain 29 Gt of C, accounting for 1.8% of the world total. Based on parameters in Table 13.3, we estimate that the net organic C accumulation in alpine tundra soils worldwide may be about 14 × 10^{12} g C yr^{-1} which comprises 3.5% of the total C stored annually by soils

TABLE 13.4
Storage and Long-Term Accumulation Rates of Organic C in Soils of
Selected Life Zones

Life zone	Mean soil C density (kg m^{-2})[a]	Long-term accumulation rate (g C m^{-2} yr^{-1})[b]
Arctic tundra, moist	11 ± 6.4	0.7 to 1.1
Alpine tundra[c]	10 ± 5.2	0.7 to 2.5
Boreal forest, moist	12 ± 8.2	0.8 to 11.7
Cool temperate forest, moist	12 ± 8.2	0.7 to 2.5
Cool temperate steppe	13 ± 9.5	2.2
Semidesert	10 ± 6.0	0.8
Tropical forest, moist	12 ± 12	2.3 to 2.5
Overall	11	2.4 ± 0.70

[a] Post et al., 1982.
[b] Schlesinger, 1990 (average for 3,000- to 10,000-year-old soils).
[c] This study (average for 2,600- to 10,000-year-old soils).

$(400 \times 10^{12}$ g C yr^{-1}; Schlesinger, 1990). The reservoir model predicts a steady-state C_e, with no net C sequestration at long times. Globally, however, alpine soils will still sequester C because not all of the soils are at equilibrium, and the model is probably incomplete.

GLOBAL WARMING AND ALPINE SOIL C POOLS

The profile C density of alpine soils increases with time of exposure (Table 13.3). However, the long-term accumulation rate of soil C depends on climate. Climatic effects are reflected by differences in rate of C input (I) and rate constant of decomposition (k) for the various chronosequences. Interval-specific C accumulation rates in the soils studied here were greatest during the late-to-middle to middle Holocene (1500–4000 years B.P.) when the climate was warmer than present; in contrast the interval specific C accumulation rates were lowest during the early Holocene to late Pleistocene (8000–10,000 years B.P.). This would seem to suggest that rates of soil organic C accumulation will increase during a climate-warming scenario. However, the effect of global warming on snowfall amount and distribution in alpine ecosystems potentially could counteract this effect.

CONCLUSIONS

Chronosequences are useful for estimating organic C pools and accumulation rates in alpine soils during time intervals of the Holocene. Data from four soil chronosequences in alpine regions were used to estimate profile storage of organic C and interval-specific C accumulation rates during subdivisions of the Holocene. Sites include the Colorado Front Range, the Wind River Range of Wyoming, and the Ben Ohau Range and Mt. Cook region of the Southern Alps in New Zealand.

Carbon storage increases with time of exposure from <1 kg C m^{-2} in late Holocene (up to several centuries old) soils to 12 ± 1.4 kg C m^{-2} in early Holocene to late Pleistocene (about 10,000 years old) soils. For soils of roughly equivalent age, C storage is dependent on topographic position in relation to snow accumulation and soil moisture, and is three times greater in wet meadow soils than in soils on exposed ridges. Carbon accumulation rates were greatest during the late-to-middle and middle Holocene (1500 to 4000 years B.P.) and lowest during the early Holocene to late Pleistocene (8000 to 10,000 years B.P.).

Based on a comparison with Holocene paleoenvironmental chronologies, these data suggest that the C accumulation rates in alpine soils were greatest during warm periods and least during cool-moist periods. Assuming that the soils examined here are representative of alpine soils world-wide, alpine soils sequester about 14×10^{12} g C yr^{-1}, which comprises about 3.5% of the total C stored annually in world soils. Carbon accumulation rates of alpine soils examined in this study may be expected to increase during global warming, suggesting that these soils will act as a sink for atmospheric CO_2. However, a major consideration is the effect of global warming on the amount and distribution of snowfall in alpine areas.

ACKNOWLEDGMENT

Dr. G.M. Marion kindly reviewed an early draft of this manuscript.

REFERENCES

Alexander, E.B., Bulk density equations for southern Alaska soils, *Can. J., Soil Sci.*, 69, 177, 1989.

Billings, W.D., Carbon balance of Alaskan tundra and taigo ecosystems: past, present, and future, *Q. Sci. Rev.*, 6, 165, 1987.

Birkeland, P.W., Holocene soil chronofunctions, southern Alps, New Zealand, *Geoderma*, 34, 115, 1984.

Birkeland, P.W., Burke, R.M., and Benedict, J.B., Pedogenic gradients for iron and aluminum accumulation and phosphorus depletion in arctic and alpine soils as a function of time and climate, *Q. Res.*, 32, 193, 1989.

Birkeland, P.W., Burke, R.M., and Shroba, R.R., Holocene alpine soils in gneissic cirque deposits, Colorado Front Range, Bull. 1590: E1-E21, U.S. Geological Survey, Washington, D.C., 1987.

Bliss, L.C., A comparison of plant development in microenvironments of arctic and alpine tundras, *Ecol. Monogr.*, 26, 303, 1956.

Burns, S.F., Alpine Soil Distribution and Development, Indian Peaks, Colorado Front Range, Ph.D. thesis, University of Colorado, Boulder, CO, 1980.

Elias, S.C., Paleoenvironmental interpretation of Holocene insect fossil assemblages from four high-altitude sites in the Front Range, Colorado, *Arctic Alpine Res.*, 17, 31, 1985.

Gennadiyev, A.N., Study of soil formation by the chronosequence method (as exemplified by the soils of the Elbrus region), *Sov. Soil Sci.*, 10, 707, 1979.

Gennadiyev, A.N. and Sokolova, T.A., Trend and rate of clay formation in some soils of the Elbrus area, *Sov. Soil Sci.*, p. 345, 1977.

Harden, J.W., Sundquist, E.T., Stallard, R.F., and Mark, R.K., Dynamics of soil carbon during deglaciation of the Laurentide sheet, *Science*, 258, 1921, 1992.

Jenny, H., Gessel, S.P., and Bingham, F.T., Comparative study of decomposition rates of organic matter in temperate and tropical regions, *Soil Sci.*, 68, 419, 1949.

Mahaney, W.C., Soil stratigraphy and genesis of Neoglacial deposits in the Arapaho and Henderson cirques, central Colorado Front Range, in *Quaternary Environments*, Geographical Monogr. 5, 197, York University Press, Downsview, Ontario, 1974.

Mahaney, W.C., Late-Quaternary stratigraphy and soils in the Wind River Mountains, western Wyoming, in *Quaternary Soils*, Geographical Abstracts, Norwich, CT, 1978, p. 223.

Marion, G.M. and Oechel, W.C., Mid- to late-Holocene carbon balance in arctic Alaska and its implications for future global warming, *Holocene*, 3, 193, 1993.

Nelson, D.W. and Sommers, L.E., Total carbon, organic carbon, and organic matter, in *Methods of Soil Analysis. II. Chemical and Microbiological Properties*, 2nd ed., Page, A.L., Ed., American Society of Agronomy, Madison, WI, 9, 539, 1982.

Oechel, W.C., Hastings, S.J., Vourlitis, G., Jenkins, M., Riechers, G., and Grulke, N., Recent change of arctic tundra ecosystems from net carbon dioxide sink to a source, *Nature*, 361, 520, 1993.

Post, W.M., Emmanuel, W.R., Zinke, P.J., and Stangenberger, A.R., Soil carbon pools and world life zones, *Nature*, 298, 156, 1982.

Rodbell, D.T., Soil-age relationships on late Quaternary moraines, Arrowsmith Range, southern Alps, New Zealand, *Arctic Alpine Res.,* 22, 355, 1990.

Schlesinger, W.H., Evidence from chronosequence studies for a low carbon storage potential of soils, *Science,* 348, 232, 1990.

Shroba, R.R. and Birkeland, P.W., Trends in late-Quaternary soil development in the Rocky Mountains and Sierra Nevada of the western U.S., in *Late-Quaternary Environments of the United States,* Vol. 1, Porter, S.C., Ed., University of Minnesota Press, Minneapolis, 1983, p. 145.

Soil Survey Staff, Soil Survey Manual, Agricultural Handbook No. 9, U.S. Government Printing Office, Washington, D.C., 1993.

Vreeken, W.J., Principle kinds of chronosequences and their significance in soil history, *J. Soil Sci.*, 26, 378, 1975.

Webber, P.J., Tundra primary productivity, in *Arctic and Alpine Environments*, Ives, J.D. and Barry, R.G., Eds., Methuen, London, 1974, p. 445.

Zwinger, A.H. and Willard, B.E., *Land Above the Trees*, Harper & Row, New York, 1972.

14 Spatial and Temporal Patterns of Soil Moisture and Depth of Thaw at Proximal Acidic and Nonacidic Tundra Sites, North-Central Alaska, U.S.

K.M. Hinkel, F.E. Nelson, J.G. Bockheim, L.L. Miller, and R.F. Paetzold

CONTENTS

INTRODUCTION

Major findings of the Arctic Flux Study in the Kuparuk River basin of north-central Alaska (Weller et al., 1995) are the abundance of moist nonacidic tundra (MNT) and the corresponding implications for arctic ecosystem dynamics and trace-gas fluxes (Walker et al., in press). Based on spectral contrasts in NDVI (normalized difference vegetation index), MNT comprises 39% of the land cover in the Kuparuk basin and covers about 14% of the Alaskan North Slope (Walker and Everett, 1991). It is also an important component of the Siberian tundra (Matveyeva, 1994).

Although physiognomically similar, MNT and moist acidic tundra (MAT) have different species compositions, biodiversity, dominant soils, and average seasonal thaw depth (Walker and Acevedo, 1987; Walker and Everett, 1991; Walker et al., 1994, 1995; Shippert et al., 1995; Bockheim et al., 1997, 1998; Nelson et al., 1997). The primary differences, as characterized by the two contrasting sites discussed in this study are summarized in Table 14.1. The greater vascular plant leaf area and moss cover in MAT provide effective thermal insulation and result in colder soils and shallower thaw depths than MNT (Walker et al., in press).

The presence of *Sphagnum*, a dominant moss at acidic sites, radically alters the thermal, geochemical, and hydrological processes. *Sphagnum* releases H ions to the soils (Sjörs, 1950;

TABLE 14.1

General Characteristics of Nonacidic and Acidic Sites;
Values Are Areal Averages, with Standard Deviation in Parentheses

	Nonacidic Tundra, Sagwon 3	Acidic Tundra, Sagwon 4
pH, surface organic layer	7.6	4.0
Organic horizon thickness (cm)	9 (1)	15 (1)
Mean active layer thickness (cm)	55 (13)	36 (8)
Cryoturbation, frost scars (%)	30	<1
OC (%) in upper soil (0–100 cm depth)	60	40

Gorham, 1956; Clymo, 1963; Andrus, 1986; Bockheim et al., 1998), decreasing the soil pH beyond that associated with normal soil leaching processes. *Sphagnum* has a high moisture-holding capacity that promotes conversion of moist sites to a waterlogged state. Decomposition of *Sphagnum*-derived peat is slow due to the high acidity, low nutrient concentrations, low decomposability of *Sphagnum*-derived organic matter, and to the cold, saturated soil conditions in northern environments.

Soils in MNT have thinner organic horizons, significantly greater pH values throughout the active layer, and greater cryoturbation than soils in MAT (Bockheim et al., 1997, 1998). They also have greater amounts of exchangeable Ca, Mg, and bases and cation-exchange capacity, and significantly lower amounts of exchangeable acidity and Al than in MAT. Tissues from forbs, sedges, and woody shrubs in MNT had two to three times as much Ca as the same or similar species in MAT (Bockheim et al., 1998).

Vegetation and soil differences between MAT and MNT have important consequences for land-atmosphere energy and gas exchange. A comparison between intensively studied MAT and MNT sites in the Arctic Foothills physiographic region revealed that the acidic sites had higher rates of evapotranspiration (ET; 13%), gross photosynthesis (63%), and ecosystem respiration (8-fold) than did nonacidic sites (Walker et al., in press). Because photosynthetic rates were higher than respiration rates, MAT experienced 23% more net carbon gain than did MNT during the 10-day measurement period. The wetter, more anaerobic soils of MAT had a five-fold greater methane flux than the MNT.

The origin of nonacidic soils in arctic Alaska is poorly understood. Their existence has been related to the distribution of calcareous loess (Carter, 1988), landscape age (Walker et al., 1989, 1995), and modern dry deposition (Walker and Everett, 1991). The latter may be related to differences in snow distribution (Evans et al., 1989). Bockheim et al. (1997, 1998) proposed the alternative hypothesis that cryoturbation exerts a strong influence on the distribution of MNT and MAT. At research sites in the Arctic Foothills, frost scars occupied 30% of the MNT research site's surface but less than 1% of the MAT site (Walker et al., in press). Cryoturbation may be inhibited in MAT by its thicker and more uniform organic mat, which insulates the soil and results in higher volumetric moisture contents and a thinner active layer (Bockheim et al., 1997). In contrast, cryoturbation in MNT soils causes mixing of the organic matter throughout the active layer and exposes the dark-colored mineral soil, maintains discontinuities in the vegetation cover, and enhances soil heat transfer and thaw-depth variability. Incorporation of organic matter assists in the development of a mollic epipedon in MNT. The active layer contains 60% of the total organic C in the upper 100 cm in the MNT and 40% in the MAT (Bockheim et al., 1997). Thus, the processes that affect development of MNT and MAT are likely to be integral to those leading to the sequestration or release of soil organic carbon.

The general conclusion reached by investigators measuring trace gases in the Kuparuk River basin is that soil moisture has opposing effects on CH_4 and CO_2 (Vourlitis et al., 1993; Reeburgh, in press; Eugster et al.,1997; King et al., in press; Oechel et al., in press). The CH_4 flux is strongly regulated by moisture, such that there is greater production in the wetter areas and in wetter years. Overall, the Kuparuk River basin is a net CH_4 source to the atmosphere. However, the Coastal Plain produces more CH_4 than the Foothills region and dry areas can be a net CH_4 sink. Although the CH_4 flux declines with soil drying, the CO_2 flux initially increases. A lowering of the water table, for example, increases soil decomposition and CO_2 release, increasing net CO_2 loss from the system.

This study presents time series of soil moisture in the active layer and upper permafrost at MAT and MNT sites. Particular attention is devoted to the summer months, when trace gas fluxes are at their maximum. Time series of soil moisture and temperature are rare in northern Alaska; technology that facilitates collection of extensive data sets in remote locations has become available only recently and may prove extremely useful for developing and evaluating models of subsurface thermal evolution. Because soil moisture content varies with time and depth, moisture impacts soil temperature through coupled heat-mass transport processes and by affecting the thermal properties of the soil materials (Kane et al., 1991; Waelbroeck et al., 1997; Hinkel, 1997). Technological advances have also facilitated collection of soil moisture measurements across relatively large areas. This study provides information about the variability of near-surface moisture at MAT and MNT sites in the northern part of the Arctic Foothills of northern Alaska.

STUDY SITES

Investigations were undertaken at the MAT and MNT sites at which most of the Flux Study's intensive work on the MAT/MNT contrasts was performed (Figure 14.1). Located in the Sagwon Upland in the northern part of the Arctic Foothills physiographic province (Wahrhaftig, 1965), the sites are separated by 4 km, occur at similar elevation, and are subject to nearly identical macroclimate. Both occupy gentle hill slopes, and the parent material (loess) is similar. The sites are very dissimilar with respect to plant diversity, soil heat flux, primary production, CO_2 uptake, methane production, soil carbon storage, and soil pH (Walker et al., in press). The site identification system adopted for the Arctic Flux Study (Weller et al., 1995) is retained in this chapter; the two Sagwon sites are referenced by the terms "MAT" or "acidic" site (Flux Study plot 95-**4**) and "MNT" or "nonacidic" site (Flux Study plot 95-**3**). A more complete enumeration of the Flux Study plots is given in Nelson et al. (1997).

The pH of the soil of the acidic site, measured in a saturated extract using distilled water, is low (4.0) and the active layer is substantially thinner than at the nonacidic site. Vegetation consists of cottongrass tussocks (*Eriophorum vaginatum*), dwarf-birch (*Betula nana* ssp. *exilis*) and other dwarf-shrub species (*Ledum palustre* spp. *decumbens*, *Vaccinium vitis-idaea* ssp. *minus*, *V. uliginosum*, *Salix planifolia* spp. *pulchra*), and *Sphagnum* moss (Walker et al., 1994). Mean thickness of the organic mat, based on 45 samples, averages 15 cm over the study area.

The nonacidic site, located 4 km north, has higher (7.6) surface pH, a substantially thicker active layer, and a thinner (9 cm) organic mat than the acidic site. Vegetation is dominated by nontussock sedges (*Carex bigelowii* and *Eriophorum triste*), a few prostrate shrubs (*Dryas integrifolia*, *Salix reticulata* ssp. *reticulata*, *S. arctica*), and brown mosses (*Tomenthypnum nitens* and *Hylocomium splendens*) beneath a relatively open canopy.

INSTRUMENTATION

To obtain time series of soil moisture and temperature, we installed Vitel Hydra® soil moisture probes at the two sites. These compact sensors determine the complex dielectric constant of soil using frequency-domain techniques based on four voltages, which are measured between four 7-cm long stainless-steel tines. A built-in thermistor is used to derive temperature-corrected soil water electrical conductivity. The sensors also measure the salinity of soil water and volumetric soil

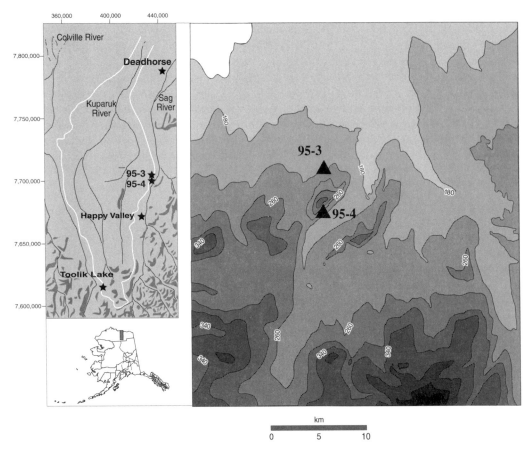

FIGURE 14.1 Maps showing location of Flux Study Plots 95-3 (nonacidic) and 95-4 (acidic) at Arctic Foothills/Arctic Coastal Plain transition in the Sagwon Uplands. Map in upper left shows location of sites in Kuparuk River basin (solid white line) and in relation to prominent geographic features in north-central Alaska.

moisture content. Comparative measurements of soil moisture using gravimetric methods and sensors based on soil dielectric properties have been evaluated by the USDA-NRCS and the U.S. Army Cold Regions Research and Engineering Laboratory (Atkins et al., 1998). Results indicate accuracy similar to that obtained with time-domain reflectometry units.

Three soil moisture probes were installed at each site in June, 1996 at depths of 7, 25, and 45 cm. Because thawing was minimal at this time, a pit was excavated using a PICO® gas-powered jackhammer. One wall of the pit was cleaned carefully and four holes drilled for the sensor tines using a drilling jig to ensure that tine spacing was maintained during installation and that a solid contact was made between the tines and frozen soil. Measurements were made hourly, and an average value recorded on a Campbell CR10 data logger every two hours. In addition to soil moisture, temperature, salinity, and electrolytic conductivity, the temperature within the logger container was also monitored; this enclosure temperature approximates the air temperature at the site.

A portable moisture monitoring device was fabricated to obtain extensive measurements over each of the 1-ha areas. The soil moisture probe was mounted on a 150-cm-long steel conduit and attached to a voltmeter. Power was provided by a 12-V lantern battery. At each measurement point, voltage from each of the four channels was recorded and later transcribed for processing. Each measurement represents the bulk volumetric soil moisture content for the upper 7 cm. The portable unit has been field tested and used to map the spatial variability of near-surface soil moisture (Hinkel et al., 1995; Miller et al., 1998; Nelson et al., in press).

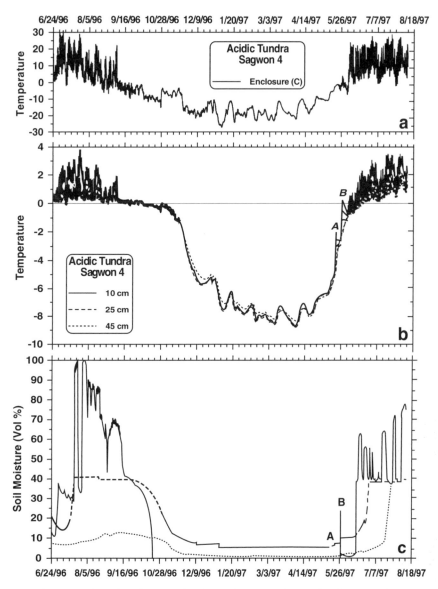

FIGURE 14.2 Time series for acidic site (Flux Study plot 95-4) showing (a) data logger enclosure temperature, (b) soil temperature, and (c) volumetric soil water content (% by volume) at 10, 25, and 45 cm; letters A and B in late May, 1997 refer to snow meltwater infiltration events.

RESULTS AND DISCUSSION

ANNUAL TEMPORAL PATTERNS OF TEMPERATURE AND SOIL MOISTURE

At the acidic site (Sagwon 4), the soil temperature traces show moderate variation related to weather conditions. Amplitude of the variation decreases with depth (Figure 14.2b). In early September, subfreezing air temperatures prevailed, followed by nearly two months in which ground temperatures remained near the freezing point. This "zero-curtain effect" occurs as a response to the release of latent heat as soil water is converted to ice, retarding penetration of the upper and lower freezing fronts and maintaining the temperature of the unfrozen layer near the freezing point. Nonconductive heat-transfer processes can also play a significant role in maintaining the effect (Outcalt et al., 1990).

Following closure of the zero curtain in mid-November, freezeback of the active layer proceeded rapidly. Subfreezing soil temperatures persisted until mid-May, when several sharp upward inflections ("A" and "B" on the figures) were observed in the traces as the soil warmed toward 0°C.

The soil moisture trace at the near-surface (10 cm) probe shows large variations during summer (Figure 14.2c). No precipitation data were collected at the Sagwon sites, but reliable records exist for Barrow. Our studies there show that rapid increases of near-surface soil moisture in summer result from infiltration of precipitation (Hinkel et al., 1997; Miller et al., 1998). Following rain events, gradual drying of the upper soil was observed near the surface while values remained nearly constant at depth. Thus, the patterns observed at the acidic site are likely the result of infiltration following precipitation.

During the period in which the zero-curtain effect dominates, soil moisture gradually freezes, primarily from the soil surface downward. During winter, the instrumental record indicates that the unfrozen soil moisture contents in the upper 45 cm are relatively constant at values less than 10%.

Spikes in the upper probe's record in mid-May ("A" and "B" in Figure 14.2b and 14.2c) are synchronous with warming indicated in the temperature record. These events represent infiltration of snow meltwater during periods with above-freezing air temperature. Soil moisture content increased at all levels, as indicated by the step-like appearance of the traces. During a 36-hr period in late May (Event "B" in Figure 14.2b and 14.2c), the temperature increased 2.7°C and soil moisture content rose from 0 to 25% at the upper probe. This rate of warming is seven times greater than the warming events observed during the winter. The temperature and soil moisture traces indicate that infiltration of snow meltwater has a strong impact on the thermal and moisture fields in the active layer and upper permafrost, and supports results of other studies that relied on measurements of soil water electrolytic conductivity (Hinkel et al., 1997).

In mid-June 1997, soil thaw reached the 10 cm level and moisture content increased rapidly to around 40% as ice was converted to soil water (Figure 14.2c). A similar pattern occurred at the 25 cm level in late June, and at the 45 cm level in late July as the thaw front penetrated to those levels.

Soil at depths greater than 10 cm appears to have remained nearly saturated at 40% following thaw at the beginning of the second summer (Figure 14.2c). The near-surface probe, however, recorded substantial variation and higher values, indicating that the soil materials were responding to infiltration of precipitation, followed by drainage and soil drying. This is likely due to the relatively thick (15 cm) organic mat at this site. The near-surface probe is embedded in the porous and permeable organic layer, and the record reflects rapid infiltration and movement of water through this layer.

Temperature patterns at the nonacidic site (Sagwon 3) are similar to those for the acidic site, although the range of temperature variations induced by short-term weather events was greater. Near-surface soil temperature was several degrees warmer in summer and several degrees cooler in winter at this site, reflecting the reduced insulating capacity of the thinner organic mat (Figure 14.3b). The progression of active layer thaw in summer 1997 is well illustrated in this figure since the effects of installation had dissipated by this time.

The soil moisture records are also generally similar between the two sites. They reflect downward soil freezing in autumn, soil moisture values less than 10% in winter, and "spikes" in spring resulting from infiltration of meltwater. During event "A" (Figure 14.3b and 14.3c), the near-surface soil temperature increased from –4.2 to –0.5°C in 36 hr, and the soil moisture increased from 5 to 9%. This rate of warming was five times faster than during winter warming events.

There was little variation in soil moisture content throughout the soil column at the nonacidic site (Figure 14.3c). Values at all probe levels were constant throughout the summer at around 40%, and there is no evidence of precipitation infiltration effects or soil drying since the upper probe is located below the organic mat.

SPATIAL PATTERNS OF THAW AND SOIL MOISTURE

Several sampling designs were implemented to examine the spatial characteristics of soil moisture over the 1-ha study areas. To measure maximum thaw depth near the end of the thaw season in

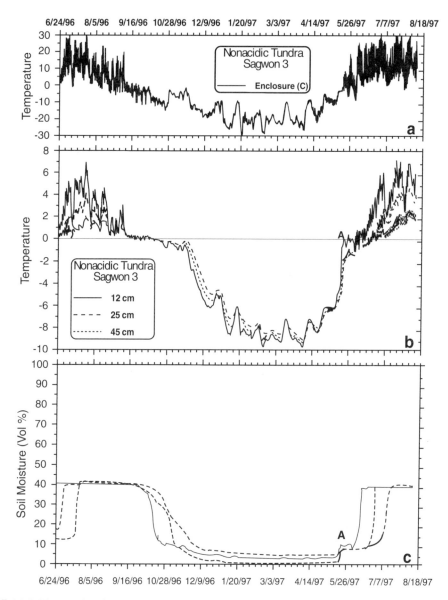

FIGURE 14.3 Time series for nonacidic site (Flux Study plot 95-3) showing (a) data logger enclosure temperature, (b) soil temperature, and (c) volumetric soil water content (% by volume) at 12, 25, and 45 cm; letter A in late May 1997 refer to snow meltwater infiltration event.

mid-August 1997, a 21 x 21 (5 m node spacing) grid was superimposed on the study areas, yielding a total of 441 grid nodes for thaw depth measurements. A 1-cm diameter steel rod, graduated in cm, was inserted into the thawed soil to the point of resistance, and the depth recorded for each grid node.

Because soil moisture readings are time consuming, an unbalanced hierarchical sampling scheme was implemented (Oliver and Badr, 1995). A total of 54 soil moisture readings were collected from each site in mid-August, 1997. Some values are missing from each data set owing to the presence of gravel. The frequency distribution of thaw depth and volumetric soil moisture content (VSM) at each site are shown in Figure 14.4, and summary statistics are given in Table 14.2. Spatial patterns are shown in Figure 14.5.

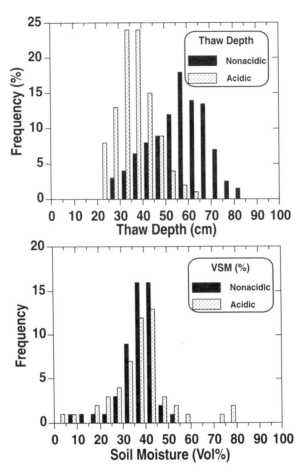

FIGURE 14.4 Frequency distribution of (a) thaw depth (n = 441), and (b) near-surface volumetric soil moisture content (n ≅ 52) at acidic and nonacidic sites.

Based on 441 measurements, the average thaw depth at the nonacidic site is substantially greater (55 cm) than at the acidic site (36 cm). The greater variability at the acidic site is readily discernible in Figure 14.4. This is likely due to the effects of the thinner organic mat and lack of *Sphagnum* at the nonacidic site, which promote penetration of heat energy into the substrate. Nonconductive effects in the organic layer at the acidic site may also play an important role (Nelson et al., 1985; Outcalt and Nelson, 1985; Hinkel et al., 1993).

Mean soil moisture values were similar (about 37%) at the two sites (Table 14.2). However, more high-frequency variation occurred in the spatial domain at the acidic site, which has a standard deviation nearly twice that at the nonacidic site. This reflects the presence of the cottongrass tussocks (*Eriophorum vaginatum*). Relatively low soil moisture values occur within the tussocks (minimum of 2%), while inter-tussock areas are often near saturation (maximum of 79%). Conversely, soil moisture values were less variable at the nonacidic site on the sampling date.

Spatial patterns of thaw and soil moisture, presented as maps in Figure 14.5, depict large variations over short distances. Examination of the maps does not reveal any apparent relation between soil moisture and thaw depth, and this impression is supported by inspection of scatter plots and results from a point-in-polygon analysis. This result is not surprising, however, given that end-of-season thaw depth represents a cumulative, time-integrated response to conditions since the onset of thawing. Conversely, near-surface soil moisture values experience high-frequency fluctuations in response to precipitation events and evaporative drying (Miller et al., 1998).

TABLE 14.2
Summary Descriptive Statistics for Thaw Depth and Near-Surface Volumetric Soil Moisture Content at Nonacidic and Acidic Sites

	Nonacidic Tundra Sagwon 3	Acidic Tundra Sagwon 4
	Thaw depth (cm)	
n	441	441
Spatial mean	55	36
Standard deviation	13	8
Minimum	19	13
Maximum	86	63
Normal distribution[a]	no	no
	Soil moisture (%)	
n	51	52
Spatial mean	36	38
Standard deviation	8	14
Minimum	7	2
Maximum	53	79
Normal distribution[a]	no	no

[a] At 95% level.

Two other results arise from this analysis:

1. Downslope-oriented lineations with relatively large depths of thaw occur at both sites, although they are best developed at the acidic site. These features appear integrated, as if forming part of a network, and may reflect the thermal influence of subsurface water flow on thaw depth. Throughflow of meteoric and meltwater, flowing on top of the frozen soil, would transport sensible heat and enhanced thermal ablation. Subsurface flow appears to follow channels that are oriented downslope, but the sampling interval of 5 m appears to be too coarse to capture the necessary detail. We are examining this phenomenon in more detail.

2. The 5 m node spacing used in this study is apparently sufficient to address at least some of the spatial pattern of thaw depth at these Foothills sites; the 100 m spacing used in an earlier analysis (Nelson et al., in press) was incapable of resolving any underlying pattern at several Foothills locations. Two-dimensional correlograms (Oden and Sokal, 1986; Jacquez, 1991), shown in Figure 14.6, show positive and statistically significant cross-slope autocorrelation at both sites. At the acidic site, autocorrelation in the up- and downslope direction is not statistically significant in many distance/direction classes (reflecting random variation) and negative in others (reflecting systematic dissimilarities along the slope). At the nonacidic site, however, positive and statistically significant autocorrelation exists at medium and long lags along an axis oriented from NNW to SSE (up- and downslope). These patterns may be related to water movement in linear pathways as discussed above, although conclusive results await application of a denser network of sampling locations, as well as process-oriented work.

SUMMARY

Near-surface soil moisture values were highly variable in summer at the acidic site, probably in response to the thick organic mat at this site. The upper probe at 10 cm is embedded within the

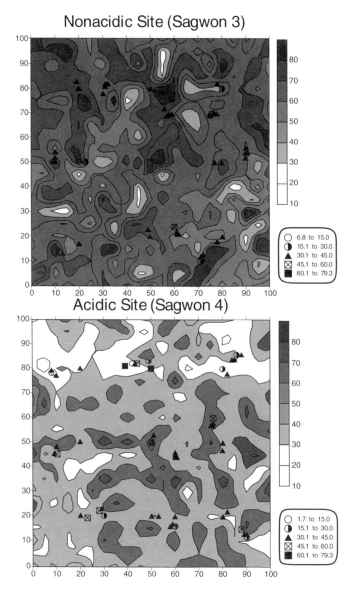

FIGURE 14.5 Maps of 1 ha study areas showing spatially interpolated thaw depth (n = 441) and near-surface soil moisture content (n ≅ 52) at nonacidic and acidic sites. Note that scales are identical at both sites; north is to the top; near-average values are shown as triangles; lower and higher values are shown as circles and squares, respectively.

porous and permeable mat, which facilitated a rapid response to precipitation events. At the nonacidic site, near-surface soil moisture was more constant. Soil moisture content in the middle and lower parts of the active layer remained nearly constant at around 40% throughout the summer at both sites.

The two sites experienced a zero-curtain effect that persisted for 6–8 weeks in early winter. Infiltration of snow meltwater in spring occurred as pulses, which affected the thermal and moisture content throughout the instrumented column.

End-of-season thaw depth is highly variable at both locations. The nonacidic site experienced substantially greater depth of thaw than the acidic site. This situation, which is repeated from year to year, is probably related to the dominant vegetation types and especially the presence of

Nonacidic Tundra
Sagwon 3

Acidic Tundra
Sagwon 4

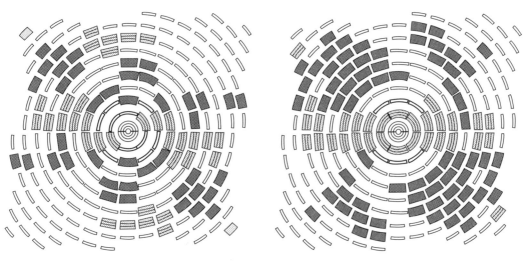

FIGURE 14.6 Two-dimensional (directional) correlograms for nonacidic and acidic sites; correlograms are based on Moran's *I* statistic; virtually all statistically significant distance/direction classes involve values of *I* less than 0.333; statistically significant positive and negative autocorrelation in specific distance/direction classes are represented as larger, light- and dark-toned rectangular boxes, respectively; smaller, unshaded boxes denote lack of statistical significance; for further details on directional correlograms, see Jacquez (1991) and Oden and Sokal (1986).

Sphagnum at the acidic site. Evidence exists for integrated subsurface flow networks where the thaw depth is greater. Both of these observations imply the thermal effects of channeled soil water throughflow, which accelerates ablation of frozen ground.

Although spatially averaged soil moisture content is similar at these sites, there is greater spatial variability at the acidic site. Presumably, this reflects the influence of the tussock vegetation.

ACKNOWLEDGMENTS

This research was supported by NSF grants OPP-9529783 and OPP-9732051 to KMH and OPP-9612647 to FEN. N.I. Shiklomanov and Linda Parrish (University of Delaware) prepared Figures 14.1 and 14.6, respectively.

REFERENCES

Andrus, R.E., Some aspects of Sphagnum ecology, *Can. J. Bot.*, 64, 416, 1986.

Atkins, R.T., Pangburn, T., Bates, R.E., and Brockett, B.E., Soil Moisture Determinations Using Capacitance Probe Methodology, Spec. Rep. 98-2, U.S. Army Cold Regions Research and Engineering Laboratory, Hanover, NH, 1998.

Bockheim, J.G., Walker, D.A., and Everett, L.R., Soil carbon distribution in nonacidic and acidic tundra of arctic Alska, in *Soil Processes and the Carbon Cycle,* Lal, R., Kimble, J.M., Follett, R,F., and Stewart, B.A., Eds., CRC Press, Boca Raton, FL, 1997, p. 143.

Bockheim, J.G., Walker, D.A., Everett, L.R., Nelson, F.E., and Shiklomanov, N.I., Soils and cryoturbation in moist nonacidic and acidic tundra in the Kuparuk River basin, arctic Alaska, U.S., *Arctic Alpine Res.,* 30, 166, 1998.

Carter, L.D., Loess and deep thermokarst basins in Arctic Alaska, in *Proc. 5th Int. Conf. Permafrost,* Vol. 1, Tapir, Trondheim, 1988, p. 706.

Clymo, R.S., Ion exchange in Sphagnum and its relation to bog ecology, *Ann. Bot. N.S.,* 27, 309, 1963.

Eugster, W., McFadden, J.P., and Chapin, F.S., III, Structurally similar arctic tundra vegetation types differ in surface energy and CO_2 fluxes, *Boundary-Layer Meteorol.,* 85, 293, 1997.

Evans, B.M., Walker, D.A., Benson, C.S., Nordstrand, E.A., and Petersen, G.W., Spatial interrelationships between terrain, snow distribution, and vegetation patterns at an arctic foothills site in Alaska, *Holarctic Ecol.,* 12, 270, 1989.

Gorham, E., On the chemical composition of some waters from the Moor House Nature Preserve, *J. Ecol.,* 44, 375, 1956.

Hinkel, K.M., Nelson, F.E., and Paetzold, R.F., Monitoring moisture in the active layer and upper permafrost: field testing and results, *Eos Transactions of the American Geophysical Union,* 76, F243, 1995.

Hinkel, K.M., Outcalt, S.I., and Nelson, F.E., Near-surface summer heat-transfer regimes at adjacent permafrost and non-permafrost sites in central Alaska, *Proc. 6th Int. Conf. Permafrost,* Vol. I, South China University of Technology Press, Wushan, Guangzhou, China, 1993, p. 261.

Hinkel, K.M., Estimating seasonal values of thermal diffusivity in thawed and frozen soils using temperature time series, *Cold Regions Sci. Technol.,* 26, 1, 1997.

Hinkel, K.M., Outcalt, S.I., and Taylor, A.E., Seasonal patterns of coupled flow in the active layer at three sites in northwest North America, *Can. J. Earth Sci.,* 34, 667, 1997.

Jacquez, G.M., *C2D: Spatial Autocorrelation in Two Dimensions,* Exeter Publishing, Setauket, NY, 1991.

Kane, D.L., Hinzman, L.D., and Zarling, J.P., Thermal response of the active layer to climatic warming in a permafrost environment, *Cold Regions Sci. Technol.,* 19, 111, 1991.

King, J.Y., Reeburgh, W.S., and Regli, S.K., Methane emission and transport by arctic sedges in Alaska: results of a vegetation removal experiment, *J. Geophys. Res.,* 103(d22), 29,083-29,092, 1998.

Matveyeva, N.V., Floristic classification and ecology of tundra vegetation of the Taymyr Peninsula in northern Siberia, *J. Veg,. Sci.,* 4, 813, 1994.

Miller, L.L., Hinkel, K.M., Nelson, F.E., Paetzold, R.F., and Outcalt, S.I., Spatial and temporal patterns of soil moisture and thaw depth at Barrow, Alaska, in Proc. 7th Int. Conf. Permafrost, Centre d'Etudes Nordique, Université Laval, Québec, Canada, 1998.

Nelson, F.E., Outcalt, S.I., Goodwin, C.W., and Hinkel, K.M., Diurnal thermal regime in a peat-covered palsa, Toolik Lake, Alaska, *Arctic,* 38, 310, 1985.

Nelson, F.E., Shiklomanov, N.I., Mueller G., Hinkel, K.M., Walker, D.A., and Bockheim, J.G., Estimating active-layer thickness over a large region: Kuparuk River basin, Alaska, U.S., *Arctic Alpine Res.,* 29, 367, 1997.

Nelson, F.E., Hinkel, K.M., Shiklomanov, N.I., Mueller, G.R., Miller, L.L., and Walker, D.A., Active-layer thickness in north-central Alaska: systematic sampling, scale, and spatial autocorrelation, *J. Geophys. Res. - Atmos.,* 103(D22)(28963), 1998.

Oden, N.L. and Sokal, R.R., Directional autocorrelation: an extension of spatial correlograms to two dimensions, *Syst. Zool.,* 35, 608, 1986.

Oechel, W.C., Vourlitis, G.L., Hastings, S.J., Ault, R.P., Jr., and Bryant, P., The effects of water table manipulation and elevated temperature on the net CO_2 flux of wet sedge tundra ecosystems, *Global Change Biol.,* 4(1) 77-90, 1998.

Oliver, M.A. and Badr, I., Determining the spatial scale of variation in soil radon concentration, *Math. Geol.,* 27, 893, 1995.

Outcalt, S.I. and Nelson, F.E., A model of near-surface coupled-flow effects on the diurnal thermal regime in a peat-covered palsa, *Arch. Meteorol. Geophys. Bioclimatol.,* A33, 345, 1985.

Outcalt, S.I., Nelson, F.E., and Hinkel, K.M., The zero-curtain effect: heat and mass transfer across an isothermal region in freezing soil, *Water Resour. Res.,* 26, 1509, 1990.

Reeburgh, W.S., King, J.Y., Regli, S.K., Kling, G.W., Auerbach, N.A., and Walker, D.A., A CH_4 emission estimate for the Kuparuk River Basin, Alaska, *J. Geophys. Res.,* 103(D22), 29,005-29,013, 1998.

Shippert, M., Walker, D.A., Auerbach, N.A., and Lewis, B.E., Biomass and leaf-area index maps derived fron SPOT images for Toolik Lake and Imnavait Creek areas, Alaska, *Polar Rec.,* 31, 147, 1995.

Sjörs, H., The relation between vegetation and electrolytes in north Swedish mires, *Oikos,* 2, 241, 1950.

Vourlitis, G.L., Oechel, W.C., Hastings, S.J., and Jenkins, M.A., The effect of soil moisture and thaw depth on methane flux from wet coastal tundra ecosystems of the North Slope of Alaska, *Chemosphere,* 26, 329, 1993.

Waelbroeck, C., Monfray, P., Oechel, W.C., Hastings, S., and Vourlitis, G., The impact of permafrost thawing on the carbon dynamics of tundra, *Geophys. Res. Lett.*, 24, 229, 1997.

Walker, D.A. and Acevedo, W., Vegetation and a LANDSAT-derived land cover map of the Beechy Point Quadrangle, Arctic Coastal Plain, Alaska, U.S. Army Cold Regions Research and Engineering Laboratory, Hanover, NH, 1987.

Walker, D.A. and Everett, K.R., Loess ecosystems of northern Alaska: regional gradient and toposequence at Prudhoe Bay, *Ecol. Monogr.*, 61, 437, 1991.

Walker, D.A., Auerbach, N.A., and Shippert, M.M., NDVI, biomass, and landscape evolution of glaciated terrain in northern Alaska, *Polar Rec.*, 31, 169, 1995.

Walker, D.A., Chapin, F.S., III, Auerbach, N.A., Bockheim, J.G., Eugster, W., King, J.Y., McFadden, J.P., Michaelson, G.J., Nelson, F.E., Ping, C.-L., and Reeburgh, W.S., Landscape age and substrate pH: controls of arctic ecosystem processes, *Nature*, 394(6692), 469, 1998.

Walker, M.D., Walker, D.A., and Auerbach, N.A., Plant communities of a tussock tundra landscape in the Brooks Range foothills, Alaska, *J. Veg. Sci.*, 5, 843, 1994.

Walker, M.D., Walker, D.A., and Everett, K.R., Wetland soils and vegetation, arctic foothills, Alaska, Biological Rep. 89(7), June, U.S. Fish and Wildlife Service, Washington, D.C., 1989.

Weller, G., Chapin, F.S., III, Everett, K.R., Hobbie, J.E., Kane, D., Oechel, W.C., Ping, C.-L., Reeburgh, W.S., Walker, D., and Walsh, J., The Arctic Flux Study: a regional view of trace gas release, *J. Biogeogr.*, 22, 365, 1995.

15 Areal Evaluation of Organic Carbon Pools in Cryic or Colder Soils of China

G.B. Luo, G.L. Zhang, and Z.T. Gong

CONTENTS

INTRODUCTION

Recently, the possibility of global climatic changes brought about by the "greenhouse" effect of CO_2 in the atmosphere, which is closely related to the carbon pool in soils, has prompted much carbon-related research. Although estimates vary widely, most people agree that about 1500 Pg (10^{15} g) organic C is stored in the soil reservoir (Eswaran, 1995). Even small changes in such large pools of carbon would be expected to have dramatic feedback on the global climate system (Schlesinger, 1995). A large portion of this pool is held in the soils of the cryic or colder ecosystems, with an estimated 350 to 455 Pg (Billings, 1987) which is equivalent to 23–30% of the total world soil organic C pool, where dramatic changes in climate are expected during the new century. That is why more and more attention is being paid to these regions of the world.

Although some CH_4 behavior in paddy lands were reported in China (Sun, 1993), organic C pool in soils of China are hard to find. This chapter is based on soil temperature regimes of each county of China which were calculated according to the definition in *Soil Taxonomy* and on the basis of climatic data. It is found that areas with cryic soil temperature regimes or colder are generally concentrated in two parts (Figure 15.1). One is near or north of latitude 40° N, and another is in the Qingzang plateau with an elevation >3000 m which mainly includes Helongjiang, Jilin, Inner Mongolia and Qinghai, Tibet, the north part of Xinjiang and a part of Gansu, Sichuan, Ningxia, Liaoning, etc. (Appendix 15.1). The former covers areas of about 1.70 million km² where Boric Luvisols (*Cryoboralfs*), Udic Isohumisols (*Cryoborolls*), Ustic Isohumisols (*Calciborolls*), and Cryic Aridisols (*Haplocryids*) spread from east to west and the latter about 1.85 million km² where Cryic (Gelic) Cambisols (*Cryumbrepts*) and Cryic Aridisols (*Haplocryids*) distribute (Figure 15.1).

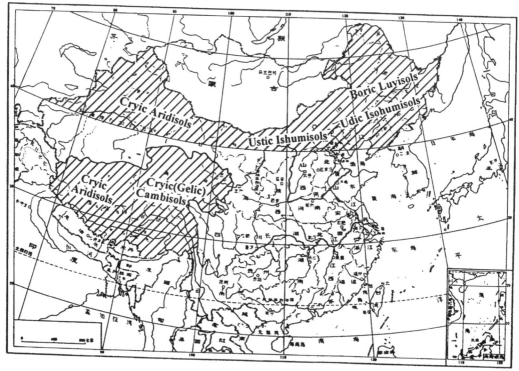

FIGURE 15.1 Sketch of Cryic or colder soil distribution in China.

METHOD AND MATERIALS

CALCULATION MODEL

In order to calculate Soil Organic Carbon (SOC) storage in soils, we must know the SOC distribution status in soils. In an ideal situation, the organic C storage (OCS) can be obtained accurately from the three-dimensional spatial integral, i.e., $OCS(x,y,z) = \iiint B*C*dxdydz$, by use of boundary conditions x from 0 to $x1$, y from 0 to $y1$, and z from 0 to $z0$ which distinguishes pedosphere from other spheres. Besides both B (soil bulk density) and C (organic carbon content) are functions of x, y, z. However, in practice, it is almost impossible because soil scientists so far have little knowledge of soil property changes in the latter direction and nobody knows the exact $B(x,y,z)$ and $C(x,y,z)$ functions. Thus, we have to simplify our concept to skip the soil changes in the horizontal direction and only consider them in the vertical direction. Therefore, in most cases we processed, i.e., $OCS = S\int B*Cdz$, S is a surface area as a constant and both B and C are functions of a depth z. The problem is to find suitable $B(z)$ and $C(z)$ function forms.

The organic C content normally decreases with the depth in the soil profile except in Histisols and some soils with alluvium parent materials. There are two types of decreasing trends (Figure 15.2). One is in Isohumisols derived from steppe land with a deep humic accumulation; as a result the organic C content in the profile is gradually decreased. It is characterized by the ratio of organic matter amount stored in 0–20 cm to that in 0–100 cm being 0.4 or less. And the other is that organic C content sharply decreases with depth as in forest soils Boric Luvisols. In both situations, however, we can consider that organic C will decrease with depth according to exponential regulation, i.e., $c(z) = c(0)*e^{-bz}$, of which $c(z)$ is organic C content in the depth z m from the surface, $c(0)$ is organic C content in the surface, and the exponent b is to determine the rate of decrease in C. By use of boundary conditions, when $z = 0$, and $c(z) = c(0)$, and when $z = 0.5$ and $c(z) = c(0.5)$, b can be obtained, i.e., $2*\ln[c(0)/c(0.5)]$. The bulk density B is considered to be a

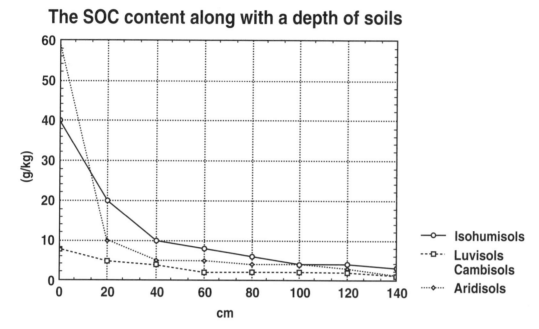

FIGURE 15.2 Organic C changes and soil depth.

linear function of the depth from the surface, i.e., $B = 0.3*z + 1.2$. Thus, in the homogeneous soil region (S), amount of stored organic C is from surface to 1 m depth (C) (because organic carbon will be near zero at this depth in most soil profiles) (Equation 15.1)

$$C(s) = S*\int c(0)*e^{-bz} (0.3z + 1.2)dz \qquad (15.1)$$

After integral from 0 to 1 m, we can get the following formula (Equation 15.2)

$$C(s) = S*c(0)*\{0.3/b^2 +1.2/b - (1.2b+0.6)/[b*(c(0)/c(0.5))]^2\} = S*k \qquad (15.2)$$

Integrally from 0 to 0.5 m leads to Equation 15.3

$$C(s) = S*c(0)*\{0.3/b^2 +1.2/b - (1.2b+0.6)/[b^2*(c(0)/c(0.5))]\} = S*k_1 \qquad (15.3)$$

Equation 15.3 is applicable in shallow soils such as those in Tibet.

Obviously, total organic C stored in soils depends on the region and k which is in relation to C(0) and C(0.5). As for a certain region, the stored C amount is characterized by k or k1 value, i.e., OC amount in unit area.

If n homogeneous parts are distinguished from cryic and colder soils in China, the total organic C stored amount can easily be obtained from the sum j = 0 to n, ΣC_j. In Equations 15.2 and 15.3, S takes a million km^2 as its unit and C(0) takes g/kg, thus C(s) will finally get Pg as a unit, of which k is in kg/m^2.

DATA COLLECTION

All soil data used are obtained from the Reports of Second National Soil Survey, including the books *Soil Series of China* (National Soil Survey Office, 1995), *Soil Survey Data of China* (National Soil Survey Office, 1996), *Soils of Helongjiang* (Soil Survey Office of Helongjiang, 1993), *Soils of Jilin* (Soil Survey Office of Jilin, 1993), *Soils of Inner Mongolia* (Soil Survey Office of Inner Mongolia, 1993), *Soils of Xinjinag* (Soil Survey Office of Xinjiang, 1993), *Soils of Tibet* (Soil Survey

TABLE 15.1
Main Soil Types and Areas of Cryic Soils in China and Their Specific Parameters

Regions	Soils	Area (10⁶ km²)	C (0)	C (0.5)	K or K1	Area*k
Helongjiang and Jilin	Boric Luvisols (*Cryoboralfs*)	0.292	41.3	8.5	16.2	4.73
	Udic Isohumisols (*Cryoborolls*)	0.205	38.5	15.5	23.6	4.84
	Boric Spodosols (*Humicryods*)	0.040	69.7	6.4	18.3	0.73
	Gleyisols (*Cryaquepts*)	0.035	53	10	19.7	0.69
	Cryic Cambisols (*Cryumbrepts*)	0.055	8.9	4.0	5.95	0.33
Inner Mongolia	Boric Luvisols (*Cryoboralfs*)	0.086	40	4.3	11.2	0.96
	Boric Spodosols (*Humicryods*)	0.053	36.4	5.6	12.1	0.64
	Ustic Isohumisols (*Calciborolls*)	0.298	20.5	8.5	12.9	3.84
	Cryic Aridisols (*Haplocryids*)	0.226	6.0	3.9	5.2	1.17
Northern Xingiang	Cryic Aridisols (*Haplocryids*)	0.47	5.9	4.1	5.2	2.44
The Qing-Zang Plateau	Udic Cambisols (Cryumbrepts)	0.50	40.7	7.8	13.9	6.95
including Sichuan	Ustic Cambisols (Cryumbrepts)	0.54	8	6.4	4.1	2.21
	Cryic Aridisols (*Haplocryids*)	0.83	9.4	2.4	3.7	3.07

Office of Tibet, 1993), *Soils of Qinghai* (Soil Survey Office of Qinghai, 1993), etc. The relevant data are collected in Table 15.1. C(0) and C(0.5) are obtained from the weighted average of soil series concerned, and k and Area*k are a calculated results.

RESULTS AND DISCUSSION

Based on the above calculations, in northern China organic C storage is about 20.4 Pg and in the Qingzang plateau it is 12.2 Pg. About 32.6 Pg organic C is stored in cryic or colder soils of China, about 2% of the world's total. It also means that in northern China there are about 11.6 kg organic C/m² whereas there are 6.5 kg C/m² in the Tibet region where the soils are weakly developed and shallow.

In northern China, the stored organic C in an unit area obviously decreases from east to west, while the soil moisture regime is from udic and ustic to desert (Figure 15.3). The same situation is observed in the Tibetan Mountains Region.

In comparison with other places of the world, the figures of organic C (kg/m²) are reasonable. In the moist cryic or colder soils in China, i.e., Helongjiang and Jilin, the stored organic C is similar to that of the Alaska region, although for the maximum value the latter is bigger. As for the arid regions, C content is similar to the desert area of southern New Mexico (Table 15.2).

In northeast China, Udic Isohumisols (*Cryoborolls*), also called Black Soils by civilians, are main soils for cultivation. However, due to changes of land use or unsuitable management practice, organic C content will decrease with years of cultivation. Thus, organic C decomposed CO_2 will flux from the pedosphere into the atmosphere. This not only degrades the soil chemical and physical properties, but also has certain effects on the environment. After 20 years of cultivation OC can decline from 68.7 to 43.8 g/kg or to 34.5 g/kg after 40 years of cultivation in the surface layer, which means almost half of C storage in the surface is lost. Meanwhile, other soil properties have also changed (Figure 15.4). According to such rates, an average of 4 kg OC/m² has been emitted from pedosphere. If 300,000 ha were cultivated for 50 years, 0.12 Pg OC would be lost, which is almost double for the OC storage in paddy soils from Suzhou to Shanghai, a part of the Yangtze delta (Luo Guobao, 1995). Such loss would have a great impact on climatic environment, although we don't know to what extent.

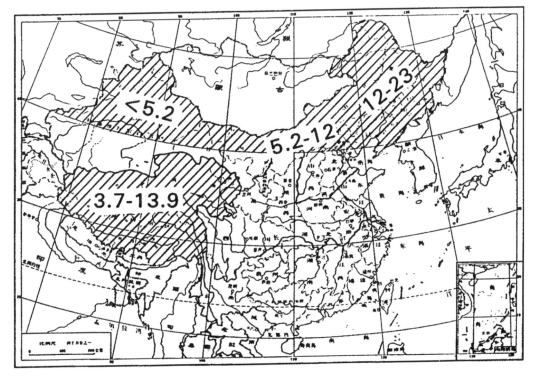

FIGURE 15.3 Sketch showing changes in the stored amount of C along west-east direction.

TABLE 15.2
Comparison of Soil Organic C in Different Regions

Authors/Regions/Soils	OC (kg/m²)
Alaska	
Shallow Entisols	16.9
Deep Entisols	32.4
Shallow Spodosols	17.4
Deep Spodosols	29.8
Cryofolists	14.2
Southern New Mexico	
Typic Haplargid, fine-loamy	3.9
Ustollic Haplargid, fine-loamy	4.8
Typic Haplargid, coarse-loamy	3.7
Ustollic Haplargid, fine-loamy	1.8
	4.1
Cryic China	
Boric Luvisols (northeast China) — *Cryoboralfs*	16.2
Udic Isohumisols (northeast China) — *Cryoborolls*	23.6
Boric Spodosols (northeast China) —*Humicryods*	18.3
Gleyisols (northeast China) — *Cryaquept*	19.7
Cryic Aridisols (northwest China) — *Haplocryids*	5.2
Udic Cambisols (Tibet) — *Cryumbrepts*	13.9
Ustic Cambisols (Tibet) — *Cryumbrepts*	4.1
Cryic Aridisols (Tibet) — *Haplocryids*	3.7

Soil degradation after cultivation years

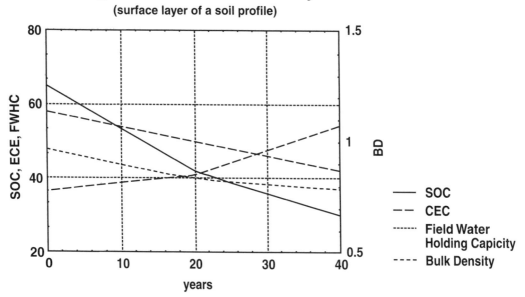

FIGURE 15.4 Soil properties change with years of cultivation.

Comparing of calculated and measured SOC data

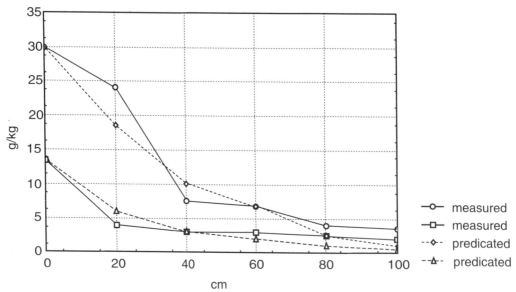

FIGURE 15.5 Comparison of actual and calculated OC values.

CONCLUSIONS

1. An exponent model can be used to estimate OC storage in cryic or colder soils in China (Figure 15.5). But this model is not suitable for some Histisols and Fluvisols and soils modified greatly by humans in which OC may distribute irregularly. Latter soils only occupy a small part in cryic or colder soils in China, and other models need to be used.

2. In the humid region of cryic China, OC storage ranges from 12–23 kg/m^2, including the alpine mountainous region. In the ustic region of cryic or colder soils in China, OC storage ranges from 5.5–11 kg/m^2 and in the arid regions of cryic soils in China, OC storage is normally less than 5.5 kg/m^2.
3. At present in cryic soils of China there are 33 Pg OC pools which may change with environmental changes including artificial and natural activities.

ACKNOWLEDGMENTS

Financial support for this research was jointly provided by National Natural Science Foundation of China and Chinese Academy of Sciences. Many thanks to Professor Zhao Qiguo for his valuable discussions and review of this manuscript.

REFERENCES

Billings, W.D., Carbon balance of Alaskan tundra and taiga ecosystems: past, present and future, *Q. Sci. Rev.*, 6, 165, 1987.

Eswaran H., van den Berg, E., Reich, P., and Kimble, J., Global soil carbon resources, in *Soils and Global Change*, Lal, R., Kimble, J., Levine, E., and Stewart, B.A., Eds., Lewis Publishers, Boca Raton, FL, 1995, p. 27.

Gong Zitong and Luo Guobao, Pedogeochemical environment and health in China, *Pedosphere*, 2, 71, 1992.

Gong Zitong, Luo Guobao, Zhang Ganlin, Spaargaren, O.C., and Kauffman, J.H., Eds., Reference Soil Profiles of the China (Field and Analytical Data). ISSAS-ISRIC Report, ISSN:1381-5571, 1995.

Gong Zitong et al., Eds., Chinese Soil Taxonomy (revised version), Chinese Agricultural Science and Technology Press, Beijing, 1996.

Henning Rodhe, Modeling Biogeochemical Cycles, in *Global Biogeochemical Cycle*, Butcher, S.S., Charlson, R.J., Orians, G.H., and Wolfe, G.V., Eds., Academic Press, NY, 1992, p. 55.

Luo Guobao, Unpublished, 1995.

National Soil Survey Office, Soil Series of China (Volume II and Volume VI). Agricultural Publishing House, Beijing, 1995.

National Soil Survey Office, General Databank of National Soil Survey of China. Agricultural Publishing House, Beijing, 1996.

Oechel, W.C. and Vourlitis, G.L., Global changes on carbon storage in cold soils, in *Soils and Global Change*, Lal, R., Kimble, J., Levine, E., and Stewart, B.A., Eds., Lewis Publishers, Boca Raton, FL, 1995, p. 117.

Grossman, R.B., Ahrens, R.J., Gile, L.H., Montoya, C.R., and Chadwick, O.A., Areal evaluation of organic and carbonate carbon in a desert area of south New Mexico, in *Soils and Global Change*, Lal, R., Kimble, J., Levine, E., and Stewart, B.A., Eds., Lewis Publishers, Boca Raton, FL, 1995.

Soil Survey Office of Helongjiang Province, Soils of Helongjiang. Agricultural Publishing House, Beijing, 1993.

Soil Survey Office of Jilin Province, Soils of Jinlin. Agricultural Publishing House, Beijing, 1993.

Soil Survey Office of Inner Mongolia, Soils of Inner Mongolia. Agricultural Publishing House, Beijing, 1993.

Soil Survey Office of Tibet, Soils of Tibet. Agricultural Publishing House, Beijing, 1993.

Soil Survey Office of Qinghai Province, Soils of Qinghai. Agricultural Publishing House, Beijing, 1993.

Soil Survey Office of Xinjiang, Soils of Xinjiang. Agricultural Publishing House, Beijing, 1993.

Soil Survey Staff, Soil Taxonomy, Agricultural Handbook No. 436, USDA, Washington, D.C., 1975.

Sun Bo, Emission of CH_4 in Soils in *Pedosphere Material Cycling in Pedosphere in Relation to Agriculture and Environment*, Zhao Qiguo, Ed., Jiangsu Science and Technology Publishing House, Nanjing, 1993.

Schlesinger, W.H., An overview of the carbon cycle, in *Soils and Global Change*, Lal, R., Kimble, J., Levine, E., and Stewart, B.A., Eds., Lewis Publishers, Boca Raton, FL, 1995, p. 9.

Xing Guangxi, Cai Zhucong and Zhao Qiguo, Soils and greenhouse effect, in *Pedosphere Material Cycling in Pedosphere in Relation to Agriculture and Environment*, Zhao Qiguo, Ed., Jiangsu Science and Technology Publishing House, Nanjing, 1993.

APPENDIX 15.1
Cryic Soil Temperature Regimes in China

Station	Province	Latitude (N)		Longitude (E)		Elevation (m)	STR	SMR
Subeiyamajie	Gansu	41°	38"	96°	53"	1963	Frigid	Arid
Tianzhu Songshan	Gansu	37°	12"	103°	32"	2727	Cryic	Arid
Tianzhu Wuqiaoling	Gansu	37°	11"	103°	05"	3045	Cryic	Udic
Tongwei Huajialing	Gansu	35°	25"	104°	50"	2451	Cryic	Udic
Yongchang	Gansu	38°	24"	102°	09"	1987	Cryic	Arid
Zhangye Qilianshan	Gansu	38°	52"	100°	12"	3023	Cryic	Ustic
Weichang	Hebei	41°	56"	117°	45"	842	Frigid	Udic
Aihui	Helongjiang	50°	15"	127°	27"	166	Cryic	Ustic
Anda	Helongjiang	46°	24"	125°	21"	151	Frigid	Ustic
Baoqing	Helongjiang	46°	19"	132°	11"	83	Frigid	Ustic
Fu Jing	Helongjiang	47°	14"	131°	59"	64	Frigid	Udic
Gannan	Helongjiang	47°	56"	123°	30"	185	Frigid	Udic
Hai Lun	Helongjiang	47°	26"	126°	58"	239	Frigid	Ustic
Harbin	Helongjiang	45°	41"	126°	37"	172	Frigid	Udic
He Gang	Helongjiang	47°	22"	130°	20"	228	Frigid	Udic
Hu Lin	Helongjiang	45°	46"	132°	58"	100	Frigid	Udic
Hua Nan	Helongjiang	46°	12"	130°	31"	182	Frigid	Udic
Huma	Helongjiang	51°	43"	126°	39"	177	Cryic	Udic
Humamohe	Helongjiang	53°	29"	122°	21"	279	Pergelic	
Ji Xi	Helongjiang	45°	17"	130°	57"	233	Frigid	Udic
Jia Yin	Helongjiang	48°	53"	130°	24"	89	Cryic	Udic
Ke Shan	Helongjiang	48°	03"	125°	53"	237	Frigid	Udic
Lin Kou	Helongjiang	45°	17"	130°	16"	312	Frigid	Udic
Luo Bei	Helongjiang	47°	34"	130°	50"	84	Frigid	Udic
Ming Shui	Helongjiang	47°	10"	125°	54"	249	Frigid	Udic
Mudanjiang	Helongjiang	44°	34"	129°	36"	241	Frigid	Udic
Nehe	Helongjiang	48°	29"	124°	51"	203	Frigid	Udic
Nenjiang	Helongjiang	49°	10"	125°	13"	222	Cryic	Udic
Qiqihar	Helongjiang	47°	23"	123°	55"	146	Frigid	Udic
Rao He	Helongjiang	46°	48"	134°	01"	55	Frigid	Udic
Shang Zhi	Helongjiang	45°	13"	127°	58"	190	Frigid	Udic
Sui Hua	Helongjiang	46°	37"	126°	58"	180	Frigid	Udic
Suifenhe	Helongjiang	44°	23"	131°	09"	470	Cryic	Udic
Sun Wu	Helongjiang	49°	26"	127°	21"	235	Cryic	Udic
Tailai	Helongjiang	46°	24"	123°	25"	149	Frigid	Arid
Tie Li	Helongjiang	46°	59"	128°	01"	211	Frigid	Udic
Tong He	Helongjiang	45°	58"	128°	44"	109	Frigid	Udic
Xun Ke	Helongjiang	49°	35"	128°	27"	112	Cryic	Udic
Yi Chun	Helongjiang	47°	43"	128°	54"	231	Cryic	Udic
Sugehe	Inner Monglia	48°	10"	121°	15"	820	Pergilic	Udic
A Er Shan	Inner Mongolia	47°	10"	119°	57"	1027	Pergilic	Udic
A La Tane Mole	Inner Mongolia	48°	40"	116°	49"	554	Frigid	Arid
Abagaqihanbeimiao	Inner Mongolia	44°	01"	114°	57"	1126	Cryic	Arid
Aihui	Inner Mongolia	50°	15"	127°	27"	166	Frigid	Udic
Alihe	Inner Mongolia	50°	35"	123°	44"	424	Cryic	Udic
Amugulang	Inner Mongolia	48°	13"	118°	16"	642	Cryic	Arid
An Ta Chan	Inner Mongolia	46°	24"	125°	21"	151	Frigid	Ustic
Bailingmiao	Inner Mongolia	41°	42"	110°	26"	1376	Frigid	Arid
Baiyinwulahaote	Inner Mongolia	44°	35"	117°	36"	996	Cryic	Arid

APPENDIX 15.1 (continued)

Station	Province	Latitude (N)		Longitude (E)		Elevation (m)	STR	SMR
Balinzuoqi	Inner Mongolia	43°	59"	119°	24"	483	Frigid	Ustic
Bauintue	Inner Mongolia	41°	51"	107°	17"	1509	Frigid	Arid
Bautzan	Inner Mongolia	41°	53"	115°	16"	1470	Cryic	
Belemiao	Inner Mongolia	43°	51"	113°	44"	1111	Frigid	Arid
Boketu	Inner Mongolia	48°	46"	121°	55"	739	Cryic	Udic
Builingmiau	Inner Mongolia	41°	42"	110°	26"	1376	Frigid	Arid
Chaogeqihalisu	Inner Mongolia	41°	27"	106°	23"	1510	Frigid	Arid
Chicihar	Inner Mongolia	47°	23"	123°	55"	147	Frigid	Ustic
Chin Haute	Inner Mongolia	43°	41"	114°	29"	1126	Cryic	Arid
Chi-Ie	Inner Mongolia	40°	58"	113°	03"	1417	Frigid	Ustic
Doehua	Inner Mongolia	42°	11"	116°	28"	1245	Cryic	
Dong Sung	Inner Mongolia	39°	50"	109°	59"	1460	Frigid	Ustic
Songsheng	Inner Mongolia	39°	50"	109°	59"	1460	Frigid	Ustic
Dongwuzhumuqinqi	Inner Mongolia	45°	31"	116°	58"	839	Cryic	Arid
Duolun	Inner Mongolia	42°	11"	116°	28"	1245	Cryic	
Elunchunqixiaoergo	Inner Mongolia	49°	12"	123°	43"	288	Cryic	Udic
Erlianhaote	Inner Mongolia	43°	39"	112°	00"	965	Frigid	Arid
Erlien Haute	Inner Mongolia	43°	39"	112°	00"	965	Frigid	Arid
Gahailemiao	Inner Mongolia	45°	44"	118°	25"	850	Cryic	Arid
Geerguanaqi	Inner Mongolia	50°	13"	120°	12"	581	Pergelic	
Genhe	Inner Mongolia	50°	41"	121°	57"	980	Pergelic	Udic
Guaizihu	Inner Mongolia	41°	11"	103°	16"	959	Mesic	Arid
Hailaer	Inner Mongolia	49°	13"	119°	45"	613	Cryic	
Hailar	Inner Mongolia	49°	13"	119°	45"	613	Cryic	
Hailiutu	Inner Mongolia	41°	34"	108°	31"	1288	Frigid	Arid
Hailiutue	Inner Mongolia	41°	40"	108°	48"	1288	Frigid	Arid
Hanaola	Inner Mongolia	44°	47"	118°	45"	1032	Cryic	
Huade	Inner Mongolia	41°	54"	114°	00"	1483	Cryic	
Huadoe	Inner Mongolia	41°	54"	114°	00"	1483	Cryic	Arid
Huhetaolagai	Inner Mongolia	44°	28"	112°	40"	1000	Frigid	Arid
Jiergalangtumiao	Inner Mongolia	44°	41"	113°	42"	1206	Frigid	Arid
Jining	Inner Mongolia	41°	02"	113°	04"	1417	Frigid	Ustic
Juzuho	Inner Mongolia	42°	24"	112°	54"	1151	Frigid	Arid
Keyoqianqiaershan	Inner Mongolia	47°	10"	119°	57"	1027	Pergelic	Udic
Keyoqianqisnelun	Inner Mongolia	46°	37"	121°	14"	499	Frigid	Udic
Kujo	Inner Mongolia	41°	02"	110°	03"	1328	Frigid	Arid
Lin Dong	Inner Mongolia	43°	59"	119°	24"	483	Frigid	Ustic
Lin Xi	Inner Mongolia	43°	36"	118°	04"	799	Frigid	Ustic
Mandoolamio	Inner Mongolia	42°	29"	110°	08"	1206	Frigid	Arid
Mandula	Inner Mongolia	42°	32"	110°	08"	1222	Frigid	Arid
Manzhouli	Inner Mongolia	49°	34"	117°	26"	667	Cryic	Arid
Mingtumiao	Inner Mongolia	44°	43"	115°	07"	1000	Cryic	Arid
Narenbaolige	Inner Mongolia	44°	37"	114°	09"	1182	Cryic	Arid
Poketu	Inner Mongolia	48°	46"	121°	55"	739	Cryic	
San He	Inner Mongolia	50°	28"	120°	06"	663	Pergelic	
Siziwangqi	Inner Mongolia	41°	33"	111°	38"	1491	Frigid	Ustic
Sunitezhouqi	Inner Mongolia	43°	50"	113°	43"	1111	Frigid	Arid
Suo Lun	Inner Mongolia	46°	37"	121°	14"	499	Frigid	Udic
Ulanhua	Inner Mongolia	41°	33"	111°	38"	1489	Frigid	Udic
Wanggaimiao	Inner Mongolia	44°	35"	117°	36"	996	Cryic	Arid

APPENDIX 15.1 (continued)

Station	Province	Latitude (N)		Longitude (E)		Elevation (m)	STR	SMR
Wo Hu tun	Inner Mongolia	43°	42"	123°	36"	183	Frigid	Udic
Wuliyasitai	Inner Mongolia	45°	31"	116°	58"	839	Frigid	Arid
Xiao'Ergou	Inner Mongolia	49°	12"	123°	43"	288	Cryic	Udic
Xiguatuqitulihe	Inner Mongolia	50°	30"	121°	28"	733	Pergelic	Udic
Xiguituqibuoketu	Inner Mongolia	48°	46"	121°	55"	739	Cryic	Udic
Xilinhaote	Inner Mongolia	43°	57"	116°	04"	990	Frigid	Arid
Xinbaerhuyouqi	Inner Mongolia	48°	40"	116°	49"	554	Frigid	Arid
Xinbaerhuzueqi	Inner Mongolia	48°	13"	118°	16"	642	Cryic	Arid
Yakeshi	Inner Mongolia	49°	16"	120°	43"	659	Pergelic	
Yenchi	Inner Mongolia	42°	53"	129°	28"	1247	Frigid	Ustic
Yiungpoomio	Inner Mongolia	43°	19"	115°	39"	1187	Cryic	Arid
Zhurihe	Inner Mongolia	42°	24"	112°	54"	1151	Frigid	Arid
Antu Tianchi	Jilin	42°	01"	128°	05"	2670	Pergelic	Perudic
Chang Chun	Jilin	43°	54"	125°	13"	237	Frigid	Udic
Changbai	Jilin	41°	21"	128°	12"	711	Cryic	Udic
Changling	Jilin	44°	15"	123°	58"	192	Frigid	Udic
Dunhua	Jilin	43°	22"	128°	12"	524	Frigid	Udic
Fuyu Shanchahe	Jilin	44°	58"	126°	00"	197	Frigid	Udic
Helong	Jilin	42°	31"	128°	58"	443	Frigid	Udic
Huadian	Jilin	42°	58"	126°	43"	266	Frigid	Udic
Hunchun	Jilin	42°	54"	131°	17"	37	Frigid	Udic
Jilin Jiuzhan	Jilin	43°	57"	126°	28"	183	Frigid	Udic
Jinyu	Jilin	42°	21"	126°	49"	549	Frigid	Udic
Ling Jiang	Jilin	41°	43"	126°	55"	333	Frigid	Udic
Qianguoerluosi	Jilin	45°	07"	124°	50"	135	Frigid	Ustic
Tonghua	Jilin	41°	41"	125°	54"	403	Frigid	Udic
Tongyu	Jilin	44°	48"	123°	04"	148	Frigid	Ustic
Wangqin Luozigou	Jilin	43°	42"	130°	15"	496	Frigid	Udic
Yan	Jilin	42°	53"	129°	28"	177	Frigid	Udic
Qingyuan	Liaoning	42°	06"	124°	57"	237	Frigid	Udic
Xiji	Ningxia	35°	58"	105°	46"	1901	Cryic	Udic
Ala Er	Qinghai	38°	13"	90°	31"	2839	Cryic	Arid
Angqian	Qinghai	32°	11"	96°	29"	3644	Cryic	Udic
Chengduoqing Shuhe	Qinghai	33°	48"	97°	08"	4415	Pergelic	Perudic
Dachaidaan	Qinghai	37°	50"	95°	17"	3173	Cryic	Arid
Darijimai	Qinghai	33°	48"	99°	48"	3968	Cryic	Perudic
Dulan	Qinghai	36°	20"	98°	02"	3191	Cryic	Arid
Dulan Nuomuhong	Qinghai	36°	22"	96°	27"	2790	Cryic	Arid
Gang Cha	Qinghai	37°	20"	100°	10"	3302	Cryic	Udic
Cheermu	Qinghai	36°	12"	94°	38"	2808	Cryic	Arid
Cheermu Tuotuohe	Qinghai	33°	57"	92°	37"	4533	Pergelic	
Geermu Wudaoliang	Qinghai	35°	17"	93°	36"	1645	Pergelic	
Geermuwutumeiren	Qinghai	26°	51"	93°	10"	2843	Cryic	Arid
Gonghe	Qinghai	36°	17"	100°	37"	2835	Cryic	
Huzhuquzhangtan	Qinghai	37°	11"	102°	09"	2871	Cryic	Udic
Jigzhi	Qinghai	33°	19"	101°	14"	3629	Cryic	Perudic
Lenghu	Qinghai	38°	50"	93°	23"	2733	Cryic	Arid
Maduo	Qinghai	34°	57"	98°	08"	4221	Pergelic	
Mangyan	Qinghai	38°	22"	90°	09"	3139	Cryic	Arid
Mengyuan	Qinghai	37°	27"	102°	00"	2708	Cryic	Udic

APPENDIX 15.1 (continued)

Station	Province	Latitude (N)		Longitude (E)		Elevation (m)	STR	SMR
Qilian	Qinghai	38°	11"	100°	18"	2787	Cryic	
Qilian Tuole	Qinghai	34°	52"	98°	22"	3360	Pergelic	Arid
Qumalai	Qinghai	35°	32"	95°	28"	4263	Pergelic	Udic
Tongde	Qinghai	35°	09"	100°	20"	3289	Cryic	Udic
Tongren	Qinghai	35°	30"	102°	05"	2488	Cryic	
Waisi	Qinghai	34°	17"	101°	35"	3414	Cryic	Udic
Wulanchaka	Qinghai	36°	47"	99°	04"	3088	Cric	Arid
Wulandelinha	Qinghai	37°	15"	97°	08"	2881	Cryic	Arid
Xinghai	Qinghai	35°	48"	99°	40"	3323	Cryic	
Yushu	Qinghai	33°	06"	96°	45"	3703	Cryic	Udic
Zadoi	Qinghai	32°	54"	95°	19"	4068	Cryic	Udic
Taishan	Shandong	36°	15"	117°	06"	1534	Frigid	Udic
Uzai	Shanxi	38°	56"	111°	49"	1400	Frigid	Udic
Youyu	Shanxi	39°	59"	112°	27"	1338	Frigid	Ustic
Aba	Si Chuan	32°	40"	102°	14"	3275	Cryic	Udic
Daocheng	Si Chuan	29°	03"	100°	11"	3728	Cryic	Udic
Emeishan	Si Chuan	29°	31"	103°	21"	3047	Cryic	Perudic
Ganzi	Si Chuan	31°	38"	99°	59"	3394	Cryic	Udic
Litang	Si Chuan	30°	00"	100°	16"	3949	Cryic	Udic
Qianning	Si Chuan	30°	30"	101°	29"	3449	Cryic	Udic
Rhu Erhe	Si Chuan	33°	20"	102°	43"	3447	Cryic	Perudic
Seda	Si Chuan	32°	25"	100°	44"	3894	Cryic	Udic
A Lo Tai	Xinjiang	47°	44"	88°	05"	735	Frigid	Arid
Altay	Xinjiang	47°	44"	88°	05"	735	Frigid	Arid
Bali Kun	Xinjiang	43°	44"	93°	04"	1638	Cryic	Arid
Burqin	Xinjiang	47°	42"	86°	52"	474	Frigid	Arid
Fuhai	Xinjiang	47°	07"	87°	30"	500	Frigid	Arid
Fuyun	Xinjiang	46°	59"	89°	30"	803	Frigid	Arid
Habahe	Xinjiang	48°	03"	86°	21"	533	Frigid	Arid
Hobokesar Mongol	Xinjiang	46°	47"	85°	43"	1292	Frigid	Arid
Jeminay	Xinjiang	47°	26"	85°	52"	1015	Frigid	Arid
Qinghe	Xinjiang	46°	40"	90°	23"	1218	Cryic	Arid
Qitai	Xinjiang	44°	01"	89°	34"	796	Frigid	Arid
Qitaibeitashan	Xinjiang	45°	22"	90°	32"	1650	Cryic	Arid
Toli	Xinjiang	45°	56"	83°	36"	1118	Frigid	Arid
Wuen quan	Xinjiang	44°	59"	81°	04"	1132	Frigid	Arid
Wulumuqi Xiaoquzi	Xinjiang	43°	34"	87°	06"	2160	Cryic	Udic
Wuqia Tuergete	Xinjiang	40°	31"	75°	24"	3505	Pergelic	Arid
Yiwu	Xinjiang	43°	16"	94°	42"	1729	Frigid	Arid
Zhaosu	Xinjiang	43°	09"	81°	08"	1849	Cryic	Udic
Ban'ge	Xizang	31°	22"	90°	01"	4700	Cryic	
Ban'gehu	Xizang	31°	48"	89°	40"	4380	Pergelic	
Dangxiong	Xizang	30°	29"	91°	05"	4200	Cryic	Udic
Deng Qing	Xizang	31°	25"	95°	36"	3873	Cryic	Udic
Dingri	Xizang	28°	35"	86°	37"	4300	Cryic	
Gaer	Xizang	32°	30"	80°	05"	4278	Cryic	Arid
Jia Li	Xizang	30°	48"	93°	17"	4317	Cryic	Perudic
Jiangzhi	Xizang	28°	55"	89°	36"	4040	Cryic	
Langkazi	Xizang	28°	58"	90°	25"	4432	Cryic	Udic
Longzi	Xizang	28°	25"	92°	28"	3900	Cryic	

APPENDIX 15.1 (continued)

Station	Province	Latitude (N)		Longitude (E)		Elevation (m)	STR	SMR
Naqu (H.H.)	Xizang	31°	29"	92°	03"	4507	Cryic	Udic
Shenzha	Xizang	30°	57"	88°	38"	4670	Cryic	
Suo Xian	Xizang	21°	54"	93°	48"	3950	Cryic	Udic
Yiadong Beli	Xizang	27°	44"	89°	05"	4300	Cryic	Udic
Deikan	Yunnan	28°	39"	99°	10"	3589	Cryic	Udic

Note: STR — soil temperature regime; SMR — soil moisture regime.

16   Temperature and Thermal Properties of Alaskan Soils

R.F. Paetzold, K.M. Hinkel, F.E. Nelson, T.E. Osterkamp, C.L. Ping, and V.E. Romanovsky

CONTENTS

INTRODUCTION

Alaska covers an area of 586,400 square miles, one fifth of it north of the Arctic Circle. Point Barrow, the northernmost point in the U.S. at 71°18' N latitude, is 2078 km (1291 mi) south of the North Pole. The range of climatic conditions is commensurately broad, giving rise to vegetation associations as disparate as temperate rainforest in the southeast and tundra in the north. Although much of the state is under the influence of the oceans and their currents, the interior of the state experiences a very continental climate with large annual temperature amplitude. The mountains of southern Alaska have the largest relief in North America, and exert a major climatic influence over much of the state. Denali (Mt. McKinley) is the highest peak in North America at 6194 m (20,320 ft). The central and western part of the Aleutian Islands chain is one of the stormiest regions in the world (Baldwin, 1973).

Mean annual air temperature ranges from 7.9°C (46.3°F) in Ketchikan to –12.6°C (9.3°F) in Barrow (Rieger et al., 1979). Although soil temperature is related to air temperature, it is strongly influenced by the microclimatic factors that determine the energy balance at the ground surface, by the nature and properties of the surface cover, and by subsurface factors such as moisture content and soil thermal properties. Topographic form is also an important consideration because it governs

FIGURE 16.1 Alaska atmospheric climatic stations. (Adapted from the Alaska Climate Research Center.)

the geometry of solar radiation receipts and local details of the vegetation and snow cover. Vegetation, soil surface color, and surface roughness determine albedo. Soil water content, organic matter content, and soil porosity characteristics greatly influence soil thermal conductivity and heat capacity. Snow affects soil temperature through its insulating characteristics and albedo.

The objectives of this chapter are to present a brief, general review of soil temperature and soil thermal properties as applicable to Alaska. Locations of some historical and some current soil temperature sites are given. Use of soil thermal properties in some models is discussed.

ALASKA CLIMATE

Air temperature data for Alaska are available through the internet (http://climate.gi.alaska.edu/history/ History.html). Total precipitation and snowfall data are available at these sites, also. Most of the stations have 30-year records; however, the network of stations is rather sparse. Most of the stations are south of the Yukon River and six of the eight stations north of the Yukon river are located on the coast. Figure 16.1 shows the locations of the atmospheric climate stations in Alaska. Figures 16.2 through 16.5 show typical data from two stations: Anchorage and Barrow.

There are several climate classifications. Two of the better known and most widely used are Köppen's and Thornthwaite's. Köppen's classification is based upon annual and monthly means of temperature and precipitation as well as "native vegetation" (Trewartha, 1954). Thornthwaite's classification is based on precipitation effectiveness which is determined by dividing the total monthly precipitation by the total monthly evaporation and summing the 12 monthly ratios to obtain a P/E index (Trewartha, 1954). A relatively general map showing a classification based primarily on temperature and precipitation is shown in Figure 16.6 (Moore et al., 1993). This map is similar to other classification schemes. The names may differ, but the basic delineations are the similar. The northern part of the state, north of the Brooks Range, is classified as the arctic zone. In other classifications, this area may be called subarctic tundra (Landsberg et al., 1965), polar tundra (Trewartha, 1954), or tundra (Trewartha, 1954). The southern coastal portion of the state is called the maritime zone, the western coastal area is the transition zone (transition between the maritime and continental zones), and the large central portion of the state is in the continental zone.

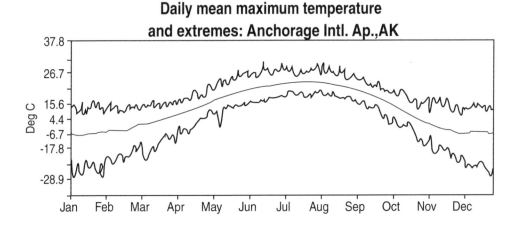

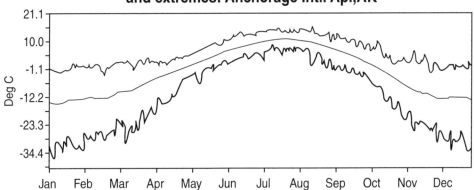

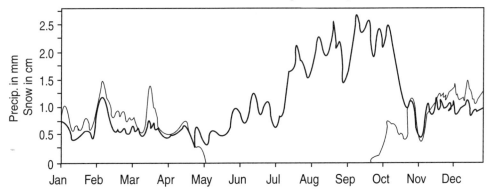

FIGURE 16.2 Air temperature and precipitation data for Anchorage, Alaska 1961 to 1990 (NOAA, NCDC). Maximum temperature 1961 to 1990: 29.4°C (85°F). Minimum temperature 1961 to 1990: –36.7°C (–34°F). Mean annual precipitation: 40.4 cm (15.9 in.). Mean annual snowfall: 171.5 cm (67.5 in.).

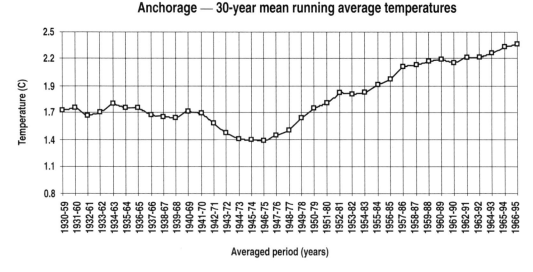

FIGURE 16.3 Anchorage, Alaska 30-year running average air temperatures. (Adapted from NOAA, NCDC.)

The area north of the Brooks Range, designated as the arctic zone in Figure 16.6, is an area of continuous permafrost as indicated in Figure 16.7. The southern coastal area is a no permafrost zone and the rest of the state is classified as discontinuous permafrost. The discontinuous permafrost zone contains areas of permafrost mixed with areas of no permafrost. The permafrost in the discontinuous permafrost zone can be destroyed fairly easily. For many uses, e.g., roads, buildings, etc., it is very important to know whether the site contains permafrost or not. Building on permafrost can cause it to melt with possibly disastrous, or at least expensive, consequences.

ALASKA SOIL CLIMATE

There are fewer soil climate classification schemes than climate classifications. Soil climate classifications generally are part of overall soil classification schemes or soil taxonomies. Two of the most important soil classification systems for cold soil regions are the U.S. Soil Taxonomy (Soil Survey Staff, 1975, 1996) and the Canadian Soil Classification System (Agriculture Canada Expert Committee on Soil Survey, 1987). Table 16.1 is a comparison of the soil temperature part of these two soil classification systems along with the proposed changes to the U.S. system.

The temperatures represent mean annual soil temperatures measured at 50 cm (20 in.). The soil temperature ranges show some overlap in various regimes. This is because the definitions of the regimes are a bit more complicated than just temperature ranges (see Appendix A for complete definitions). The ranges, however, are sufficient for comparison purposes.

Figure 16.8 shows the distribution of soil temperature regimes based on the current USDA Soil Taxonomy. The regimes closely follow the permafrost distribution as shown in Figure 16.7.

Measured soil temperature data are relatively scarce in the state. Locations of known soil temperature monitoring sites are presented in Figure 16.9. The sites shown are those for which latitude and longitude are known. Other measurement sites are known to exist, but could not be plotted on the map. Most soil temperature measurements are from short duration efforts ranging from one to five or six years. Many of the sites plotted in Figure 16.9 are from Chang (1958). Other data sources include Ping (1987), Rieger (1973), Alexander (1991), DaMore (1998), Hinkel (1998), and Nelson (1998).

The mean annual soil temperatures for the sites shown in Figure 16.9 were used to construct a soil temperature regime map based on the proposed USDA Soil Taxonomy criteria. The map is

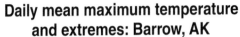

Daily mean maximum temperature and extremes: Barrow, AK

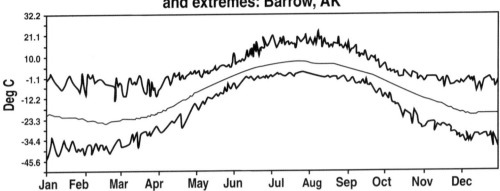

Daily mean minimum temperature and extremes: Barrow, AK

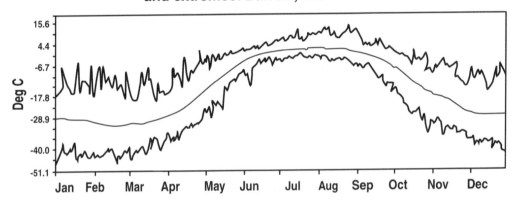

Daily mean precipitation and snowfall: Barrow, AK

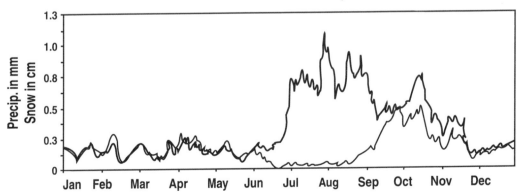

FIGURE 16.4 Air temperature and precipitation data for Barrow, Alaska 1961 to 1990 (NOAA, NCDC). Maximum temperature 1961 to 1990: 24.4°C (76°F). Minimum temperature 1961 to 1990: –47.8°C (–54°F). Mean annual precipitation: 11.4 cm (4.5 in.). Mean annual snowfall: 73.2 cm (28.8 in.).

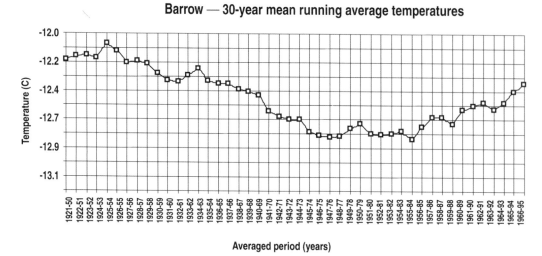

FIGURE 16.5 Barrow, Alaska 30-year running average air temperatures. (Adapted from NOAA, NCDC.)

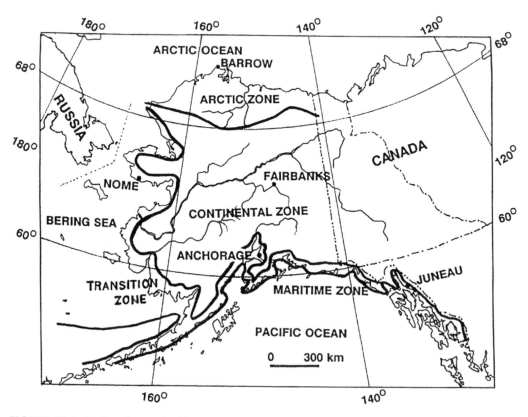

FIGURE 16.6 Alaska climate classification based primarily on air temperature and precipitation. (Adapted from Moore, J.P., Swanson, D.K., Fox, C.A., and Ping, C.L., International Correlation Meeting on Permafrost-Affected Soils, NRCS National Soil Survey Center, Lincoln, NE, 1993.)

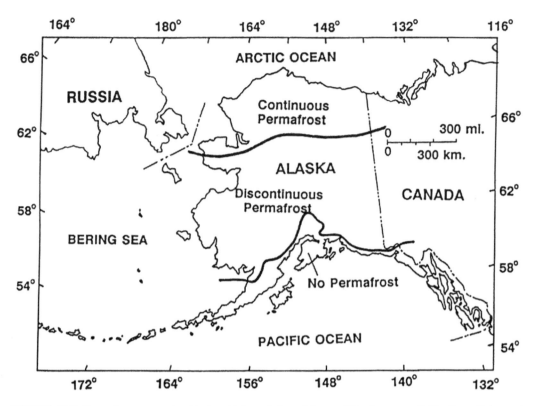

FIGURE 16.7 Permafrost zones in Alaska. (Adapted from Moore, J.P., Swanson, D.K., Fox, C.A., and Ping, C.L., International Correlation Meeting on Permafrost-Affected Soils, NRCS National Soil Survey Center, Lincoln, NE, 1993.)

TABLE 16.1
Soil Temperature Regime Classifications

Soil Taxonomy		Soil Taxonomy Proposed		Canadian Soil Classification System	
Regime	Temperature Range	Regime	Temperature Range	Regime	Temperature Range
		Hypergelic	<–10°C	Extremely Cold	<–7°C
Pergelic	<0°C	Pergelic	–10°C to –4°C	Very Cold	–7°C to –2°C
Cryic	0°C to 8°C	Subgelic	–4°C to +1°C	Cold	–8°C to +2°C
Frigid	<8°C	Frigid	<8°C	Cool	5°C to 8°C
Mesic	8°C to 15°C	Mesic	8°C to 15°C	Mild	8°C to 15°C
Thermic	15°C to 22°C	Thermic	15°C to 22°C		
Hyperthermic	>22°C	Hyperthermic	>22°C		

presented in Figure 16.10. Note that this is significantly different from the map (Figure 16.8) based on the current USDA Soil Taxonomy definitions. The pergelic areas are roughly the same. In Figure 16.10 there are two areas of subgelic and a large area of frigid. One subgelic area forms a transition between the pergelic and the warmer frigid areas. The other subgelic area intrudes from the eastern part of the state to denote an area of colder soil temperature. The proposed USDA Soil Taxonomy criteria do not appear to adequately denote the areas of no permafrost and discontinuous permafrost. Figure 16.8 was developed from estimates of soil temperature based on soil morphology,

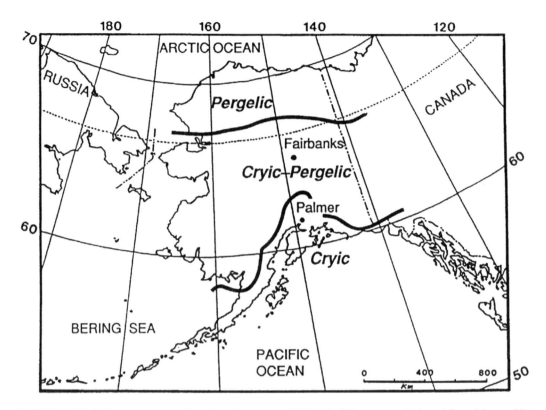

FIGURE 16.8 Soil temperature regimes based on current USDA Soil Taxonomy. (Adapted from Moore, J.P., Swanson, D.K., Fox, C.A., and Ping, C.L., International Correlation Meeting on Permafrost-Affected Soils, NRCS National Soil Survey Center, Lincoln, NE, 1993.)

vegetation, permafrost observations, and other field clues. Because of this development procedure, it closely approximates the permafrost zones in Alaska. A map based on the measured mean annual temperatures of the same stations used in the construction of Figure 16.10 gives a somewhat different soil temperature regime map (Figure 16.11).

The sparse network of measured soil temperature data, along with the short periods of record, contribute to the uncertainty of the soil temperature regime distribution map. Some additional soil temperature data are available; however, latitude and longitude are not available. In general, soil temperature data are not readily available. Some soil temperature data sets are inaccessible. Even when published, soil temperature data are limited in usefulness due to lack of information on site location, elevation, slope, aspect, vegetation, and other details. In addition, there are very few, if any, long-term soil temperature data sets available.

SOIL TEMPERATURE

As heat is removed from water, the temperature lowers until the freezing point, 0°C, is reached. At this point, the temperature remains constant until the liquid water solidifies, or changes to ice, at which time the temperature will once again start to drop. In soils, however, water does not all freeze at the same temperature. Some freezes at or near 0°C, while some freezes at much lower temperatures. No unique freezing point exists for soil water (Koopmans and Miller, 1966). There are two reasons for this. One, soil water contains solutes or salts which lower the freezing point. As the soil water freezes, the salts become more concentrated due to exclusion from the ice. And two, the "structure" of soil water becomes increasingly ordered closer to soil particles. This changes

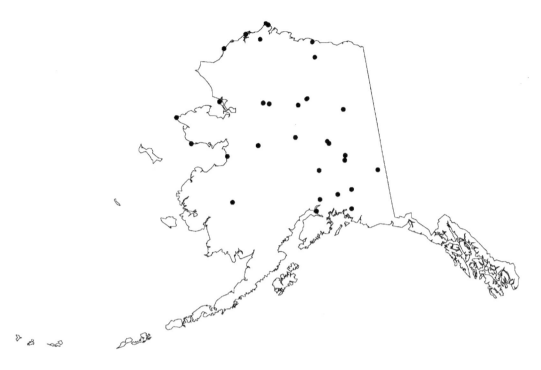

FIGURE 16.9 Alaska soil temperature monitoring sites.

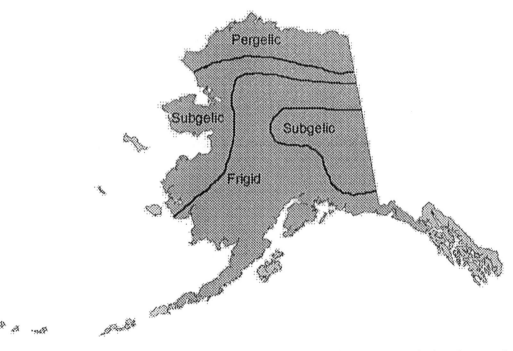

FIGURE 16.10 Soil temperature regime map based on proposed USDA Soil Taxonomy criteria and measured soil temperature.

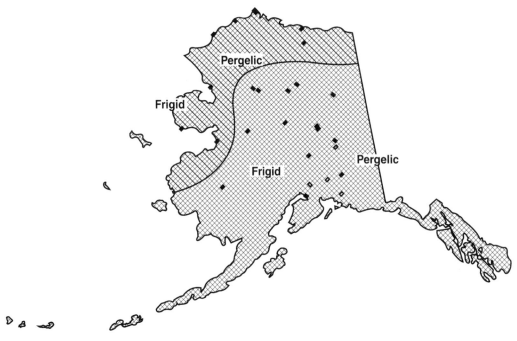

FIGURE 16.11 Soil temperature regime map based on current soil taxonomy criteria and measured soil temperature.

the free energy of the soil water and lowers the freezing point (Williams et al., 1989). A similar phenomenon can be observed in capillary tubes (Edlefsen and Anderson, 1943).

Anderson and Tice (1972) give the following relationship among soil unfrozen water content, temperature, and soil specific surface area.

$$\ln \theta_v = 0.2618 + 0.5519 \ln S - 1.449 \, S^{-0.264} \ln T \qquad (16.1)$$

where: θ_v = unfrozen water content (g g^{-1}),
S = specific surface area (m^2 g^{-1}), and
T = temperature in degrees below 0°C.

The soil specific surface area is controlled primarily by the clay content and clay mineralogy (Baver et al., 1972). In some soils, organic matter content exerts a major influence. While sand and silt have specific surface areas of less than 1 m^2 g^{-1}, the specific surface of clay ranges from 5 to 800 m^2 g^{-1}. Organic matter can have an effective specific surface of up to 1000 m^2 g^{-1}.

Thermal properties of soils depend greatly on the soil water content (Crawford, 1950) and in frozen soils, the proportion of unfrozen water and ice. Because the unfrozen water content of soil depends on the temperature, characterization of the soil thermal properties is complicated at temperatures at and just below 0°C.

For a given soil material the amount of unfrozen water present is known to depend principally upon temperature and to a lesser extent upon pressure; and, except for very low water contents, it is virtually independent of the total water content. At a given temperature and pressure, the principal factors governing the unfrozen water content are the specific surface area of the soil, its chemical and mineralogical character, the number and kind of the exchangeable ions, and the species and concentrations of soluble substances present (Anderson, 1966). Determinations of the amount of unfrozen water in frozen soil have been accomplished by: the dilatometer method, the calorimetric method, an X-ray diffraction method, and with nuclear magnetic resonance (NMR). The dilatometer method yields values that are probably too high (the clay-adsorbed water has a density less than

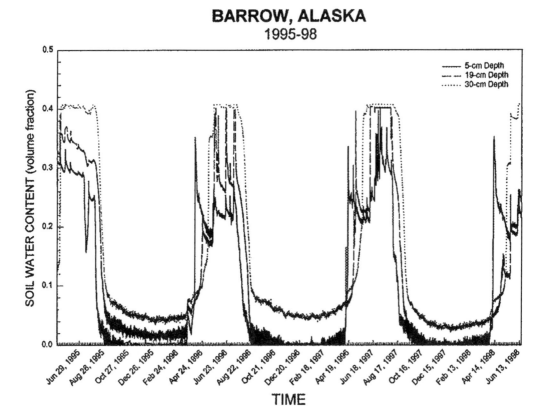

FIGURE 16.12 Unfrozen soil water content near Barrow, Alaska.

that of pure bulk water). The calorimetric method yields values that probably are too high (calculations require assumptions that the specific heat and latent heat of freezing the clay-adsorbed water are the same as for pure bulk water). The X-ray method is applicable only to mixtures with water of 2:1 expanding lattice clay minerals and cannot account for the existence of unfrozen water located outside the interlamellar space, thus it underestimates the unfrozen water content. NMR is relatively new and expensive technology, but gives the best results in determination of unfrozen soil water content (Tice et al., 1982).

Figure 16.12 shows unfrozen soil water content patterns in a soil near Barrow, Alaska. The unfrozen soil water content increases with depth, corresponding to temperature increases with depth.

THERMAL PROPERTIES OF SOILS

Soil temperature is influenced by air temperature, incoming solar radiation, slope, aspect, latitude, elevation, vegetation, soil moisture, and thickness of a surface organic layer. Snow is an effective insulator that affects winter soil temperatures. Thermal energy is transferred between air and soil through the process of convection. In addition to the temperature difference between the air and the ground, the wind speed profile and elevation (air density) are important factors governing this process. Vegetation can have considerable influence on the wind speed profile close to the ground.

Thermal energy is transferred from the sun to the soil through the process of radiation. The amount of energy reaching the soil surface depends on albedo, vegetation, cloudiness, latitude, slope, and aspect.

Thermal energy moves within soil largely through the process of conduction, although convection and radiation also occur. Conduction depends on the thermal conductivity of the medium and the thermal gradient. Under transient conditions, the rate of temperature change depends on the heat capacity of the medium as well as the thermal conductivity and temperature gradient. In soil, the thermal conductivity and heat capacity are largely functions of the soil water content.

DEFINITIONS OF SOME COMMON THERMAL TERMS

Heat is energy in transit from one mass to another because of a temperature difference between the two. Whenever a force of any kind acts through a distance, work is done. Like heat, work is also energy in transit. The heat capacity of a material is the amount of heat transferred to raise a unit mass of a material one degree in temperature. The specific heat of a material is the ratio of the heat capacity of a material to that of water. For most purposes, heat capacity may be assumed numerically equal to specific heat. Sensible heat is a change in the internal energy of a material. Addition or removal of heat that results in changes in the internal energy of a substance can be sensed through temperature changes of the material. For pure substances, the heat effects accompanying changes in state at constant pressure are known as latent heat, because no temperature change is evident. Heat of fusion, vaporization, sublimation, and change in crystal form are examples of latent heats. The thermal conductivity is the proportionality constant relating heat flux to the thermal gradient (see Equation 16.3).

Thermal diffusivity, α, (not to be confused with the absorptivity α) is the thermal conductivity, k, divided by the volumetric heat capacity, ρc_p, where the volumetric heat capacity is the product of the density, ρ, and the specific heat, c_p. The volumetric heat capacity is a measure of the ability of a material to store thermal energy. The units of the thermal diffusivity, α, are m^2/s. The units for volumetric heat capacity, ρc_p, are J/m^3 × K. Thermal diffusivity, α, measures the ability of a material to conduct thermal energy relative to its ability to store thermal energy. Materials of large α will respond quickly to changes in their thermal environment, while materials of small α will respond more sluggishly, taking longer to reach a new equilibrium condition.

$$\alpha = \frac{k}{\rho \cdot c_p} \qquad (16.2)$$

Heat may be transferred by three methods: radiation, conduction, and convection. The heat transfer mechanisms of mass transfer and phase change (latent heats) generally are included in convection.

The following definitions are taken from Incropera and DeWitt, 1990:

Conduction Equations

Heat conduction is heat transfer due to molecular or atomic activity-transfer of energy from more energetic to less energetic particles of a substance due to interactions between particles.

Rate Equation (one dimensional):

$$q''_x = -k\frac{dT}{dx} \qquad (16.3)$$

where: q''_x = heat flux (W/m^2) in the x direction per unit area, A, perpendicular to the direction of transfer,

   k = thermal conductivity (Wm^{-1} × K), and

   dT/dx = temperature gradient (Km^{-1}).

Note: q_x = heat transfer rate (W) = q''_x × A.

The basic tool for heat conduction analysis is the heat equation (three-dimensional):

$$\frac{\partial}{\partial x}\left(k\frac{\partial T}{\partial x}\right) + \frac{\partial}{\partial y}\left(k\frac{\partial T}{\partial y}\right) + \frac{\partial}{\partial z}\left(k\frac{\partial T}{\partial z}\right) + \dot{q} = \rho c_p \frac{c \partial T}{\partial t} \tag{16.4}$$

where: t = time,

ρ = density,

c_p = heat capacity,

$\rho c_p \dfrac{\partial T}{\partial t}$ = the time rate of change of the internal (thermal) energy of the medium per unit volume, and

$\dot{q}$ = rate of energy generation per unit volume of the medium (Wm^{-3}).

Convection

The convection heat transfer mode is comprised of two mechanisms: diffusion (random molecular motion) and bulk motion of the fluid. Note: advection refers to heat transport due solely to bulk fluid motion.

Rate Equation:

$$q'' = h(T_s - T_\infty) \tag{16.5}$$

where: h = convection heat transfer coefficient ($Wm^{-2} \times K$),

T_s = surface temperature, and

T_∞ = fluid temperature.

The convection heat transfer coefficient, h, depends on conditions in the boundary layer, which are influenced by surface geometry, the nature of the fluid motion, and an assortment of fluid thermo-dynamic and transport properties. The convection heat flux is presumed to be positive if heat is transferred from the surface ($T_s > T_\infty$) and negative if heat is transferred to the surface.

Typically, the energy that is being transferred is the sensible, or internal thermal, energy of the fluid. However, there are convection processes for which there is, in addition, latent heat exchange. This latent heat exchange is generally associated with a phase change between the liquid and vapor states of the fluid. An example is the condensation of water vapor on the outer surface of a cold water pipe.

Radiation

Thermal radiation is energy emitted by matter that is at a finite temperature. Emission may occur from solid surfaces, liquids, and gases. The energy of the radiation field is transported by electro-magnetic waves. Radiation transfer occurs most efficiently in a vacuum. The maximum flux (W/m^2) at which radiation may be emitted from a surface is given by the Stefan-Boltzmann law:

$$q'' = \sigma T_s^4 \tag{16.6}$$

where: T_s = absolute temperature (K) of the surface, and

σ = Stefan-Boltzmann constant

= 5.67×10^{-8} $W/m^2 \times K^4$.

Equation 16.6 is for an ideal radiator or blackbody.

The heat flux emitted by a real surface is less than that of a blackbody and is given by:

$$q'' = \varepsilon \sigma T_s^4 \qquad (16.7)$$

where: ε = emissivity of the surface.

The emissivity, ε, of a surface ranges from 0 to 1 and indicates how efficiently the surface emits compared to an ideal radiator. Conversely, if radiation is incident upon a surface, a portion will be absorbed, and the rate at which energy is absorbed per unit surface area may be evaluated from knowledge of a surface radiative property, termed the absorptivity α.

$$q''_{abs} = \sigma q''_{inc} \qquad (16.8)$$

where α is between 0 and 1.

Determination of the net rate at which radiation is exchanged between surfaces is more complicated. A special case that occurs frequently in practice involves the net exchange between a small surface and a much larger surface that completely surrounds the smaller one. The surface and the surroundings are separated by a gas that has no effect on the radiation transfer. Assuming the surface to be one for which $\alpha = \varepsilon$ (a gray surface), the net radiation heat exchange between the surface and its surroundings, expressed per unit area of the surface is

$$q'' = \frac{q}{A} \varepsilon \sigma (T_s^4 - T_{sur}^4) \qquad (16.9)$$

where: A = surface area

ε = the emissivity of the surface, and

T_{sur} = the temperature of the surroundings.

For this special case, the area and emissivity of the surroundings do not influence the net heat exchange rate.

The surface within the surroundings may also simultaneously transfer heat by convection to the adjoining gas. The total rate of heat transfer from the surface is then the sum of the heat rates due to the two modes. That is

$$q = q_{conv} + q_{rad} \qquad (16.10)$$

or

$$q = hA(T_s - T_\infty) + \varepsilon A \sigma (T_s^4 - T_{sur}^4) \qquad (16.11)$$

NOTE: For heat transfer to occur, there must be a temperature gradient, hence thermodynamic nonequilibrium. The first law of thermodynamics (the law of conservation of energy) states that the amount of thermal and mechanical energy which enters a control volume, plus the amount of thermal energy which is generated within the control volume, minus the amount of thermal and mechanical energy which leaves the control volume, must equal the increase in the amount of energy stored in the control volume.

$$\frac{dE_{cv}}{dt} = \dot{Q}_{cv} - \dot{W}_{cv} + \sum_i \dot{m}_i \left(h_i + \frac{V_i^2}{2} + gz_i \right) - \sum_e \dot{m}_e \left(h_e + \frac{V_e^2}{2} + gz_e \right) \qquad (16.12)$$

where: E_{cv} = the energy stored in the control volume,
Q_{cv} = thermal energy going into the control volume,
W_{cv} = mechanical energy leaving the control volume,
m = mass of the material,
h = enthalpy of the material (on a unit mass basis, h = u + pv, where u is the internal energy, p is the pressure, and v is the volume) not to be confused with the convection heat transfer coefficient,
V = velocity of the material,
g = the gravitational constant, and
z = elevation.

A thermal energy generation term, $\dot{E}_g$, may be added to Equation 16.12 to account for the rate of energy conversion from some other energy form (chemical, electrical, electromagnetic, or nuclear) to thermal energy within the control volume.

A control surface, having no mass or volume, may be used in some cases instead of a control volume. In this case the conservation requirement holds for both steady-state and transient conditions. $\dot{E}_{in} - \dot{E}_{out} = 0$ and $q''_{cond} - q''_{conv} - q''_{rad} = 0$, where q_{cond} is heat conduction to the control surface and q_{conv} and q_{rad} are heat convection and radiation from the control surface.

Heat Conduction

If one assumes one-dimensional heat transfer and no energy generation, Equation 16.4 reduces to:

$$\frac{\partial}{\partial x}\left(k\frac{\partial T}{\partial x}\right) = \rho c_p \frac{\partial T}{\partial t} \tag{16.13}$$

For steady-state conditions, the right side of Equation 16.13 reduces to zero. For transient conditions with surface convection in a semi-infinite solid:

$$-k\frac{\partial T}{\partial x}\bigg|_{x=0} = h[T_\infty - T(0, t)] \tag{16.14}$$

The closed-form solution to Equation 16.14 is given by:

$$\frac{T(x, t) - T_i}{T_\infty - T_i} = erfc\left(\frac{x}{2\sqrt{\alpha t}}\right) - \left[\exp\left(\frac{hx}{k} + \frac{h^2\alpha t}{k^2}\right)\right]\left[erfc\left(\frac{x}{2\sqrt{\alpha t}} + \frac{h\sqrt{\alpha t}}{k}\right)\right] \tag{16.15}$$

where: T_i = the initial temperature at the point x, and
T_∞ = the temperature at a point within the solid that remains constant.

This is the general equation used for heat transfer in soils. The convection heat transfer coefficient, h, is not easy to evaluate. It depends on conditions in the boundary layer, which are influenced by surface geometry, the nature of the fluid motion, and an assortment of fluid thermodynamic and transport properties. In addition, radiation effects must be incorporated into the solution.

Equation 16.14 is the starting point for most mechanistic models of soil temperature. There are many types of models. Some other types of soil temperature models are stochastic, empirical, and deterministic. Some models are combinations or employ other indicators of soil temperature. While mechanistic models use material properties such as thermal conductivity, heat capacity, density, etc., stochastic models usually don't. Some models use inputs such as wind speed, air temperature, and solar radiation which must either be monitored or estimated from nearby historical data.

TABLE 16.2
Thermal Properties[a] of Selected Materials

Material	k (W m⁻¹ K⁻¹)	C_p (J kg⁻¹ K⁻¹)	$\alpha \times 10^6$ (m² s⁻¹)
Air	0.026	1007	22.5
Water	0.62	4200	0.14
Ice	2.0	2000	1.0
Snow	0.05–2.25	—	—
Soil	0.1–1.3	800–1800	1.0

[a] Properties generally vary with temperature and density.

TABLE 16.3
Thermophysical Properties[a] of Air at Atmospheric Pressure

Composition	Temperature (K)	ρ (kg m⁻³)	C_p (kJ kg⁻¹ K)	k 10³ (W m⁻¹ K)	α 10⁶ (m² s⁻¹)	μ 10⁷ (N s m⁻²)	ν 10⁶ (m² s⁻¹)	Pr
Air	250	1.3947	1.006	22.3	15.9	159.6	11.44	0.720
	300	1.1614	1.007	26.3	22.5	184.6	15.89	0.707
	350	0.9950	1.009	30.0	29.9	208.2	20.92	0.700

[a] ρ(kg m⁻³) is density; ρ(kg m⁻³) is specific heat at constant pressure; C_p(kJ kg⁻¹ K) is thermal conductivity; k 10⁶ (W m⁻¹ k) is thermal diffusivity; α 10⁶ (m² s⁻¹) is viscosity (same as kg s⁻¹ m); ν 10⁶ (m² s⁻¹) is kinematic viscosity; Pr (dimensionless) is Prandtl number.

Source: Adapted from Incropera, F.P. and DeWitt, D.P., *Fundamentals of Heat and Mass Transfer*, John Wiley & Sons, New York, 1990. With permission.)

When exposed to a temperature gradient, water in a soil will move from points of high temperature to points of low temperature. As the water moves, some heat is carried with it, which changes the temperature gradient. Furthermore, heat can be transferred by conduction through the solid matrix as well as by conduction and convection through the water and air phases. Significant variation in soil temperature may result from latent heat as water changes state. As water is evaporated, energy for latent heat is required. This heat is partially supplied by the moist soil and the soil surface thus gradually cools. The latent heat for evaporation of pure water at room temperature is approximately 2.43×10^6 J kg⁻¹ (580 cal g⁻¹). Latent heat required to transfer soil water to the vapor phase may exceed this value by as much as 0.84×10^6 J kg⁻¹ (200 cal g⁻¹) (Nielsen et al., 1972). The drier the soil, the greater the latent heat released or required for condensation and evaporation. Condensation of water vapor on soil can cause a temperature rise. NOTE: Isothermal conditions are almost always assumed when water movement in soil is analyzed. Thermal properties of selected materials are shown in Table 16.2

The effect of impurities in ice is to increase its apparent specific heat. Table 16.3 illustrates the dependence of the thermal properties of air on temperature. Specific heats for various materials often found in soils are given in Table 16.4. Other sources for soil thermal property data include Shul'gin (1957), Gol'tsberg and Davitaya (1971), Carson, (1961), Baver et al. (1972), and various soil physics and engineering texts.

MODELING

This section is a brief introduction to some of the current arctic soil temperature-modeling efforts. No attempt is made to describe the models or results in any detail, only to provide introductory

TABLE 16.4
Mean Specific Heats of Various Solids
(32 to 212°F, 273 to 373 K)

Material	C (kJ kg^{-1} K^{-1})	Material	C (kJ kg^{-1} K^{-1})
Alumina	0.77	Limestone	0.91
Gneiss	0.75	Marble	0.88
Granite	0.84	Quartz	0.96
Graphite	0.84	Sand	0.82
Gypsum	1.10	Sandstone	0.92
Hornblende	0.84	Serpentine	1.05
Humus (soil)	1.80	Silica	0.80
Kaolin	0.94	Talc	0.87

Source: Adapted from Avallone, E.A. and Baumeister, T., *Marks' Standard Handbook for Mechanical Engineers*, 9th ed., McGraw-Hill, New York, 1987. With permission.

information on some of the work being done in this area. And this is certainly not a complete list of arctic soil temperature-modeling efforts.

Hinzman, et al. (1997) have a thermal model that simulates thermal processes at the surface of the tundra, within the active layer, and in the underlying permafrost. They tested it using data from seven meteorological stations within the Kuparuk River Watershed. Their model uses air temperature, adjusted for elevation, and short-wave solar radiation, adjusted for slope, to estimate surface temperature. The surface temperature then is used to estimate, using finite element analysis, temperature profile and depth of thaw. The model needs subsurface thermal properties and works better as a regional model than as a site-specific model.

Romanovsky and Osterkamp (1997b) used measured soil temperature data (0.01–1 m, 1993-1995) from Barrow to calibrate models (Goodrich model and Guymon model). The Goodrich model used air temperature and snow cover thickness as inputs. The Barrow active layer average thickness is 0.35 m (ranging from 0.25 to 0.49 m). The mean annual ground temperature ranges from –7.3°C to –12.8°C, and the mean annual permafrost surface temperature ranges from –8 to –13.2°C. They also looked at unfrozen soil water content at Barrow and Prudhoe Bay.

Osterkamp and Romanovsky (1997) and Romanovsky and Osterkamp (1997a) documented upward freezing from the bottom of the active layer at West Dock, Deadhorse, and Franklin Bluffs sites. They used a model for unfrozen water content in active layer and near surface permafrost. Analytical calculations of the apparent thermal diffusivity showed that its observed variations with temperature were due to the latent heat effects of the unfrozen water. They note that the physical and mechanical properties of frozen ground and permafrost become strongly dependent on temperature when unfrozen water is present. Freeze-up dates for the active layer were taken to be the dates when the temperatures at all depths began to decrease as the zero curtain disappeared. In general, freeze-up dates depend on air temperatures, active layer thickness, soil water content, thermal properties of the soil, mean annual permafrost surface temperatures, the dates for the start of freezing at the ground surface, and the timing of the snow cover and its rate of accumulation. Freeze concentration of the pore solutes due to expulsion of the impurities from the growing ice and ice lenses can, under specific conditions, produce a significant shift in the freezing point and phase equilibrium temperatures. They note that a freezing point of –0.2°C would require a soil water salinity of more than three parts per thousand assuming that all the freezing point depression is attributable to salts derived from sea water, which is a large value for this area. Unfortunately, there were no data available about soil water salinity for these layers. Slow cooling of the active layer after freeze-up was attributed to significant amounts of unfrozen water in the active layer.

Many soils in the Arctic exhibit a degree of frost churning or cryoturbation. Also, they often contain ice lenses. These properties result in very nonuniform soil profiles, which, in turn, create difficulties in modeling soil behavior. Modelers, therefore, usually opt to use average or apparent values for soil properties. Due to soil variability and inconsistency in published data, it is always better to determine soil thermal properties or the apparent thermal properties in the location or area of interest. This is especially true when modeling. Some models are more sensitive to variations in soil thermal properties than others.

Soil temperature modelers need measured soil temperature data in order to calibrate and verify their models. If a model requires soil thermal properties, it is best that they be determined for each particular situation. Usually these will be "effective" properties that account for the high variability of Arctic soils. The great localized variability, caused by cryoturbation, in these soils makes it impractical to measure soil thermal properties or even use averages.

SUMMARY

Alaska is a large and sparsely populated state and as such, has relatively few climatic stations. Soil temperature monitoring stations are few and most have been recently initiated. Soil climate regime definitions, in the various soil classification schemes, are complex and different. Maps of soil climate regimes based on observations of vegetation, permafrost, air temperature, etc. do not match very well with those based on soil temperature measurements. Soil temperature is influenced by many factors, including air temperature, solar radiation, slope, aspect, latitude, elevation, vegetation, soil moisture, and thickness of a surface organic layer. Some factors affect energy input and others influence heat conduction. Various mathematical models use different approaches to estimate soil temperatures. Some models attempt to correlate soil temperature with other, more easily measured parameters, such as air temperature. Others use a mechanistic approach and try to model soil temperature based on thermal inputs and soil thermal properties. Soil thermal properties are largely determined by soil water content. Measured values of soil temperature are necessary to calibrate and verify mathematical models.

REFERENCES

Agriculture Canada Expert Committee on Soil Survey, The Canadian System of Soil Classification. 2nd ed. Agric. Can. Publ. 1646, Agricultural Canada, Ottawa, 1987.

Alaska Climate Research Center, Histormental climate, *History for Alaska*, University of Fairbanks, Alaska, online, http://climategi.alaska.edu/.

Alexander, E.B., Soil temperatures in forest and muskeg on Douglas Island, Southeast Alaska, *Soil Survey Horizons*, Winter, 1991, p. 108.

Anderson, D.M., Phase composition of frozen Montmorillonite-water mixtures from heat capacity measurements, *Soil Sci. Soc. Am. Proc.*, 30, 670, 1966.

Anderson, D.M. and Tice, A.R., Predicting unfrozen water contents in frozen soils from surface area measurements, *Highway Res. Rec.*, 393, 12, 1972.

Avallone, E.A. and Baumeister, T., III, *Marks' Standard Handbook for Mechanical Engineers*, 9th ed., McGraw-Hill, New York, 1987.

Baldwin, J.L., Climates of the United States. U.S. Department of Commerce, NOAA, Washington, D.C., 1973.

Baver, L.D., Gardner, W.H., and Gardner, W.R., *Soil Physics*, 4th ed., John Wiley & Sons, New York, 1972.

Carson, J.E., Soil Temperature and Weather Conditions, U.S. Atomic Energy Commission, Argonne National Laboratory ANL-6470, Argonne, IL, 1961.

Crawford, C.B., Soil Temperatures (A Review of Published Records). National Research Council of Canada, Preprint of a paper presented to the 30th ann. meet. Highway Research Board in Washington, D.C., January, 1951.

Chang, J.H., *Ground Temperature*, Vol. I & II, Harvard University, Cambridge, MA, 1958.

DaMore, D., U.S. Forest Service, Juneau, AK, personal communication, 1998.

Edlefsen, N.E. and Anderson, A.B.C., Thermodynamics of soil moisture, *Hilgardia*, 15, 31, 1943.

Gol'tsberg, I.A. and Davitaya, F.F., Eds., *Soil Climate*, Gidrometeorologicheskoe Press, Leningrad. Translated from Russian, Published for the Soil Conservation Service, USDA and the National Science Foundation, Washington, D.C., Amerind Publishing, New Delhi, 1980.

Hinkel, K., University of Cincinnati, personal communication, 1998.

Hinzman, L., Li, S., Lilly, E., Goering, D., and Kane, D., Distributed thermal modeling of the surface energy balance on the north slope of Alaska, Arctic System Science, Land-Atmosphere-Ice Interactions, Science Workshop, March 27-29, 1997, Seattle, WA, 1997.

Incropera, F.P. and DeWitt, D.P., *Fundamentals of Heat and Mass Transfer*, John Wiley & Sons, New York, 1990.

Koopmans, R.W.R. and Miller, R.D., Soil freezing and soil water characteristic curves, *Soil Sci. Soc. Am. Proc.*, 66, 680, 1966.

Landsberg, H.E., Lippmann, H., Paffen, K.H., and Troll, C., *World Maps of Climatology*, 2nd ed., Springer-Verlag, New York, 1965.

Moore, J.P., Swanson, D.K., Fox, C.A., and Ping, C.L., International Correlation Meeting on Permafrost Affected Soils. Guide Book — Alaska Portion, USDA NRCS National Soil Survey Center, Lincoln, NE, 1993.

NCDC, National Climatic Data Center, on-line, http://www.ncdc.noaa.gov/.

NOAA, Climate Diagnostic Center, on-line, http://www.cdc.noaa.gov/, 1999.

Nelson, F., University of Delaware, personal communication, 1998.

Nielsen, D.R., Jackson, R.D., Cary, J.W., and Evans, D.D., Eds., *Soil Water*, American Society of Agronomy, Madison, WI, 1972.

Osterkamp, T.E. and Romanovsky, V.E., Freezing of the active layer on the coastal plain of the Alaskan Arctic, *Permafrost Periglacial Processes*, 8, 23, 1997.

Ping, C.L., Soil temperature profiles of two Alaskan soils, *Soil Sci. Soc. Am. J.*, 51, 1010, 1987.

Rieger, S., 1973. Temperature regimes and classification of some well-drained alpine soils in Alaska, *Soil Sci. Soc. Am. Proc.*, 37, 806, 1973.

Rieger, S., Schoephorster, D.B., and Furbush, C.E., Exploratory Soil Survey of Alaska, USDA Soil Conservation Service, Washington, D.C., 1979.

Romanovsky, V.E. and Osterkamp, T.E., Thawing of the active layer on the coastal plain of the Alaskan Arctic, *Permafrost Periglacial Processes*, 8, 1, 1997a.

Romanovsky, V.E. and Osterkamp, T.E., Long-term (1949-1996) unfrozen water dynamic in the Barrow frozen active layer and its potential importance for CO_2 fluxes and other environmental characteristics, Arctic System Science, Land-Atmosphere-Ice Interactions, Science Workshop, March 27-29, Seattle, WA, 1997b.

Shul'gin, A.M., The Temperature Regime of Soils, GIMIZ Gidrometeorologicheskoe Izdatel'stvo, Leningrad. Translated from Russian, Israel Program for Scientific Translations, Jerusalem, 1965.

Tice, A.R., Oliphant, J.L., Nakamo, Y., and Jenkins, T.F., Relationship between the ice and unfrozen water phases in frozen soil as determined by pulsed nuclear magnetic resonance and physical data, CRREL Rep. 82-15, U.S. Army Cold Regions Research and Engineering Laboratory, Hanover, NH, 1982.

Trewartha, G.T., *An Introduction to Climate*, 3rd ed., McGraw-Hill, New York, 1954.

U.S. Department of Agriculture, Soil Conservation Service, Soil Survey Staff, Soil Taxonomy: A basic system of soil classification for making and interpreting soil surveys, USDA Handbook 436, U.S. Government Printing Office, Washington, D.C., 1954.

U.S. Department of Agriculture, Natural Resources Conservation Service, Soil Survey Staff, Keys to Soil Taxonomy, 7th ed., USDA NRCS, U.S. Government Printing Office, Washington, D.C., 1996.

Watson, C.E., Climates of the states: Alaska, U.S. Dept. Commerce, Weather Bureau, Washington, D.C., 1959.

Williams, P.J. and Smith, M.W., *The Frozen Earth*, Cambridge University Press, Cambridge, U.K., 1989.

APPENDIX A1

SOIL TAXONOMY: SOIL TEMPERATURE REGIMES.

Classes Of Soil Temperature Regimes

The following soil temperature regimes are used in defining classes at various categoric levels in the taxonomy.

Pergelic (L. per, throughout in time and space, and L. gelare, to freeze; meaning permanent frost) — Soils with a pergelic temperature regime have a mean annual temperature lower than 0°C. These are soils that have permafrost if they are moist, or dry frost if there is no excess water. It seems likely that the moist and the dry pergelic regimes should be defined separately, but at present we have only fragmentary data on the dry soils of very high latitudes. Ice wedges and lenses are normal in such soils in the U.S.

Cryic (Gr. kryos, coldness; meaning very cold soils) — Soils in this temperature regime have a mean annual temperature higher than 0°C but lower than 8°C.

1. In mineral soils, the mean summer soil temperature (June, July, and August in the northern hemisphere and December, January, and February in the southern hemisphere) either at a depth of 50 cm from the soil surface or at a densic, lithic, or paralithic contact, whichever is shallower, is as follows:

   a. If the soil is not saturated with water during some part of the summer and
      (1) If there is no O horizon: lower than 15°C; or
      (2) If there is an O horizon: lower than 8°C; or
   b. If the soil is saturated with water during some part of the summer and
      (1) If there is no O horizon: lower than 13°C; or
      (2) If there is an O horizon or a histic epipedon: lower than 6°C.

2. In organic soils, the soil is either:

   a. Frozen in some layer within the control section in most years 2 months after the summer solstice; i.e., the soil is very cold in winter but warms up slightly in summer; or
   b. Not frozen in most years below a depth of 5 cm from the soil surface; i.e., the soil is cold throughout the year but, because of marine influence, does not freeze in most years.

Cryic soils that have an aquic moisture regime commonly are churned by frost. All isofrigid (see below) soils without permafrost are considered to have a cryic temperature regime.

Frigid — The concept of the frigid soil temperature regime and other soil temperature regimes listed below are used chiefly in defining classes of soils in the low categories. A soil with a frigid regime is warmer in summer than a soil with a cryic regime, but its mean annual temperature is lower than 8°C, and the difference between mean summer and mean winter soil temperatures (June-July-August and December-January-February) is more than 5°C either at a depth of 50 cm from the soil surface or at a densic, lithic, or paralithic contact, whichever is shallower.

Mesic (Gr. mesos, intermediate) — The mean annual soil temperature is 8°C or higher but lower than 15°C, and the difference between mean summer and mean winter soil temperatures is more than 5°C either at a depth of 50 cm from the soil surface or at a densic, lithic, or paralithic contact, whichever is shallower.

Thermic — The mean annual soil temperature is 15°C or higher but lower than 22°C, and the difference between mean summer and mean winter soil temperatures is more than 5°C either at a depth of 50 cm from the soil surface or at a densic, lithic, or paralithic contact, whichever is shallower.

Hyperthermic — The mean annual soil temperature is 22°C or higher, and the difference between mean summer and mean winter soil temperatures is more than 5°C either at a depth of 50 cm from the soil surface or at a densic, lithic, or paralithic contact, whichever is shallower.

If the name of a soil temperature regime has the prefix iso (Gr. isos, equal), the mean summer and mean winter soil temperatures for June, July, and August and for December, January, and February differ by less than 5°C at a depth of 50 cm or at a densic, lithic, or paralithic contact, whichever is shallower.

Isofrigid — The mean annual soil temperature is lower than 8°C.
Isomesic — The mean annual soil temperature is 8°C or higher but lower than 15°C.
Isothermic — The mean annual soil temperature is 15°C or higher but lower than 22°C.
Isohyperthermic — The mean annual soil temperature is 22°C or higher.

SOIL TEMPERATURE CLASSES

Soil temperature classes, as named and defined here, are used as family differentiae in both mineral and organic soils. The names are used as family modifiers unless the criteria for a higher taxon carry the same limitation. Thus frigid is implied in all boric and cryic suborders and cryic great groups and subgroups, and would be redundant if used in the names of families within these classes of soils.

The Celsius (centigrade) scale is the standard. It is assumed that the temperature is that of a soil that is not being irrigated.

Control Section for Soil Temperature

The control section for soil temperature is either at a depth of 50 cm from the soil surface or at the upper boundary of a root-limiting layer, i.e., a duripan, a fragipan, a petrocalcic, petrogypsic, or placic horizon, or continuous ortstein; or at a densic, lithic, paralithic, or petroferric contact, whichever is shallower. The soil temperature classes, defined in terms of the mean annual soil temperature and difference between mean summer and mean winter temperature, are determined using the following key:

Key to soil temperature classes
   A. Lower than 8°C (47°F); Frigid, or
   B. 8° (47°F) to 15°C (59°F); Mesic, or
   C. 15° (59°F) to 22°C (72°F); Thermic, or
   D. 22°C (72°F) or higher; Hyperthermic

All other soil thats have mean annual soil temperature, as follows:
   1. Lower than 8°C (47°F); Isofrigid, or
   2. 8° (47°F) to 15°C (59°F); Isomesic, or
   3. 15° (59°F) to 22°C (72°F); Isothermic, or
   4. 22°C (72°F) or higher; Isohyperthermic

The soil temperature classes of Histosols are determined using the same key and definitions as those used for mineral soils. The modifier frigid, however, would be redundant in the family names of boric and cryic great groups and cryic and pergelic subgroups and is therefore omitted.

APPENDIX A2

SOIL TAXONOMY: PROPOSED SOIL TEMPERATURE FAMILIES

Soil Temperature Classes

Soil temperature classes, as named and defined here, are used as family differentiae in both mineral and organic soils. The names are used as family modifiers unless the criteria for a higher taxon carry the same limitation. Thus frigid is implied in all cryic suborders and cryic great groups and subgroups, and would be redundant if used in the names of families within these classes of soils.

The Celsius (centigrade) scale is the standard. It is assumed that the temperature is that of a soil that is not being irrigated.

Control Section for Soil Temperature

The control section for soil temperature is either at a depth of 50 cm from the soil surface or at the upper boundary of a root-limiting layer, i.e., a duripan, a fragipan, a petrocalcic, petrogypsic, or placic horizon, or continuous ortstein; or at a densic, lithic, paralithic, or petroferric contact, whichever is shallower. The soil temperature classes, defined in terms of the mean annual soil temperature and difference between mean summer and mean winter temperature, are determined using the following key.

Key to soil temperature classes:
  A. Soils in the order of Gelisol and that have a mean annual soil temperature as follows:
      1. –10°C or lower;   **Hypergelic,** or
      2. –4°C to –10°C   **Pergelic,** *or*
      3. +1°C to –4°C   **Subgelic.**
  B. Soils that have a difference in soil temperature of 5°C or more between mean summer (June, July, and August in the northern hemisphere) and mean winter (December, January, and February in the northern hemisphere) and a mean annual soil temperature of:
      1. Lower than 8°C (47°F);   **Frigid,** or
      2. 8° (47°F) to 15°C (59°F);   **Mesic,** or
      3. 15° (59°F) to 22°C (72°F);   **Thermic,** or
      4. 22°C (72°F) or higher.   **Hyperthermic**
  C. All other soil that have mean annual soil temperature, as follows:
      1. Lower than 8°C (47°F);   **Isofrigid,** or
      2. 8° (47°F) to 15°C (59°F);   **Isomesic,** or
      3. 15° (59°F) to 22°C (72°F);   **Isothermic,** or
      4. 22°C (72°F) or higher.   **Isohyperthermic.**

APPENDIX A3

CANADIAN SOIL TEMPERATURE CLASSES*

Extremely Cold
    MAST <–7°C.**
    Continuous permafrost usually occurs below the active layer within 1 m of the surface.
    No significant growing season, <15 days >5°C.
    Remains frozen within the lower part of the control section.
    Cold to very cool summer, MSST*** <5°C.
    No warm thermal period >15°C.

Very Cold
    MAST -7-2°C.**
    Discontinuous permafrost may occur below the active layer.
    Soils with Aquic regimes usually remain frozen within part of the control section.
    Short growing season, <120 days >5°C.
    Degree-days >5°C are <550.
    Moderately cool summer, MSST 5-8°C.
    No warm thermal period >15°C.

---

* Agriculture Canada Expert Committee on Soil Survey, 1987, p. 121.
** MAST: Mean Annual Soil Temperature.
*** MSST: Mean Summer Soil Temperature.

Cold

    MAST 2-8°C.*

    No permafrost.

    Undisturbed soils are usually frozen in some part of the control section for a part of the
        dormant season.

    Soils with Aquic regimes may remain frozen for part of the growing season.

    Moderately short to moderately long growing season, 140-220 days >5°C.

    Degree-days >5°C are 550-1250.

    Mild summer, MSST 8-15°C.

    An insignificant or very short, warm thermal period, 0-50 days >15°C.

    Degree-days >15°C are <30.

Cool

    MAST 5-8°C.*

    Undisturbed soils may or may not be frozen in part of the control section for a short part of
        the dormant season.

    Moderately short to moderately long growing season, 170-220 days >5°C.

    Degree-days >5°C are 1250-1700.

    Mild to moderately warm summer, MSST** 15-18°C.

    Significant very short to short warm thermal period, >60 days >15°C.

    Degree-days >15°C are 30-220.

Mild

    MAST 8-15°C.*

    Undisturbed soils are rarely frozen during the dormant season.

    Moderately long to nearly continuous growing season, 200-365 days >5°C.

    Degree-days >5°C are 1700-2800.

    Moderately warm to warm summer, MSST** 15-22°C.

    Short to moderately ware thermal period, 90-180 days >15°C.

    Degree-days >15°C are 170-670.

---

* MAST: Mean Annual Soil Temperature.
** MSST: Mean Summer Soil Temperature.

Section IV

Recommendations and Conclusions

17 The Fate of C in Soils of the Cold Ecoregions

R. Lal and J.M. Kimble

CONTENTS

INTRODUCTION

Approximately 50% of earth's land mass is frozen at sometime during the winter, with 20% of it underlain by permafrost (Sharratt et al., 1997). Similar to soils of other ecosystems, processes and properties of the soils of the cold regions are also to be understood and quantified to ensure stability and minimize adverse effects of natural or anthropogenic perturbations. These soils also contain as much as 40% of the glacial soil C pool, and are now a net sink for atmospheric carbon. Will these soils continue to be a sink even with severe anthropogenic perturbation in the Boreal Forest and Alpine Grassland biomes, and projected global warming of 1 to 4°C with in the next 50 to 100 years (Bengtsson, 1994; IPCC, 1995)? These soils could be a potential source of atmospheric C as CO_2 and CH_4. Because of their high productivity, the natural ecosystems of the Boreal Forest biome are being converted to managed ecosystems; yet, the impact of such a conversion on soil C pool and fluxes is not very well understood. Assessment of the impact of such a conversion necessitates study of the processes at the soil-atmosphere interface, and within the active layer above the permafrost.

CHARACTERISTICS OF FROZEN SOILS

Visually, soils of the cold regions have two identifiable layers: the active (thawed) layer and the permafrost. The depth of freezing may range from a few centimeters to several meters, depending on the location and landscape characteristics. The depth of the active layer undergoes drastic seasonal changes including frost heave and structural dynamics. Freezing of the active layer alters its macro- and microstructure, and the magnitude of alteration depends on the water content and on the proportion of unfrozen water in frozen soils. These processes can have a drastic impact on moisture retention and transmission properties (Burt and Williams, 1976; Chamberlain and Gow, 1979). The mechanics of frozen soils, thermal properties, and hydrologic characteristics need to be studied (Williams, 1997). It is also these properties that determine the gas exchange between soil and the atmosphere, a process with the direct bearing on the accelerated greenhouse effect.

MAJOR ISSUES

The data on total soil C pool for soils of high latitudes and altitudes are limited and need improvements in view of large variability and knowledge gaps in the existing information. Consequently, several important issues which need to be addressed include the following:

(a) *Assessment of C at Depth*: Most available data on soil C pool are based on the top 1 m depth. Yet, a large proportion of C exists in the subsoil up to 3 m of depth. Therefore, precise assessment of the total soil C pool requires assessment of the C at depth. In addition to its importance in the total pool, C at depth may also influence C emission through degradative processes, e.g., erosion. The C at depth may also influence C emission after alteration of the permafrost depth due to potential global warming.

(b) *Soil Inorganic Carbon*: Most available information on soil C pools is based on SOC content. However, soils derived from calciferous parent material also contain SIC, which must also be quantified and included in the total C pool. Quantification of SIC content is also relevant to understand the importance of secondary carbonates in C sequestration.

(c) *Alpine Ecoregions*: Soils of the alpine ecoregions are important to the global C cycle. Yet, C pools in these soils have not been widely assessed in relation to land use and management.

(d) *Total System Carbon*: The SOC content and its dynamics are closely linked to C in other components of the ecosystem. The SOC content is closely linked to aboveground plant biomass, soil fauna, C in (living and dead) microbial biomass, and particulate and dissolved organic and inorganic carbon in aquatic ecosystems. Soil C pool and fluxes must also be linked to other pertinent biogeochemical cycles, e.g., N, P, and S.

(e) *Data Variability and Reliability*: The available data on soil C pool and fluxes are highly variable, and the accuracy and reliability are also questionable. These problems are attributed to numerous factors including the following:

   (i)   Nonuniform terminology used for soil, climate, and ecosystems;
   (ii)   Different soil classification systems used;
   (iii) Differences in components of the total system C assessed, e.g., SOC, SIC, vegetation biomass, variable soil depth, etc;
   (iv)   Different methods of soil sampling and analyses, and mapping; Lack of accepted systems of linking national procedures with international standards;
   (vi)   Different ways of data expression and interpretation (median rather than mean value for highly variable data set); and
   (vii) High variability in SOC and SIC content due to differences in parent material, slope, vegetation, and microclimate.

   Data accuracy and reliability can only be improved with due consideration to these and other factors.

(f) *Soil Degradation*: Soil degradation, a decline in functional attributes of soil within an ecosystem, is a severe problem even within Arctic and Subarctic ecoregions. Soil degradation may be caused by several processes (e.g., erosion, compaction, change in the depth of active soil layer) and factors (e.g., fire, grazing, agricultural activities). Causes of soil degradation may be natural and anthropogenic. Soil degradative processes impact the C pool and its dynamics through direct and indirect influences, and the cause-effect relationships need to be established between soil C pools and fluxes and degradation-induced changes in soil properties.

(g) *Methods and Approaches*: A wide range of methods and approaches are used to obtain soil samples, conduct laboratory analyses, and interpret data on soil C pool and fluxes. It is important to harmonize conceptual bases, analytical procedures, and data interpretation standards. Several considerations to standardize procedures include the following:

(i)   Sampling within a defined pedon may minimize the error and reduce variability;
(ii)   Developing appropriate methods of measuring bulk density of organic and frozen soils may be useful for improving the database;
(iii)   Land use effects on soil C pool can only be evaluated when compared with that of an undisturbed pedon;
(iv)   Data accuracy and reliability may be improved through adoption of multiple approaches, e.g., analytical, modeling, and isotopic measurements;
(v)   Establishment of pedotransfer functions, relating SOC and SIC pools and fluxes to other soil properties, may be a valuable approach to establishing the cause-effect relationship, and to data extrapolation and interpretation;
(vi)   Regional and national maps should be validated through ground truthing and informed opinion surveys; and
(vii)   There exists the potential of using some remotely sensed information in improving the database on soil C and vegetational C pool and fluxes.

(h) *Principle C Pools in Soils of Cold Ecoregions*: There are three principal pools of soil C in cold ecoregions that need to be carefully assessed. These are

(i)   Wetlands which constitute a major SOC pool in cold ecoregions. Yet, C pool (both SOC and SIC) in wetlands (Plate 17.1) soils should not be ignored.
(ii)   Soil C below the 1 m depth constitutes a significant part of the total C pool.
(iii)   The SIC is an important component of the total pool.

Management of wetlands in cold ecoregions is a major consideration toward management of the global C pool.

(i) *Fire*: Fire plays an important role in C dynamics in soils of the Boreal forest ecoregion (Plate 17.2). The impact of natural fires on total system C and its dynamics is not widely understood, especially in relation to direct flux and C loss due to post-fire emissions. Total system C pool (soils + vegetation) need to be studied in relation to fire intensity, frequency of fire cycle, and temporal changes in C pool and fluxes caused by differences in intensity and frequency of disturbances due to fire.

(j) *Relationship among Various Pools*: There is a lack of understanding about the relationship among various C pools, especially with regards to the following:

(i)   Relationship between gross primary productivity, net primary productivity, and biomass C with SOC and SIC pools;
(ii)   Relationship between aquatic and terrestrial C pools (e.g., dissolved organic carbon vs. soil organic carbon); and
(iii)   Relationship between the residence time and characteristics of C in different pools (above and below ground).

KNOWLEDGE GAPS AND RESEARCHABLE PRIORITIES

The total C pool in soils of the cold ecoregion is large, and plays an important role in the global C cycle. Considerable progress has been made in assessing C pool and dynamics for soils of these ecoregions (Tarnocai and Smith, 1992; Whalen et al., 1996). Yet, the workshop identified numerous knowledge gaps that need to be filled. Soils of these ecoregions have been a net sink of C in the past. Because of the potential climatic change and natural or anthropogenic disturbances, these

PLATE 17.1a Wetlands are a major component of the landscape in cold ecoregions: coastal plains on the North Slope of Alaska showing extreme wetness.

PLATE 17.1b Wetlands are a major component of the landscape in cold ecoregions: most of the land between the ponds contains organic soils (Histels).

PLATE 17.1c Wetlands are a major component of the landscape in cold ecoregions: wetlands on the North Slope are thaw ponds and a sink for C.

PLATE 17.2a Fire is an important factor in the Boreal and Taiga biomes: forest fire in the Tundra in Far Eastern Russia on the lower Kolyma River.

PLATE 17.2b Fire is an important factor in the Boreal and Taiga biomes: naturally occurring forest fires are very common in the Taiga region of Far Eastern Russia.

PLATE 17.2c Fire is an important factor in the Boreal and Taiga biomes: a large forest fire in the Taiga Tundra in Far Eastern Russia along the Kolyma River.

PLATE 17.3 Climate station near the base of the Wilson-Piedmont Glacier in Antarctica.

soils can be a major source of emissions of greenhouse gases or a lesser sink than they are now. In this regard, several researchable priorities that need to be pursued include the following:

(i) *How Deep is Deep*? Precise measurement of the C pool requires the information on C depth in the pedosphere to be determined. For most soils, the depth of C assessment should be 2 or 3 m or to a lithic or paralithic contact.

(ii) *Multiple Approaches to C Pool and Flux Assessment*: Soil C pools and fluxes need to be assessed for benchmark soils in principal ecoregions by several approaches. The technique using stable isotopes (O_2, C, H) is very useful and should be calibrated with other approaches, e.g., chronosequence, models, and C balance.

(iii) *Standardization of Methods*: There is a need to identify appropriate methods of assessment of the C pool and fluxes, and to standardize/internalize these methods. Reference samples should be tested by all laboratories conducting C analyses using a common scale and depth of measurements.

(iv) *Laboratory Studies and Field Flux Rate*: Measurements of gaseous fluxes in the field (Plate 17.3) should be related to laboratory measurements of soil respiration rates based on incubation studies.

(v) *Regional Maps of Soil C Pool*: Multinational efforts are needed to develop reliable maps of the C pool of different ecoregions. These maps need to be based on standardized terminology (soils and ecoregions), methods (soil sampling and analysis), and scaling procedures. It is also important to develop the C impact map with relation to land use of the climate change.

(vi) *Pedotransfer Functions*: Developing pedotransfer functions is a useful strategy for relating soil C pool and fluxes to other soil properties and processes. Some useful pedotransfer functions include the following:

PLATE 17.4 Oil pipeline in the foothills of the Brooks Range in Northern Alaska.

- Total system C vs. biomass C, clay content, slope gradient, etc.;
- Soil bulk density vs. texture and SOC content;
- Gaseous emissions vs. soil temperature and moisture regimes;
- Active soil depth above the frost layer vs. soil temperature regime, moisture regimes, etc.;
- Old vs. new C pool.

(vii) *Soil Inorganic and Secondary Carbonates*: Inorganic or carbonate-C constitutes an important component of the total C pool, and must be quantified, especially for soils developed from calcareous parent material. Further, formation of secondary carbonates may play an important role in sequestration of C in soils and needs to be assessed.

(viii) *Modeling*: Modeling is an important strategy to identify the missing links. The purpose of modeling is not to get to reality, but to identify the missing link to understand the broader picture. It is important, however, to identify what models are the most appropriate for soils and environments of the cold ecoregions. Soil and ecoregion specific data are needed to run and validate various models. Some specific issues that need to be addressed through modeling are

- The impact of potential environmental change on the total C pools and fluxes from soils of the cold ecoregions;
- The relation between total C pool and different components, e.g., SOC, SIC, vegetation (above and below ground biomass), microbial biomass, etc.; and
- Impact of land use, soil management, and natural/anthropogenic disturbances on soil C pool and fluxes.

(ix) *Cryoturbation*: The impact of cryoturbation on soil C pools and their dynamics are not well known, and need to be improved for principal soils of the Arctic, Subarctic, and Boreal regions.

(x) *Alpine Soils*: Soil C pool and dynamics of alpine soils (e.g., the Himalayan-Tibetan, Andean, Alps, and other soils of high altitudes) need to be studied in relation to land use and management.

(xi) *Soil Degradation*: It is important that the nature of the "disturbance" is understood. The term disturbance implies "interruption of natural processes." There may be several types of disturbances that can have a drastic impact on soil C pool and fluxes. Some commonly observed disturbances include:

- Local disturbance, e.g., oil exploration (Plate 17.4);
- General disturbance, e.g., grazing;
- Intense disturbance, e.g., fire;
- Systemic disturbance, e.g., potential climatic change.

Assessing the impact of such disturbances on soil quality is important to understanding the dynamic of soil C. Knowledge of the system's resilience or its ability to recover following a disturbance is important to understand the ease of restoration following the disturbance (e.g., grazing, fire). What are the direct and indirect effects of such disturbances, and how long does it take for the system to attain steady state following disturbance?

(xii) *Agriculture in the Boreal Forest Ecoregions*: Agriculture intensification in the Boreal ecoregion can have a drastic impact on C pool and fluxes, and this impact must be assessed. Important considerations for this ecoregion are

- Aforestation and its impact on C dynamics;
- Soil management and its impact on hydrological balance especially with regards to erosion, transport of dissolved organic C, and C redistribution in the aquatic ecosystems;
- Land use impact on biological productivity; and
- Assessment of C pool and dynamics in upland, non-peat soils of these regions.

(xiii) *Energy Balance and the C Pool*: Natural and anthropogenic disturbances impact an ecoregion's energy balance and must be quantified. The energy balance impacts soil moisture and temperature regimes, percent snow cover, biomass productivity and the canopy cover, and the ultraviolet light emissions.

(xiv) *Soil Respiration*: Soil respiration needs to be measured for benchmark soils of the region in relation to land use and ecoregional characteristics.

(xv) *C Credit Procedures*: Because of the large variability in the database due to different procedures used at regional and national levels, it is important to establish international procedures for evaluating C credits due to adoption of recommended land use and soil management practices.

(xvi) *Soil Erosion and C Dynamics*: Similar to other ecoregions, erosional impacts on C pool and dynamics are not adequately understood. It is important to study the C pool in relation to topography, and the fate of C dislocated by erosional processes, e.g., buried C, dissolved organic C, C exposed due to disruption of aggregates, and C redistributed over the landscape.

RECOMMENDATIONS

(a) Establish multination workgroups/task force to do the following:
   (i) Standardize methods of soil sampling, analyses, mapping, scaling, modeling, and evaluating C credit,
   (ii) Produce reliable/credible C pool(s), C fluxes, and C impact maps for different depths at ecoregional and national levels,

(iii)   Develop a common databank, and
(iv)   Record the depth of soil at which the precise measurement of the C pool is determined.
(b)   Develop long-term monitoring sites for evaluating SOC and SIC pools and their dynamics in relation to land use and management, soil properties, and climatic factors.
(c)   Establish benchmark sites for monitoring temporal changes in organic soil surface dynamics in relation to soil moisture and temperature regimes.

REFERENCES

Bengtsson, L., Climate of the 21st Century, *Agric. For. Meteorol.*, 72, 3, 1994.

Burt, T.P. and Williams, P.J., Hydraulic conductivity in frozen soils, *Earth Surf. Process.*, 1, 349, 1976.

Chamberlain, E.J. and Gow, A.J., Effect of freezing and thawing on the permeability and structure of soils, *Eng. Geol.*, 13, 73, 1979.

IPCC, *Climate Change 1995*, Working Group 1, Oxford University Press, Oxford, U.K., 1995.

Koopmans, R.W.R. and Miller, R.D., Soil freezing and soil water characteristic curves, *Soil Sci. Soc. Am. Proc.*, 3, 680, 1966.

Sharratt, B.S., Radke, J.K., Hinzman, L.D., Iskandar, I.K., and Gruenevelt, P.H., Physics, chemistry, and ecology of frozen soils in managed ecosystems: an introduction, in *Proc. Int. Symp. Physics, Chemistry, Ecology of Seasonally Frozen Soil*, Tech. Spec. Rep. 97-10, U.S. Army Cold Regions Research and Engineering Laboratory, Hanover, NH, June 10-12, 1997, p. 1.

Tarnocai, C.S. and Smith, C.A.S., The formation and properties of soils in the permafrost region of Canada, First Int. Conf. on Cryopedol. Workshop Proc., Puschino, Russia, 1992, p. 21.

Whalen, S.C., Reeburg, W.S., and Reimers, C.E., Control of tundra methane emission by microbial oxidation, in *Landscape Function and Disturbance in Arctic Tundra: Ecological Studies 120*, Reynolds, J.F. and Tenhunen, J.D., Eds., Springer-Verlag, Berlin, 1996, p. 257.

Williams, P.J., The seasonally frozen layer: geotechnical significance and needed research, in *Proc. Int. Symp. Physics, Chemistry, Ecology of Seasonally Frozen Soils*, Iskandar, I.K., Wright, E.A., Radke, J.K., Sharratt, B.S., Gruenvelt, P.H., and Hinzman, L.D., Eds., Tech. Spec. Rep. 97-10, U.S. Army Cold Regions Research and Engineering Laboratory, Hanover, NH, June 10-12, 1997.

Index

A

DATE DUE

GAYLORD   #3522PI   Printed in USA